W0231919

Drying of Polymeric and Solid Materials
Modelling and Industrial Applications

Jean-Maurice Vergnaud

Drying of Polymeric and Solid Materials

Modelling and Industrial Applications

With 216 Figures

Springer-Verlag
London Berlin Heidelberg New York
Paris Tokyo Hong Kong
Barcelona Budapest

Jean-Maurice Vergnaud, PhD
Université de Saint-Etienne, Laboratoire de Chimie des Matériaux et
Chimie Industrielle, Faculté de Sciences et Techniques, 23 rue du
Docteur Paul Michelon, 42023 Saint-Etienne Cedex 2, France

Cover illustration: Ch.15, Fig.11. Surface concentration–time histories during
the process of drying in a finite volume of air. (Adaptation.)

ISBN-13:978-1-4471-1956-2   e-ISBN-13:978-1-4471-1954-8
DOI: 10.1007/978-1-4471-1954-8

British Library Cataloguing in Publication Data
Vergnaud, Jean-Maurice
  Drying of Polymeric and Solid Materials: Modelling and Industrial
  Applications.
  I. Title
  660
  ISBN-13:978-1-4471-1956-2

Library of Congress Cataloging-in-Publication Data
Vergnaud, J. M.
  Drying of polymeric and solid materials: modelling and industrial
applications / J. M. Vergnaud.
    p.   cm.
  Includes index.
  ISBN-13:978-1-4471-1956-2
  1. Drying. I. Title.
TP363.V44 1992   91-42551
660'.28426–dc20   CIP

Apart from any fair dealing for the purposes of research or private study, or criticism or
review, as permitted under the Copyright, Designs and Patents Act 1988, this publication
may only be reproduced, stored or transmitted, in any form or by any means, with the
prior permission in writing of the publishers, or in the case of reprographic reproduction
in accordance with the terms of licences issued by the Copyright Licensing Agency.
Enquiries concerning reproduction outside those terms should be sent to the publishers.

© Springer-Verlag London Limited 1992
Softcover reprint of the hardcover 1st edition 1992

The publisher makes no representation, express or implied, with regard to the accuracy of
the information contained in this book and cannot accept any legal responsibility or
liability for any errors or omissions that may be made.

Typeset by Keytec Typesetting Ltd, Bridport Dorset
69/3830-543210 Printed on acid-free paper

Preface

There are several objectives in this book devoted to the fundamentals of drying processes. The first is to consider the basic process of drying, which is very often controlled not only by the stage of evaporation but also by the stage of diffusion. In fact, two types of diffusion appear: "internal diffusion" of the liquid within the solid; and "external" diffusion of the vapour in the surrounding atmosphere through the grains of the solid. A certain emphasis is placed upon this "internal" diffusion through the solid itself, while various shapes for the grains of solids are examined: planes, cylinders of various lengths, spheres, and parallelepipeds.

The parameters of concern are: the shape of the solid; the diffusivity; the rate of evaporation; and the surrounding atmosphere. The diffusivity may either be constant or may depend on the concentration of the diffusing substance. The role played by the surrounding atmosphere is described, and the value of the liquid concentration which is at equilibrium with this atmosphere is considered. The volume of the atmosphere can either be infinite or finite. The effect of the ratio of the volumes of the solid and of the atmosphere is examined in various cases, and especially in Chap. 15. Generally, no change in dimension is considered during the drying process, except in Chap. 16, where the shrinkage following the drying of polymer beads is examined.

The second objective is to develop various models capable of describing the process of drying, by considering the stages of diffusion and evaporation. Analytical solutions are given following classical mathematical treatment, and special attention is focused on the assumptions which are made to obtain them. Particular consideration is given to the modelling of the process and to the latest application of computer simulation, by developing numerical models with finite differences. As they can easily take all the factors into account, they have a very wide use. In fact, the problems do not arise specifically from the numerical model but rather from the factual knowledge of the process.

As some assumptions are generally needed to build the models, it is thus understandable that the best person to make these assumptions and so to establish the model is the person who is responsible for the study of the drying process. In order to help all people interested in this task, various mathematical and numerical models are described for various shapes of the solid.

The third objective is to show how a method based essentially on experiments and modelling can be of interest. Experiments are necessary for at least the following three reasons:

- To establish a deep and precise knowledge of the process, by varying the operational conditions of the experiments.
- To determine all data, such as the diffusivity and its concentration dependence with liquid concentration, the rate of evaporation and the total amount of liquid which can evaporate.
- To test the validity of the model with long tests on real samples carried out under different conditions.

In order to determine the data, short tests are performed by using small samples of simple shapes, e.g., a thin plane sheet or a sphere, as the time for diffusion to proceed is proportional to the square of the dimension (thickness for the sheet, radius for the sphere).

Modelling the drying process with either an analytical solution or a numerical analysis followed by computerisation can be very powerful. On the one hand, the analytical solution results from classical mathematical treatment which can be achieved in only a few simple cases: where the diffusivity is constant, the initial concentration of the liquid in the sample is uniform, the material is isotropic, and where the shape and the boundary conditions are simple. On the other hand, numerical models can be built for all cases, by taking into account all the known facts, especially the following conditions: the concentration dependency of the diffusivity, the various initial and boundary conditions, the nature of the surrounding atmosphere, and even the change in dimension of the solid which can accompany the process of drying.

Mathematical and numerical models are of great help for the user, and they are capable of bringing the following advantages:

- Building the model and drawing the process on a flowsheet requires all the various steps of the process to be clearly distinguished.
- The role played by each parameter can be determined by simulating the process, the optimisation of the conditions of the process enabling the user to attain the best values for these parameters.
- A simulation of the process with the help of the model is generally of great help for industrial applications. From experiments performed on a small scale, or a laboratory scale, the results on a larger industrial scale can be evaluated by calculating. Short tests with small or thin samples of the same material as the industrial sample, are able to give data within a short period of time. By using the models and these data, highly time-consuming experiments with the large real samples can thus be avoided.

- Modelling is generally capable of reducing the number of experiments. It is also of help to determine the plan of the experiments which have to be performed to accomplish the work.

The book is divided into three parts. The first part presents an overview of the mathematical treatment of drying, considering the liquid diffusion within the solid and the evaporation from the surface. Various shapes are considered for the solid: thin plane sheets, cylinders and spheres. In order to help the reader, some emphasis is placed on the conditions in which an analytical solution is available. For people wanting to improve their background knowedge of the mathematical treatment of diffusion–evaporation, various examples are developed in a didactic way in the first four chapters:

- In Chap. 1, general equations of diffusion and evaporation are given for transient conditions.
- In Chap. 2, various plane sheets of solid are dried under transient conditions.
- In Chap. 3, cylinders of infinite or finite lengths are dried under transient conditions. Cases of drying under stationary conditions, with hollow cylinders are also described, as they are of interest for tubes.
- In Chap. 4, isotropic spheres are dried in transient conditions, as well as hollow spheres in stationary conditions.

The second part is devoted to numerical treatment, in order to accustom the users to this new and efficient way of working. Essentially explicit numerical methods are developed, because of their easy use with microcomputers. This part is divided into four chapters.

- In Chap. 5, plane sheets are considered and typical examples of numerical analysis are developed in these simple cases: the diffusivity of the liquid within the solid is either constant or concentration dependent, while various values of the rate of evaporation are given.
- Chap. 6 deals with the numerical treatment of the process of drying with solid cylinders or with hollow cylinders, the length being either finite or infinite. The diffusivity can be either constant or concentration dependent, with various values for the rate of evaporation.
- In Chap. 7, the numerical treatment is described in the case of solid spheres or of hollow spheres, with various cases for the diffusivity and the rate of evaporation.
- Chap. 8 is devoted to the numerical treatment of diffusion–evaporation taking place through and out of parallelepipedic solids. Isotropic media are considered, as well as anisotropic ones with three principal axes of diffusion and with three principal diffusivities.

The third part examines various approaches to many industrial problems with practical purposes. Each of these different cases are

discussed in Chaps. 9 to 16, working through the difficulties encountered with experiments or calculation, and finally leading to industrial
applications.

- Chap. 9 deals with the problem of drying classical paints, these
  paints consisting essentially of polymers and various additives as
  dyes, and of a solvent which must evaporate. A difficulty appears
  with the concentration dependency of diffusivity which makes
  complete drying difficult if not impossible. Moreover, the particu-
  larly interesting case of the drying of multi-layer paints is also
  considered, and the time of drying for the first layer is shown to be
  of concern.

- Chap. 10 discusses the question of how humid earth dries, especially when this earth is the essential component of a wall. This
  case may also be of practical interest not only for walls used in
  developing countries, but also for walls built in industrial countries
  with concrete.

- Chap. 11 focuses on the problems of rubbers, when they are in
  contact from time to time with a liquid such as a hydrocarbon, and
  when it is necessary for the liquid, which has been previously
  absorbed, to evaporate. Very often, industrial samples cannot be
  tested, these tests of absorption and desorption being highly time
  consuming and destructive for the material. The short test technique used with a thin plane sheet, coupled with the modelling of
  the process, is thus capable of predicting the kinetics of drying of
  large industrial samples as well as the profiles of concentration of
  liquid developed within various parts of interest of this solid.
  Various shapes for the solid, such as sheets, cylinders, tube and
  annuli, are considered in succession, and various models are built
  and tested in all these cases.

- Chap. 12 shows the complexity of the matter transfers which take
  place within a plasticised PVC, when this polymer is put into
  contact with various liquids such as food, blood and solvents. Two
  matter transfers are observed, the liquid entering the PVC and
  enabling the plasticiser to leave the polymer, while the diffusivity
  for the liquid depends on the concentrations of liquid and plasticiser. In spite of this high complexity, modelling and short experiments prove to be powerful, and they are especially efficient at
  determining the best operational conditions of temperature and
  time necessary for the majority of the liquid to evaporate out of
  the polymer.

- Chap. 13 concentrates on the rather complex problem of the drying
  of wood, when the moisture content is below or beyond the fibre
  saturation point. Three different cases are considered: one-dimensional transport of water through the thickness of thin sheets of
  wood, which is of help for determining the values of each principal
  diffusivity along each principal axis of diffusion; two-dimensional
  transport which can be observed when a dimension is very large
  compared with the other two; and finally the common three-
  dimensional transport with three principal diffusivities (longitu-

dinal, radial and tangential) through the solid, and evaporation
from the surface. Numerical models are estalished and successfully
tested for various shapes, such as plane sheets and beams, and for
various values of the moisture content below and above the fibre
saturation point.

- In Chap. 14, the drying of thermoset coatings made of pure epoxy
  resin is examined, essentially at temperatures ranging from 50 to
  100°C, when water is the evaporating substance.

- Chap. 15 is devoted to the important problem of drying of dosage
  forms made of a polymer and drug. These dosage forms are
  capable of delivering the drug into the patient's stomach at a
  controlled rate. A good way for preparing these forms consists of
  making a paste with the mixture of polymer and drug in powder
  form by addition of a liquid, and then of making this liquid
  evaporate as completely as possible. The effect of two parameters,
  the temperature and volume of the surrounding atmosphere, on
  the process, is specifically determined. Moreover, the numerical
  model is also used for predicting the kinetics of drying as well as
  the profiles of concentration developed through the spherical form,
  when the vapour pressure of the liquid in the atmosphere is
  controlled.

- Chap. 16 examines the difficult problem of drying when a change
  in dimension is taken into account during the process. Polymer
  beads of ethylene vinyl acetate are selected for this purpose, and
  the new numerical model describes the process of drying and
  shrinkage with high accuracy. Comparisons made between this new
  model and classical models neglecting this change in dimension of
  the solid show that the new way of modelling must be used when
  the extent of evaporating substance is higher than 20%. This case
  is often encountered when the solid to be dried is a polymer or
  rubber and the evaporating substance a hydrocarbon or other
  chemical agent.

Acknowledgements

Many people working in industrial firms have influenced this work through various industrial contracts, and many colleagues and students have supported my efforts. Eminent among them is my good friend and colleague, Jean Bouzon, Professor at the University of Saint-Etienne.

I am grateful for the collaboration of my colleagues: A. Accetta, J. C. David, J. L. Taverdet, and of my students: A. Aboutaybi, J. Y. Armand, H. P. Blandin, M. H. David, A. Droin-Josserand, M. El Kouali, Y. Khatir, S. Laoudi, H. Mounji and N. Laghoueg-Derriche.

My deepest appreciation to my students A. Aboutaybi for his drawings and A. El Brouzi for his help in calculation.

Special thanks to my colleague M. Norton who read the initial drafts.

Many grateful thanks to C. Cervantes, F. Faverot and D. Berthet for their competent typing of the manuscript.

October 1991   J.-M. Vergnaud

Contents

Chapter 1

Principles and General Equations

The general problem of the drying of solids can be complex as three types of transport exist for liquid and vapour (Fig. 1.1):

1. The transport of liquid within a solid.
2. The evaporation of liquid from the surface of the solid.
3. The transport of vapour through the grains of the solid.

It is of interest to consider these various means of transport separately.

1.1 Transport of Liquid Through a Solid

Very often, the transport of the liquid through the solid is controlled by diffusion with a constant or concentration-dependent diffusivity. This transport can be performed either under stationary conditions (when the concentration of the liquid varies with position) and under transient conditions (when the concentration of the liquid varies with position and time).

1.1.1 Polymers

The diffusion of liquid through a polymer can differ strongly according to the nature of the polymer. Polymers usually have a wide spectrum of relaxation times associated with their structural changes and especially with the motion of the polymer segments. All of these relaxation times decrease as temperature or the concentration of the liquid in the polymer is increased, and the motion of the polymer segments is enhanced. The diffusion of liquids in polymers is thus associated with the finite rates at which the polymer structure changes in response to the motion of the liquid. An important feature of polymers is their glass transition temperature. Above this temperature, the polymers are in a rubbery state, and below this temperature they are in a glassy state.

The diffusion behaviour of many polymers in a rubbery state is described by

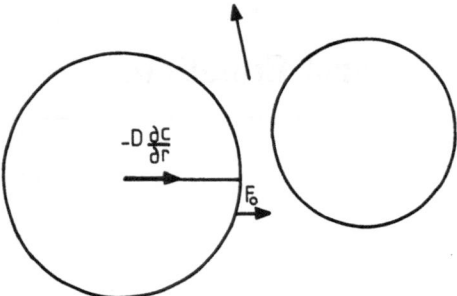

Fig. 1.1. Scheme of the three types of transport in the process of drying: internal diffusion of liquid; evaporation; external diffusion of vapour.

Fick's laws with constant or concentration-dependent diffusivity. The rate of diffusion is much less than the rate of relaxation, and the polymer chains adjust so quickly to the presence of the liquid that they do not cause diffusion anomalies. The amount of liquid transferred at time t, Q_t, as a fraction of the total amount of liquid transferred at infinite time, Q_∞, is very often expressed as a function of time by the relationship

$$\frac{Q_t}{Q_\infty} = kt^n \tag{1.1}$$

with $n = \frac{1}{2}$ when the diffusion is Fickian (case 1).

The properties of a glassy polymer tend to be time dependent, as the stress is slow to decay after such a polymer has been stretched. In this case, diffusion is very rapid compared with the rate of the relaxation process. The liquid advances with a constant velocity; this advancing front marks the innermost limit of penetration of the liquid, and is the boundary between a swollen gel and the glassy part free of liquid. These processes are characterised by $n = 1$ in Eq. (1.1) (case 2). Non-Fickian diffusion occurs when the relaxation and diffusion rates are comparable. This anomalous diffusion lies between case 1 and case 2, and n takes an intermediate value between $\frac{1}{2}$ and 1.

1.1.2 Wood Materials

Two schools of thought can be recognised in the matter of moisture (called also bound-water) transport in wood, according to the nature given to the driving force of the liquid in Eq. (1.2):

$$\text{Rate of transport} = \text{Conductance} \times \text{Driving force} \tag{1.2}$$

One school of thought maintains that bound-water diffusion takes place in response to a vapour pressure gradient and that the potential which drives moisture is a gradient of pressure. The other school, representing the vast majority of wood scientists, believes that the transport is described by diffusion and obeys Fick's laws, the concentration gradient of moisture playing the role of the driving force. The main difficulty in wood transport is that wood is an anisotropic medium which has different diffusion properties in different directions. Fortunately, there are three principal axes of diffusion with three principal diffusivities, and wood beams and wood planes are usually cut along these axes.

1.1.3   Basic Equations of Diffusion (Fick's Laws)

As shown in the case of polymers, diffusion is the process by which the liquid is transported from one part of the solid to another as a result of random molecular motions. As a result, there is a transport of diffusing molecules from the region of higher to that of lower concentration of these molecules, because there are more molecules able to diffuse in the region of higher concentrations.

Transfer of heat by conduction is due to random molecular motions; this analogy between the two processes was recognised by Fick in 1855 who adopted the mathematical equation of heat conduction derived previously by Fourier in 1822. The rate of transfer of the diffusing substance through unit area of a section of an isotropic substance is proportional to the concentration gradient measured normal to the section:

$$\mathrm{Ra} = -D\,\frac{\partial C}{\partial x} \tag{1.3}$$

where Ra is the rate of transfer per unit area of section, C the concentration of diffusing substance, x the space coordinate measured normal to the section and D the diffusivity (expressed in $\mathrm{cm}^2\ \mathrm{s}^{-1}$).

1.2   Evaporation of Liquid From the Surface

The liquid evaporates from the surface of the solid, and it is of interest to know the rate of evaporation from the surface. Two different cases can be considered, according to the boundary conditions and the value of the rate of evaporation.

1.2.1   Infinite Rate of Evaporation

When the rate of evaporation is very high, so as to be considered infinite, the following simple assumption can be made: the surface of the solid reaches equilibrium with the surrounding atmosphere instantaneously when evaporation starts. Of course, when the atmosphere is free of vapour the concentration at the surface falls immediately to zero. A strong agitation of the atmosphere is needed, and the volume of this external atmosphere must be very much larger than that of the solid.

1.2.2   Finite Rate of Evaporation

The rate of evaporation from the surface is proportional to the difference between the actual concentration on the surface of the solid at any time, C_s, and the concentration which would be in equilibrium with the vapour pressure remote from the surface, C_{ext}, the coefficient of proportionality F_0 being the rate of evaporation of the pure liquid in the same conditions

$$F = F_0(C_s - C_{ext}) \tag{1.4}$$

Some evaporative processes are controlled by diffusion in the sense that the rate of evaporation depends largely on the rate at which the diffusing liquid is supplied by internal diffusion to the evaporating surface. This is taken to be

$$-D\left(\frac{\partial C}{\partial x}\right)_s = F_0(C_s - C_{ext}) \tag{1.5}$$

where $(\partial C/\partial x)_s$ is the gradient of concentration next to the surface and the diffusivity D corresponds to the concentration C_s.

Of course, the rate of evaporation from the surface increases with the rate of evaporation of the pure liquid. As the rate of evaporation of a pure liquid is proportional to its vapour pressure, the rate of evaporation from the surface increases with temperature by following the Clausius–Clapeyron equation:

$$\ln\frac{F_{T_1}}{F_{T_2}} = -\frac{\Delta H_v}{R}\left(\frac{1}{T_1} - \frac{1}{T_2}\right) \tag{1.6}$$

where F_{T_1} and F_{T_2} are the rates of evaporation of the pure liquid at temperatures T_1 (K) and T_2 respectively, ΔH_v is the enthalpy of vaporisation of the liquid and R the constant of ideal gas.

A similar condition holds for concentration at equilibrium with the surrounding atmosphere, and the rate of evaporation from the surface is increased by decreasing the value of the concentration C_{ext} and the external vapour pressure.

1.3   Diffusion of Vapour Outside the Solid

The diffusion of vapour outside the solid, which can be called external diffusion in contrast with the internal diffusion within the solid, also plays a very important role in the rate of the process in industrial applications.

This external diffusion is dependent on the rate at which the vapour is eliminated from the solid, and it is very often the limitation on the rate of the drying process.

If this external diffusion could be high enough, the concentration C_{ext} shown in Eqs (1.4) and (1.5) would be maintained at the theoretical value which is very low when the surrounding atmosphere is almost free of vapour. But of course, the evaporation of the liquid from the surface is responsible for an increase in the vapour pressure. Moreover, the presence of gradients of temperature in the surrounding atmosphere may be responsible for vapour condensation and the evaporation of this condensate retards the process of drying by maintaining the value of the external concentration C_{ext} at a high accidental value.

A strong circulation of the external atmosphere is then necessary to eliminate the vapour from the surface of the solid as soon as it is produced, and thus to improve the process of drying. Moreover, using an atmosphere at a temperature higher than that of the solid can also be of help for this purpose.

1.4 Effect of Parameters

It is of interest to discuss the effect of some parameters on the process of drying by using relationships. Two equations are worth noting. One is concerned with the diffusion of the liquid through the sample. A dimensionless number often appears either in the analytical equations expressed in the form of a trigonometrical series (see Chaps 2–4) or in the numerical relations (see Chaps 5–8) for calculating the kinetics of the matter transferred and the profiles of concentration:

$$\frac{Dt}{L^2} \quad \text{or} \quad \frac{Dt}{R^2} \tag{1.7}$$

where D is the diffusivity, L is the thickness for a plane sheet, and R is the radius for a solid sphere.

The other relation expresses the rate of evaporation of matter from the surface in terms of the concentration of liquid on the surface C_s and of the concentration necessary to maintain equilibrium with the surrounding atmosphere, C_{ext}:

$$F = F_0(C_s - C_{ext}) \tag{1.4}$$

1.4.1 Dimensions of the Grain

Equation (1.7) shows that the time t necessary for a given amount of matter to be transferred is directly proportional to the square of the main dimension, e.g., the thickness L of a sheet, the radius R of a sphere. The rate of evaporation of the liquid from the surface is proportional to the surface area of the material. As a result, it is shown that the dimensions of the grains of the material play an important role, both on diffusion and on evaporation, as small grains form a larger area of solid at the external surface.

1.4.2 Nature of the Solid and Liquid

The diffusivity is a parameter related to both the nature of the liquid and the solid. This parameter is of great importance as shown in Eq. (1.7), where the time necessary for a given amount of matter to be transferred is inversely proportional to the diffusivity. The diffusivity is generally the same throughout bulk polymers except when the polymer is stretched. In the case of wood, three principal directions of transport are observed with three main diffusivities.

The nature of the liquid, and especially its volatility, predominates for the rate of evaporation in Eq. (1.4).

Moreover, more complex phenomena sometimes occur, especially when the diffusivity is concentration dependent. By following the principle that the lower the concentration of liquid, the lower the diffusivity, it is possible that a part of the liquid would remain trapped in the solid while the surface of this solid would be almost perfectly dry. This is observed in special polymers such as plasticised PVC where the distribution of plasticiser is modified during the immersion of the polymer in the liquid (see Chap. 12), and also for the drying of paints (Chap. 9) and the drying of thermosetting polymers previously soaked in a liquid (Chap. 14).

1.4.3 Shape of the Solid

The shape of the solid influences the process of drying rather strongly, especially when the solid is anisotropic. Drying a wood beam is quite different from drying a plane sheet of the same material, because the longitudinal diffusivity of the moisture is considerably greater than the other main diffusivities (see Chap. 13).

1.4.4 Surrounding Atmosphere

The surrounding atmosphere plays a very important role in the drying process. The vapour pressure in this atmosphere is directly related to the value of the concentration at equilibrium C_{ext} shown in Eq. (1.4). Of course, the lower the vapour pressure in the air, the higher the rate of evaporation.

In industrial applications on a large scale, the surrounding atmosphere can also be non-uniform.

1.4.5 Temperature

Temperature influences the drying of a solid in the following ways. Firstly, an increase in temperature provokes an increase in the rate of evaporation of the pure liquid F_0, by following the Clausius–Clapeyron equation (Eq. (1.6)). Secondly, the diffusivity of a liquid in a solid, and especially in polymers, increases with temperature following an Arrhenius law. Lastly, an increase in temperature is responsible for a decrease in the relative humidity of the surrounding atmosphere, and so reduces the value of the concentration of liquid on the surface at equilibrium with this atmosphere C_{ext}, shown in Eq. (1.4).

1.5 General Equations

The following three shapes for the solid are considered:

- Plane sheets with parallel surfaces
- Cylinders of infinite and finite length
- Spheres

1.5.1 Operational Conditions

Two cases can be considered for the rate of evaporation of the liquid from the external surface of the solid:

- The rate of evaporation is so high that the concentration of liquid on the solid surface falls to the equilibrium value as soon as the process starts.
- The rate of evaporation is finite, which is a common case. The rate of evaporation of the liquid on the solid surface is proportional to the difference

between the actual concentration of liquid on the surface and this concentration which is at equilibrium with the surrounding atmosphere, the coefficient of proportionality being the rate of evaporation of the pure liquid under the same conditions of temperature and vapour pressure.

Of course, the second case is more representative of experiments, but the first case is described by very simple boundary equations which are of help for integrating the equations of transport.

Usually, the drying process is carried out under transient (or non-steady-state) conditions with the following relationships: the concentration of the liquid within the solid varies with position and time, and the rate of evaporation on the surface decreases continuously with time. In special cases, in a tube for instance, the presence of a large amount of liquid at the inner part allows the liquid to diffuse through the solid with a constant rate. The liquid is transported under stationary conditions with the two relationships: the concentration of the liquid depends only on position, and the rate of evaporation is constant.

The process of drying can be performed either in an atmosphere of finite volume, or in a very large volume which can be considered as infinite. These two cases must be considered.

1.5.2 Differential Equation of Diffusion

The rate of transfer of a substance diffusing through a unit area of a section is proportional to the gradient of concentration normal to the section. Thus the fundamental differential equation of diffusion in an isotropic medium is as follows:

$$\mathrm{Ra} = -D\frac{\partial C}{\partial x} \tag{1.3}$$

where Ra is the rate of transfer per unit area of section, C the concentration of diffusing substance, x the space coordinate normal to the section and D the diffusivity (or diffusion coefficient).

Diffusion in a Thin Isotropic Sheet

An element of volume in the form of a thin plane sheet of thickness Δx whose sides are perpendicular to the axis of diffusion is shown in Fig. 1.2. The rate of increase of diffusing substance in the element $A\mathrm{d}x$ is calculated by considering the rates of transfer of substance entering through the area A of the plane x and leaving through the same area of the plane $x + \mathrm{d}x$:

$$A(\mathrm{Ra}_x - \mathrm{Ra}_{x+\mathrm{d}x}) = -A\frac{\partial \mathrm{Ra}}{\partial x}\mathrm{d}x \tag{1.8}$$

The rate at which the amount of diffusing substance in the element increases is also given by

$$A\mathrm{d}x\frac{\partial C}{\partial t} \tag{1.9}$$

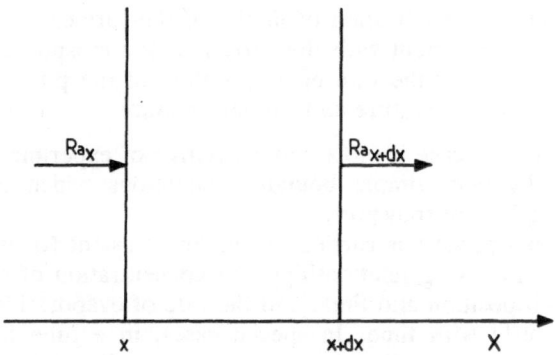

Fig. 1.2. Diffusion in one dimension through the element of volume $A\,dx$.

and from the equality of these expressions of the rate, we have

$$\frac{\partial C}{\partial t} = \frac{\partial}{\partial x}\left(D\frac{\partial C}{\partial x}\right)$$

(1.10)

When the diffusivity is constant, this equation reduces to

$$\frac{\partial C}{\partial t} = D\frac{\partial^2 C}{\partial x^2}$$

(1.11)

Diffusion in an Isotropic Cylinder

In the general case where the diffusion is radial and longitudinal with the same diffusivity, the rate of increase of the concentration of diffusing substance in the elements (Fig. 1.3) is

$$\frac{\partial C}{\partial t} = \frac{\partial}{\partial z}\left(D\frac{\partial C}{\partial z}\right) + \frac{1}{r}\frac{\partial}{\partial r}\left(rD\frac{\partial C}{\partial r}\right)$$

(1.12)

When the cylinder is very long, the longitudinal diffusion is neglected, and diffusion is everywhere radial, as shown in the circular cross-section:

$$\frac{\partial C}{\partial t} = \frac{1}{r}\frac{\partial}{\partial r}\left(rD\frac{\partial C}{\partial r}\right)$$

(1.13)

This equation becomes when the diffusivity is constant

$$\frac{\partial C}{\partial t} = D\left[\frac{\partial^2 C}{\partial r^2} + \frac{1}{r}\frac{\partial C}{\partial r}\right]$$

(1.14)

Diffusion in a Sphere

In this case the diffusion is radial, and the general equation for diffusion takes the form (Fig. 1.4):

$$\frac{\partial C}{\partial t} = \frac{1}{r^2}\frac{\partial}{\partial r}\left[r^2 D\frac{\partial C}{\partial r}\right]$$

(1.15)

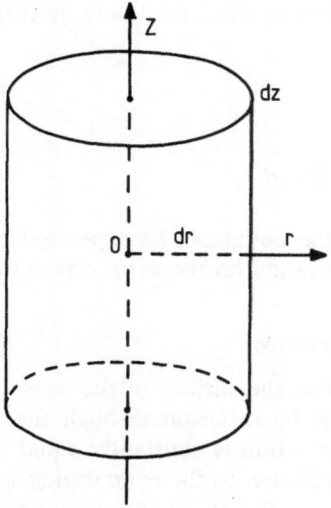

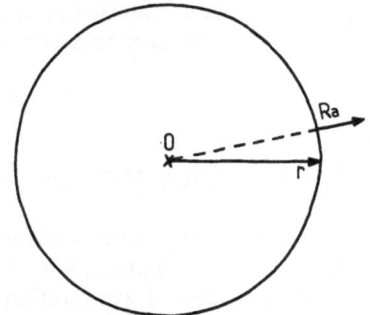

Fig. 1.4. Radial diffusion through the spherical element.

Fig. 1.3. Diffusion through a cylindrical element of volume: radial and longitudinal.

and for constant diffusivity it becomes

$$\frac{\partial C}{\partial t} = D\left[\frac{\partial^2 C}{\partial r^2} + \frac{2}{r}\frac{\partial C}{\partial r}\right] \tag{1.16}$$

or

$$\frac{\partial C}{\partial t} = \frac{D}{r^2}\frac{\partial}{\partial r}\left(r^2\frac{\partial C}{\partial r}\right) \tag{1.16'}$$

1.5.3 Equation of Evaporation

General Equation of Evaporation

The general equation of evaporation expresses the rate of evaporation of the substance in terms of the concentration of this substance on the evaporating surface:

$$F = F_0(C_s - C_{ext}) \tag{1.4}$$

where F_0 is the rate of evaporation of the pure liquid when the concentration is 1, C_s the concentration of the liquid on the surface and C_{ext} the concentration of the liquid on this surface when at equilibrium with the vapour in the surrounding atmosphere.

Following the respective values of these concentrations, absorption or evaporation takes place:

$$C_s > C_{ext} \text{ (evaporation)} \tag{1.17}$$

$$C_s < C_{ext} \text{ (absorption)} \tag{1.18}$$

When the vapour in the surrounding atmosphere is eliminated very quickly, the concentration C_{ext} reduces to a very low value:

$$F = F_0 C_s \qquad (1.19)$$

Boundary Condition at the Surface of the Solid

The case of a very high rate of evaporation of the liquid can be expressed by the following boundary condition, where the concentration on the surface falls to zero as soon as the process of evaporation starts:

$$C_s = 0 \text{ (evaporating surface)} \qquad (1.20)$$

In the general case, the liquid evaporates out of the surface of the solid, while the liquid is brought to the evaporating surface by diffusion through the solid. The boundary condition is that the rate of evaporation is constantly equal to the rate at which the liquid is supplied by internal diffusion to the evaporating surface (per unit surface).

$$-D\frac{\partial C}{\partial x} = F_0(C_s - C_{ext}) \qquad (1.4)$$

This condition can be used whatever the shape of the surface. Many cases are described with plane surfaces, cylinders and spheres.

The value of F_0 can be determined in two ways:

1. By evaporating the pure liquid and following the constant rate of evaporation.
2. By evaporating the liquid out of the solid, and determining the rate of evaporation at the beginning of the process of drying when the concentration on the surface is equal to the initial concentration. Of course in this case, $F_0 C_{in}$ is obtained.

When the solid is porous with very thin pores, or when there is a bond between the liquid and the solid, the values obtained for F_0 by the first method can differ from those obtained by the second.

Symbols

A	Area through which the diffusion takes place
C	Concentration of diffusing substance
C_{ext}	Concentration on the surface necessary to maintain equilibrium with the surrounding atmosphere
C_s	Concentration on the surface
D	Diffusivity ($cm^2\ s^{-1}$)
F_0, F	Rate of evaporation of the pure liquid, and of the liquid when concentration is C_s, respectively ($cm\,s^{-1}$).
ΔH_v	Enthalpy of vaporisation
k	Constant
L	Thickness of the sample

n	Integer
Q_t, Q_∞	Amount of substance transferred after time t, and after infinite time, respectively
r	Radial position coordinate
Ra	Rate of transfer of diffusing substance
t	Time
T	Temperature (kelvin)
x	Position coordinate
z	Position coordinate

Chapter 2

Thin Plane Sheet

Various cases of one-dimensional diffusion in a solid bounded by two parallel planes, with evaporation on the surfaces, are considered. The plane sheet is so thin that all the liquid evaporates through the plane surfaces and a negligible amount through the edges, and the diffusivity D is constant.

2.1 Non-steady State With Infinite Rate of Evaporation

2.1.1 Infinite Atmosphere With Uniform Initial Concentration of Liquid

The most simple case is encountered when the initial concentration of liquid in the solid is uniform.

The equation for diffusion in one dimension when the diffusivity D is constant is given by Fick's second law:

$$\frac{\partial C}{\partial t} = D\frac{\partial^2 C}{\partial x^2} \tag{2.1}$$

Initital conditions: $t = 0,$ $0 < x < L,$ $C = C_{in}$ (sheet)   (2.2)

Boundary conditions: $t > 0,$ $x = 0,$ $C = 0$ (surfaces)   (2.3)

$$x = L$$

The region $0 < x < L$ is initially at a uniform concentration C_{in}, and the two surfaces are kept at a constant concentration (zero in this case) (Fig. 2.1). The solution for the concentration is given in the form of a trigonometrical series by using the method of separation of variables (Apps 2.A and 2.B):

$$\frac{C_{x,t}}{C_{in}} = \frac{4}{\pi}\sum_{n=0}^{\infty}\frac{1}{2n+1}\sin\frac{(2n+1)\pi x}{L}\exp\left(-\frac{(2n+1)^2\pi^2}{L^2}Dt\right) \tag{2.4}$$

The amount of matter which has left one surface of the sheet of area A at time t is defined by Fick's first law:

$$Q_t = A \int_0^t D \left| \frac{\partial C}{\partial x} \right| dt \qquad (2.5)$$

$\partial C/\partial x$ being the gradient of concentration at $x = 0$ or $x = L$.

At infinite time, the amount of matter which has left the surface of area A is

$$Q_\infty = C_{in} LA \qquad (2.6)$$

It is shown that (App. 2B)

$$\frac{Q_\infty - Q_t}{Q_\infty} = \frac{8}{\pi^2} \sum_{n=0}^{\infty} \frac{1}{(2n + 1)^2} \exp \left(-\frac{(2n + 1)^2 \pi^2}{L^2} Dt \right) \qquad (2.7)$$

For long times, when the ratio $0.5 < Q_t/Q_\infty < 1$, the series can be reduced to the first term, and the well-known equation is written in the logarithm form:

$$\ln \frac{Q_\infty - Q_t}{Q_\infty} = -\frac{\pi^2}{L^2} Dt + \ln \frac{8}{\pi^2} \qquad (2.7')$$

Eq. (2.7') is often used for determining the diffusivity from the slope of the straight line obtained by plotting the term on the left-hand side as a function of time.

The half-life of the desorption process is easily observed experimentally. When the diffusivity is constant, the value of this constant D can be determined from Eq. (2.8) (App. 2.C).

$$\frac{Dt}{L^2} = 0.0489 \quad \text{with} \quad \frac{Q_t}{Q_\infty} = \tfrac{1}{2} \qquad (2.8)$$

By considering that (App. 2.D)

$$C = \frac{Q}{\sqrt{t}} \exp \left(-\frac{x^2}{4 Dt} \right) \qquad (2.9)$$

is a solution of Eq. (2.1), another solution of the problem comprises a series of error functions (App. 2.E):

$$\frac{C_{in} - C_{x,t}}{C_{in}} = \sum_{n=0}^{\infty} (-1)^n \text{erfc} \frac{(2n + 1)L - 2x}{4\sqrt{Dt}}$$

$$+ \sum_{n=0}^{\infty} (-1)^n \text{erfc} \frac{(2n + 1)L + 2x}{4\sqrt{Dt}} \qquad (2.10)$$

and

$$\frac{Q_t}{Q_\infty} = \frac{4}{L} \sqrt{Dt} \left\{ \frac{1}{\sqrt{\pi}} + 2 \sum_{n=1}^{\infty} (-1)^n \text{ierfc} \frac{nL}{2\sqrt{Dt}} \right\} \qquad (2.11)$$

Eq. (2.11) is very useful for determining the diffusivity when the process is carried out for short times.

When $Q_t/Q_\infty < 0.5$, Eq. (2.11) reduces to

$$\frac{Q_t}{Q_\infty} = \frac{4}{L} \sqrt{\frac{Dt}{\pi}} \qquad (2.12)$$

The diffusivity is found from the slope obtained by plotting Q_t/Q_∞ as a function of the square root of time.

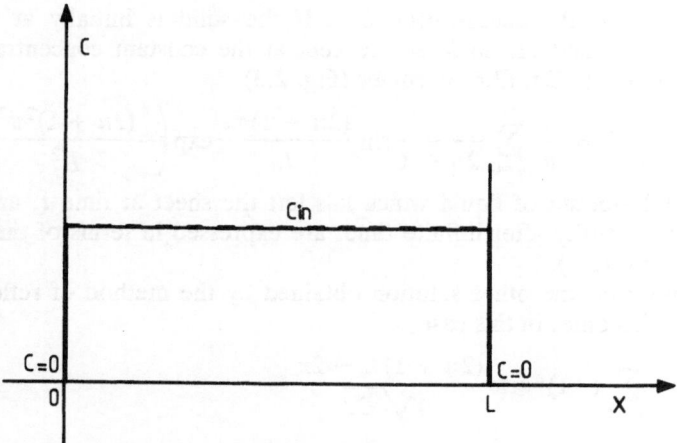

Fig. 2.1. Substance evaporating at an infinite rate from a plane sheet initially at uniform concentration, with constant (zero) concentration in the surrounding atmosphere.

The kinetics of desorption of a substance with an infinite rate of evaporation from a sheet of thickness L immersed in a surrounding atmosphere of infinite volume are shown in Fig. 2.2, by using the dimensionless coordinates

$$100 \frac{Q_t}{Q_\infty} \quad \text{and} \quad \frac{Dt}{L^2}$$

Eq. (2.7) does not converge very much for short times and a great value is needed for the integer n.

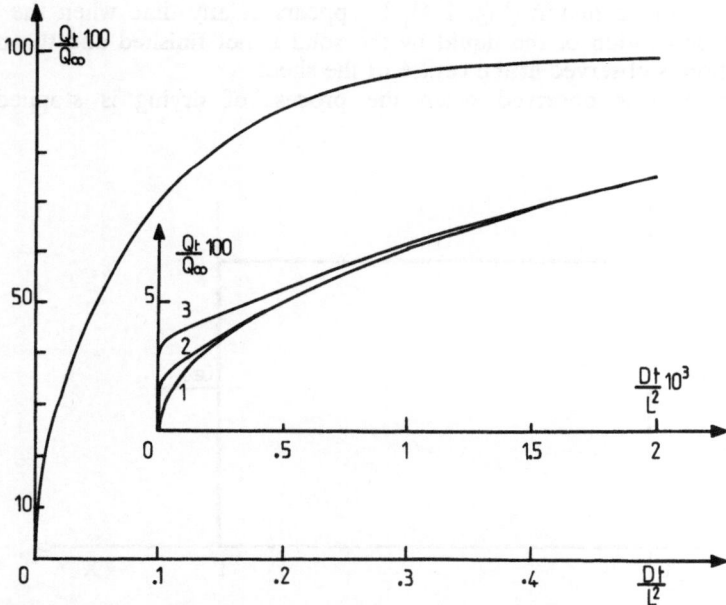

Fig. 2.2. Kinetics of evaporation from a plane sheet in a surrounding atmosphere of infinite volume, with infinite rate of evaporation. L is the thickness of the sheet.

Remark 2.1: External Concentration C_{ext}. If the solid is initially at a uniform concentration C_{in}, and the surfaces are kept at the constant concentration C_{ext}, the solution shown in Eq. (2.4) becomes (Fig. 2.3)

$$\frac{C_{ext} - C_{x,t}}{C_{ext} - C_{in}} = \frac{4}{\pi} \sum_{n=0}^{\infty} \frac{1}{2n + 1} \sin\frac{(2n + 1)\pi x}{L} \exp\left(-\frac{(2n + 1)^2\pi^2}{L^2}Dt\right) \quad (2.4')$$

Q_t, the total amount of liquid which has left the sheet at time t, and Q_∞ the corresponding quantity after infinite time, are expressed in terms of time and the diffusivity by Eq. (2.7).

In the same way, the other solution obtained by the method of reflection and superposition, becomes in this case

$$\frac{C_{in} - C_{x,t}}{C_{in} - C_{ext}} = \sum_{n=0}^{\infty} (-1)^n \text{erfc}\frac{(2n + 1)L - 2x}{4\sqrt{Dt}}$$

$$+ \sum_{n=0}^{\infty} (-1)^n \text{erfc}\frac{(2n + 1)L + 2x}{4\sqrt{Dt}} \quad (2.10')$$

with the external concentration C_{ext}.

The kinetics of the liquid leaving the sheet are expressed by Eq. (2.11) in this case.

Remark 2.2: Surface Concentrations Constant. Initial Distribution $f(x)$.

$$\text{Initial conditions:} \quad t = 0, \quad 0 < x < L, \quad C = f(x)$$

$$\text{Boundary conditions:} \quad t > 0, \quad x = 0, \quad C = C_{ext}$$

$$x = L$$

This case is very common (Fig. 2.4). It appears at any time when the previous process of absorption of the liquid by the solid is not finished and the minimum concentration is observed at the centre of the sheet.

Another case is observed when the process of drying is stopped before

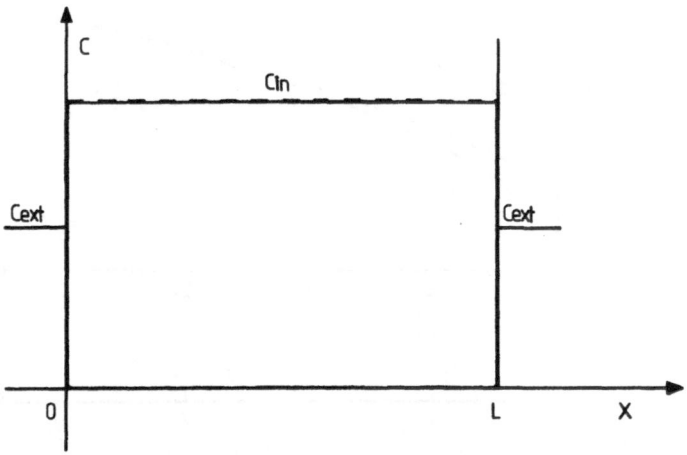

Fig. 2.3. Substance evaporating at an infinite rate from a plane sheet initially at uniform concentration, with constant concentration in the surrounding atmosphere.

completion with a maximum of the concentration of the liquid at the midplane of the sheet. The solution in the form of a trigonometrical series is

$$C_{x,t} - C_{\text{ext}} = \frac{2C_{\text{ext}}}{\pi} \sum_{n=1}^{\infty} \frac{\cos n\pi - 1}{n} \sin \frac{n\pi x}{L} \exp\left(-\frac{n^2\pi^2}{L^2}Dt\right)$$

$$+ \frac{2}{L} \sum_{1}^{\infty} \sin \frac{n\pi x}{L} \exp\left(-\frac{n^2\pi^2}{L^2}Dt\right) \int_0^L f(x') \sin \frac{n\pi x'}{L} dx'$$

$$(2.13)$$

Remark 2.3. In all the previous cases, the problem is symmetrical, the midplane being the plane of symmetry.

Remark 2.4: Determination of the Diffusivity. In the simple case where

- The diffusivity is constant
- The volume of the surrounding atmosphere is infinite
- The rate of evaporation is infinite

the diffusivity can be determined by using the following three methods:

1. For short times and $Q_t/Q_\infty < 0.5$

$$\frac{Q_t}{Q_\infty} = \frac{4}{L}\sqrt{\frac{Dt}{\pi}} \qquad (2.12)$$

2. For the half-life of the process, $Q_t/Q_\infty = 0.5$

$$\frac{Dt}{L^2} = 0.0489 \qquad (2.8)$$

3. For long times and $Q_t/Q_\infty > 0.5$

$$\ln \frac{Q_\infty - Q_t}{Q_\infty} = -\pi^2 \frac{Dt}{L^2} + \ln \frac{8}{\pi^2} \qquad (2.7')$$

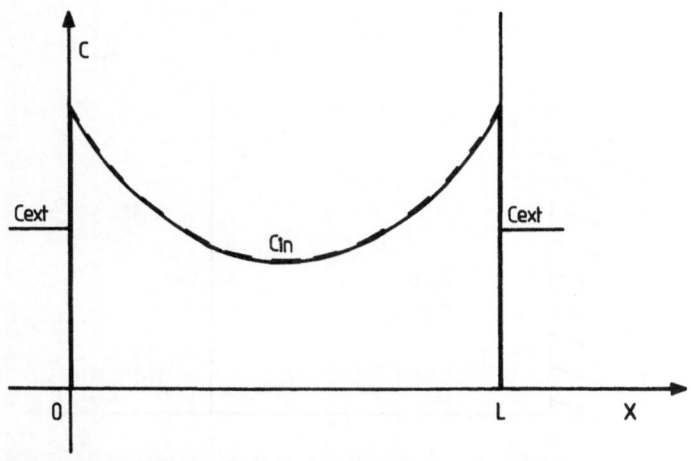

Fig. 2.4. Substance evaporating at an infinite rate from a plane sheet initially at non-uniform concentration, with constant concentration in the surrounding atmosphere.

2.1.2 Layer of Diffusing Substance Deposited on a Permeable Sheet

As the rate of evaporation of the diffusing substance is considered as infinite, the concentration on the two surfaces is constantly equal to zero (Fig. 2.5). The initial and boundary conditions are

$$t = 0, \qquad 0 < x < h, \qquad\qquad C = C_{in} \quad \text{(layer)}$$
$$\qquad\qquad h < x < L, \qquad\qquad C = 0 \quad\;\; \text{(solid)} \qquad (2.14)$$
$$t > 0, \qquad C_{0,t} = C_{L,t} = 0 \qquad\qquad\quad \text{(surfaces)} \qquad (2.15)$$

The equation for diffusion in one dimension with constant diffusivity is

$$\frac{\partial C}{\partial t} = D\frac{\partial^2 C}{\partial x^2} \qquad (2.1)$$

The solution for the concentration is given by using the method of separation of variables (App. 2.F).

$$C_{x,t} = \frac{4C_{in}}{\pi} \sum_{n=1}^{\infty} \frac{1}{n} \sin^2\left(\frac{n\pi h}{2L}\right) \sin\left(\frac{n\pi x}{L}\right) \exp\left(-\frac{n^2\pi^2}{L^2}Dt\right) \qquad (2.16)$$

The amount of diffusing substance remaining in the sheet and layer after time t is given by

$$Q'_t = \int_0^L C_{x,t}\,dx \qquad (2.17)$$

where $C_{x,t}$ is given in Eq. (2.16).

After infinite time, all the diffusing substance is evaporated, and this amount is equal to the initial amount of this substance in the layer:

$$Q_\infty = C_{in}h \qquad (2.18)$$

It is shown that

$$\frac{Q'_t}{Q_\infty} = \frac{8L}{\pi^2 h} \sum_{n=0}^{\infty} \frac{1}{(2n+1)^2} \sin^2\left(\frac{(2n+1)\pi h}{2L}\right) \exp\left(-\frac{(2n+1)^2\pi^2}{L^2}Dt\right) \;(2.19)$$

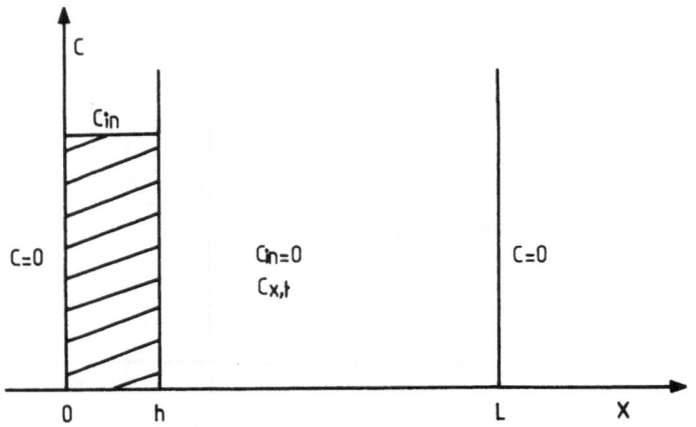

Fig. 2.5. Plane sheet with a layer of diffusing substance on one surface, with infinite rate of evaporation.

The amount of substance which has left the layer and sheet after time t, Q_t, is obviously given by

$$Q_\infty = Q_t + Q_{t'} \qquad (2.20)$$

Remark 2.5: The series in Eq. (2.19) converges very slowly.

2.1.3 Layer of Diffusing Substance Located Between Two Permeable Sheets

The case where the diffusing substance is located between two permeable sheets corresponds with a sandwich (Fig. 2.6). The initial conditions can be written as follows:

$$t = 0, \qquad \frac{L}{2} - h < x < \frac{L}{2} + h, \qquad C = C_{in} \quad \text{(layer)}$$

$$0 < x < \frac{L}{2} - h, \qquad C = 0 \quad \text{(sheet)}$$

$$\frac{L}{2} + h < x < L, \qquad C = 0 \quad \text{(sheet)} \qquad (2.21)$$

As the rate of evaporation is considered infinite, the boundary conditions are

$$t > 0, \quad x = 0, \, C = 0 \quad \text{(surfaces)}$$

$$x = L \qquad (2.22)$$

The equation for diffusion in one dimension with constant diffusivity,

$$\frac{\partial C}{\partial t} = D \frac{\partial^2 C}{\partial x^2} \qquad (2.1)$$

can be solved by using the method of separation of variables:

$$C_{ext} = \frac{4C_{in}}{\pi} \sum_{n=1}^{\infty} \frac{(-1)^{n+1}}{2n - 1} \sin\left(\frac{(2n - 1)\pi h}{L}\right) \sin\left(\frac{(2n - 1)\pi x}{L}\right)$$

$$\exp\left(-\frac{(2n - 1)^2 \pi^2}{L^2} Dt\right) \qquad (2.23)$$

When the layer of diffusing substance is thinner than the permeable sheet, this above equation reduces to

$$C_{x,t} = \frac{4C_{in}h}{L} \sum_{n=1}^{\infty} (-1)^{n+1} \sin\left(\frac{(2n - 1)\pi x}{L}\right) \exp\left(-\frac{(2n - 1)^2 \pi^2}{L^2} Dt\right) \qquad (2.24)$$

The amount of diffusing substance remaining in the unit area of the sandwich after time t can be calculated by

$$Q_t' = \int_0^L C_{ext} \, dx \qquad (2.17)$$

As the total amount of diffusing substance per unit area which can leave the sandwich after infinite time is

$$Q_\infty = 2hC_{in} \qquad (2.25)$$

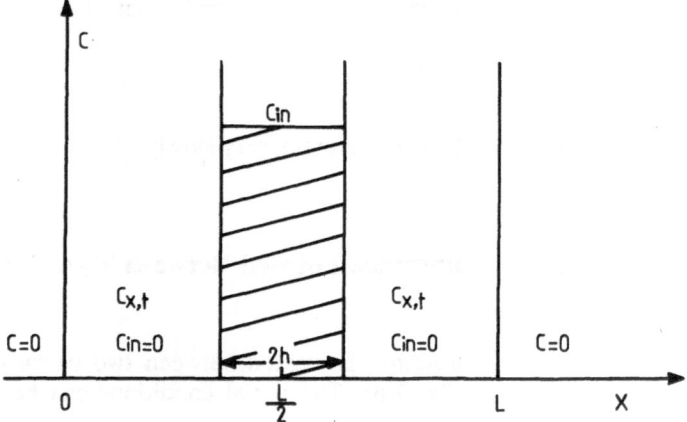

Fig. 2.6. Sandwich form, with the diffusing substance located between two permeable sheets, with infinite rate of evaporation.

the solution is given in the form of a trigonometrical series by using the method of separation of variables:

$$\frac{Q'_t}{Q_\infty} = \frac{4L}{\pi^2 h} \sum_{n=1}^{\infty} \frac{(-1)^{n+1}}{(2n-1)^2} \sin\left(\frac{(2n-1)\pi h}{L}\right) \exp\left(-\frac{(2n-1)^2 \pi^2}{L^2} Dt\right) \quad (2.26)$$

The amount of diffusing substance which has left the sandwich after time t, Q_t, is obviously given by

$$Q_\infty = Q_t + Q'_t \quad (2.20)$$

When the layer of diffusing substance is thinner than the sheet ($h \ll L$), the result is more simple

$$\frac{Q_\infty - Q_t}{Q_\infty} = \frac{4}{\pi} \sum_{n=1}^{\infty} \frac{(-1)^{n+1}}{2n-1} \exp\left(-\frac{(2n-1)^2 \pi^2}{L^2} Dt\right) \quad (2.27)$$

2.1.4 Finite Atmosphere with Constant Diffusivity

If a plane sheet initially full of liquid is suspended in a volume of air so large that the amount of liquid which evaporates from the sheet is a negligible fraction of the amount that this air can receive, and the air is well stirred, then the concentration in the air remains very small and negligible.

If, however, there is only a limited volume of air, the concentration of vapour in the air increases as the liquid evaporates. If the air is well stirred the concentration of vapour in the air depends only on time.

Two conditions are necessary for the problem. Firstly, the concentrations in the surrounding atmosphere and in the plane sheet are determined essentially by the obvious condition that the total amount of diffusing (and evaporating) substance in the air and plane sheet remains constant as diffusion–evaporation proceeds.

Following the above condition, the rate at which the liquid evaporates from the sheet is always equal to the rate at which it enters the surrounding atmosphere over the sheet surfaces.

The diffusing substance is initially uniformly distributed through the sheet and

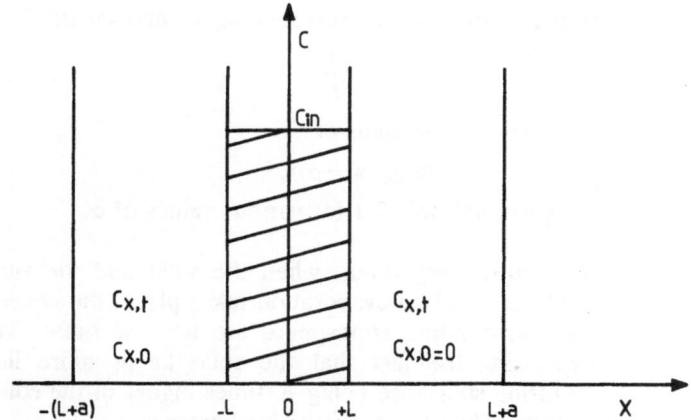

Fig. 2.7. Evaporation at an infinite rate from a sheet in a surrounding atmosphere of finite volume. The thickness of the sheet is $2L$ and the equivalent thickness of air is $2a$.

evaporates out into the well-stirred atmosphere. The unit area sheet occupies the space $-L \leqslant x \leqslant L$, while the surrounding atmosphere of limited volume in contact with this unit area occupies the space $-a - L \leqslant x \leqslant -L$ and $L \leqslant x \leqslant L + a$. The concentration of the vapour in the air is initially zero and is always uniform.

The diffusion equation in one dimension through the sheet of thickness $2L$ is

$$\frac{\partial C}{\partial t} = D\frac{\partial^2 C}{\partial x^2} \tag{2.1}$$

The initial conditions are (Fig. 2.7)

$$t = 0, \quad -L \leqslant x \leqslant L, \qquad\qquad C = C_{\text{in}} \quad \text{(sheet)}$$

$$-L - a < x < -L, \qquad C = 0 \quad \text{(air)}$$

$$L < x < L + a, \qquad\qquad C = 0 \quad \text{(air)} \tag{2.28}$$

while the boundary condition expresses the fact that the rate at which the vapour enters the atmosphere is constantly equal to the rate at which the liquid is supplied to the evaporating surface of the sheet by internal diffusion:

$$t > 0, \; a\frac{\partial C}{\partial t} = \pm D\frac{\partial C}{\partial x}, \; x = \mp L \; \text{(surfaces)} \tag{2.29}$$

The solution of this equation is obtained by using the Laplace transform [1].

The concentration within the sheet is given by the expression

$$\frac{C_{x,t} - C_\infty}{C_{\text{in}} - C_\infty} = \sum_{n=1}^{\infty} \frac{2(1 + \alpha)}{1 + \alpha + \alpha^2 q_n^2} \frac{\cos\dfrac{q_n x}{L}}{\cos q_n} \exp\left(-\frac{q_n^2}{L^2}Dt\right) \tag{2.30}$$

where C_∞ is the uniform concentration in the sheet after infinite time, and $C_{x,t}$ the concentration in the sheet at position x and time t.

The amount of diffusing substance which has left the sheet at time t, Q_t, is expressed as a fraction of the corresponding quantity after infinite time, Q_∞:

$$\frac{Q_\infty - Q_t}{Q_\infty} = \sum_{n=1}^{\infty} \frac{2\alpha(1 + \alpha)}{1 + \alpha + \alpha^2 q_n^2} \exp\left(-\frac{q_n^2}{L^2}Dt\right) \tag{2.31}$$

where α is the ratio of the volumes of the surrounding air and sheet:

$$\alpha = \frac{a}{L} \tag{2.32}$$

and the q_ns are the non-zero positive roots of

$$\tan q_n = -\alpha q_n \tag{2.33}$$

Roots of Eq. (2.33) are given in Table 2.1 for various values of α.

Remark 2.6: Partition Factor. Very often, when the solid and the surrounding atmosphere are at equilibrium and no evaporation takes place, the concentrations in the solid and in the surrounding atmosphere are not the same. There is a partition factor K expressing the fact that the solid keeps more liquid, the concentration of the diffusing substance being K times higher in the solid than in the surrounding atmosphere. Then the coefficient α becomes

$$\alpha = \frac{a}{KL} \tag{2.34}$$

The amount of liquid which evaporates after infinite time as a fraction of the amount of liquid located in the sheet is then expressed in terms of α:

$$\frac{M_\infty}{2LC_{\text{in}}} = \frac{\alpha}{1 + \alpha} \tag{2.35}$$

The kinetics of evaporation of a substance with an infinite rate of evaporation are shown in Fig. 2.8 for various values of the ratio α of the volumes of the surrounding atmosphere and sheet, by using the dimensionless coordinates

$$\frac{Q_t}{Q_\infty}100 \quad \text{and} \quad \frac{Dt}{L^2}$$

L being half the thickness of the sheet.

2.2 Non-steady State with a Finite Rate of Evaporation

This problem is very common, and it can be encountered in the following two cases:

1. When the volume of the surrounding atmosphere is considerably larger than that of the sheet, and the amount of matter which evaporates does not change the concentration of the vapour.
2. When the volume of the surrounding atmosphere is not large with regard to that of the sheet, but the concentration of vapour is kept constant, for instance by condensation at constant temperature.

The equation for diffusion in one dimension through the sheet of thickness $2L$ with constant diffusivity is

$$\frac{\partial C}{\partial t} = D\frac{\partial^2 C}{\partial x^2} \tag{2.1}$$

Table 2.1. Roots of $\tan q_n = -\alpha q_n$

α	q_1	q_2	q_3	q_4	q_5	q_6	q_7	q_8	q_9	q_{10}
9.0000	1.6385	4.7359	7.8681	11.0057	14.1451	17.2852	20.4258	23.5667	26.7077	29.8489
4.0000	1.7155	4.7648	7.8857	11.0183	14.1549	17.2933	20.4326	23.5726	26.7129	29.8535
2.3333	1.8040	4.8014	7.9081	11.0344	14.1674	17.3036	20.4413	23.5801	26.7196	29.8595
1.5000	1.9071	4.8490	7.9378	11.0558	14.1841	17.3173	20.4529	23.5902	26.7285	29.8674
1.0000	2.0288	4.9132	7.9787	11.0856	14.2075	17.3364	20.4692	23.6043	26.7409	29.8786
0.6667	2.1746	5.0037	8.0385	11.1296	14.2421	17.3649	20.4934	23.6253	26.7595	29.8953
0.4286	2.3521	5.1386	8.1334	11.2010	14.2990	17.4119	20.5335	23.6602	26.7904	29.9229
0.2500	2.5704	5.3540	8.3029	11.3349	14.4080	17.5034	20.6120	23.7289	26.8514	29.9778
0.1111	2.8363	5.7172	8.6587	11.6532	14.6870	17.7481	20.8283	23.9218	27.0250	30.1354
0.0000	3.1416	6.2832	9.4248	12.5664	15.7080	18.8496	21.9912	25.1328	28.2744	31.4160

The initial conditions are very simple, as the concentrations are uniform in the sheet, as well as in the surrounding atmosphere:

$$t = 0, \quad -L > x > L, \qquad C = C_{in} \qquad \text{(sheet)}$$

$$x < -L \text{ or} \qquad C = C_{ext} \qquad \text{(air)}$$

$$x > L \tag{2.36}$$

The boundary conditions express the fact that the rate of evaporation is constantly equal to the rate at which the liquid is supplied to the evaporating surface by internal diffusion. The rate of evaporation is generally proportional to the difference between the actual concentration of liquid on the surface and the concentration on the surface necessary to maintain equilibrium with the surrounding atmosphere, the coefficient of proportionality being the rate of evaporation of the pure liquid (Fig. 2.9).

$$D \left| \frac{\partial C}{\partial x} \right| = F_0(C_0 - C_{ext}) \tag{2.37}$$

where C_0 (or C_L) is the concentration of liquid on the surface, C_{ext} the concentration on the surface necessary to maintain equilibrium with the surrounding atmosphere and F_0 the rate of evaporation expressed in terms of the volume of pure liquid which evaporates per unit time per unit area of liquid.

If the law of evaporation shown in Eq. (2.37) holds on both surfaces of the sheet, the solution is given by [1]:

$$\frac{C_{ext} - C_{x,t}}{C_{ext} - C_{in}} = \sum_{n=1}^{\infty} \frac{2S \cos \dfrac{\beta_n x}{L}}{(\beta_n^2 + S^2 + S) \cos \beta_n} \exp\left(-\frac{\beta_n^2}{L^2} Dt\right) \tag{2.38}$$

where the β_ns are the positive roots of

$$\beta \tan \beta = S \tag{2.39}$$

with the dimensionless number S:

$$S = \frac{F_0 L}{D} \tag{2.40}$$

Roots of Eq. (2.40) are given in Table 2.2 for various values of S.

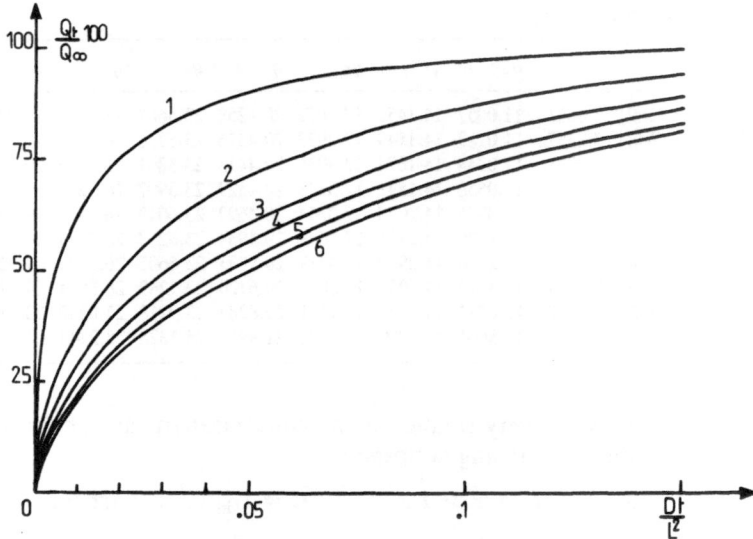

Fig. 2.8. Effect of the ratio α of the volumes of the surrounding atmosphere and sheet on the kinetics of evaporation ($2L$ represents the thickness of the sheet). 1: $\alpha = 0.25$; 2: $\alpha = 1$; 3: $\alpha = 2.33$; 4: $\alpha = 4$; 5: $\alpha = 9$; 6: $\alpha = \infty$.

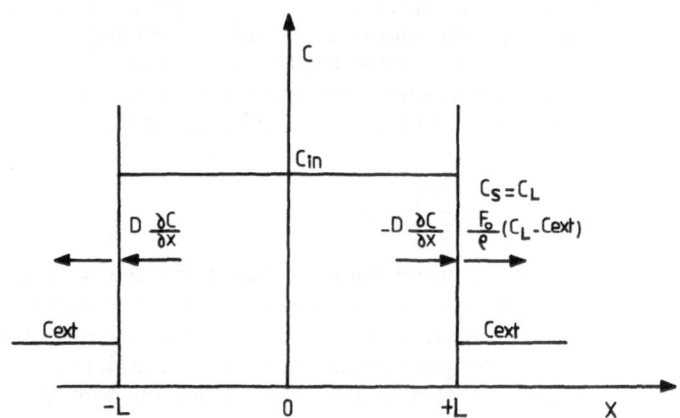

Fig. 2.9. Evaporation of substance from a sheet at a finite rate.

The total amount of liquid evaporating from the sheet up to time t is expressed as a fraction of the corresponding quantity after infinite time:

$$\frac{Q_\infty - Q_t}{Q_\infty} = \sum_{n=1}^{\infty} \frac{2S^2}{\beta_n^2(\beta_n^2 + S^2 + S)} \exp\left(-\frac{\beta_n^2}{L^2}Dt\right) \qquad (2.41)$$

Of course, depending on whether the concentration of liquid on the surface C_s is greater or less than the concentration at equilibrium with the surrounding atmosphere, the evaporation or condensation takes place.

$$C_s > C_{ext} \text{ (evaporation from the sheet)} \qquad (2.42)$$

$$C_s < C_{ext} \text{ (condensation on the sheet)} \qquad (2.43)$$

Table 2.2. Roots of $\beta \tan \beta = S$

S	β_1	β_2	β_3	β_4	β_5	β_6	β_7	β_8	β_9	β_{10}
0.000	0.0000	3.1416	6.2832	9.4248	12.5664	15.7080	18.8496	21.9911	25.1327	28.2743
0.010	0.0998	3.1448	6.2848	9.4258	12.5672	15.7086	18.8501	21.9916	25.1331	28.2747
0.100	0.3111	3.1731	6.2991	9.4354	12.5743	15.7143	18.8549	21.9957	25.1367	28.2779
0.200	0.4328	3.2039	6.3148	9.4459	12.5823	15.7207	18.8602	22.0002	25.1407	28.2814
0.500	0.6533	3.2923	6.3616	9.4775	12.6060	15.7397	18.8760	22.0139	25.1526	28.2920
1.000	0.8603	3.4256	6.4373	9.5293	12.6453	15.7713	18.9024	22.0365	25.1724	28.3096
2.000	1.0769	3.6436	6.5783	9.6296	12.7223	15.8336	18.9547	22.0815	25.2119	28.3448
5.000	1.3138	4.0336	6.9096	9.8928	12.9352	16.0107	19.1055	22.2126	25.3276	28.4483
10.00	1.4289	4.3058	7.2281	10.200	13.2142	16.2594	19.3270	22.4108	25.5064	28.6106
100.0	1.5552	4.6658	7.7764	10.887	13.9981	17.1093	20.2208	23.3327	26.4450	29.5577
900.0	1.5691	4.7072	7.8453	10.983	14.1215	17.2596	20.3977	23.5358	26.6739	29.8120

The kinetics of evaporation of different substances from a sheet immersed in a surrounding atmosphere of infinite volume are shown in Fig. 2.10 for various rates of evaporation, by using the dimensionless coordinates

$$\frac{Q_t}{Q_\infty} 100 \quad \text{and} \quad \frac{Dt}{L^2}$$

L being the thickness of the sheet and S the dimensionless number shown in Eq. (2.40) expressed in terms of the rate of evaporation F_0.

Remark 2.7: Case of a Very Low Concentration in the Surrounding Atmosphere.
When the vapour pressure in the surrounding atmosphere is very low, the concentration on the solid surface necessary to maintain equilibrium with this atmosphere is also very low. The modifications to Eq. (2.38) for $C_{ext} = 0$ are obvious.

2.3 Membrane Suspended in an Infinite Atmosphere

The following two cases are considered for a membrane with one face kept at a constant concentration C_i and the other at another constant concentration C_{ext}:

1. The interval of time during which the liquid is transferred and evaporates under transient conditions.
2. When the steady-state condition is set up.

In both cases, the membrane is initially at a uniform concentration which is generally zero. The solution can be drawn when the initial uniform concentration is not zero, C_{in} for instance.

2.3.1 Transient Conditions with Infinite Rate of Evaporation

One surface $x = 0$ of the membrane is kept at a constant concentration C_i and the other $x = L$ at zero, while the membrane is initially free of diffusing substance (Fig. 2.11).

26

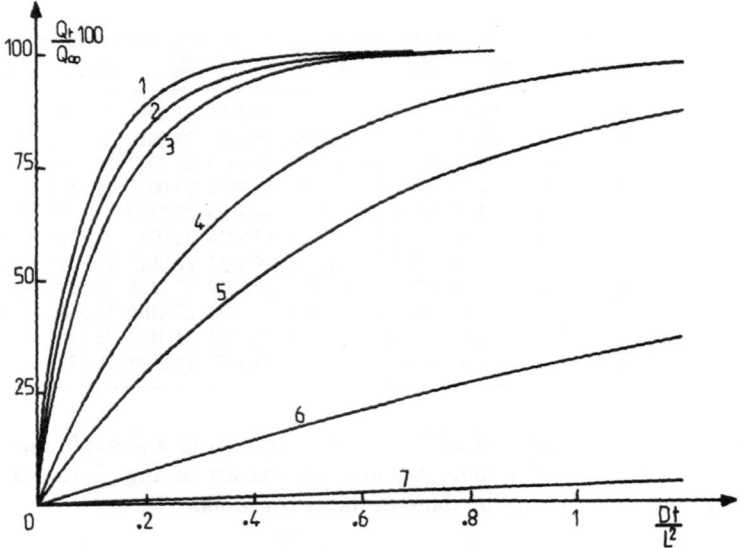

Fig. 2.10. Effect of the ratio S on the kinetics of evaporation, in surrounding atmosphere of infinite volume (L is the thickness of the sheet). 1: $S = 100$; 2: $S = 10$; 3: $S = 5$; 4: $S = 1$; 5: $S = 0.5$; 6: $S = 0.1$; 7: $S = 0.01$.

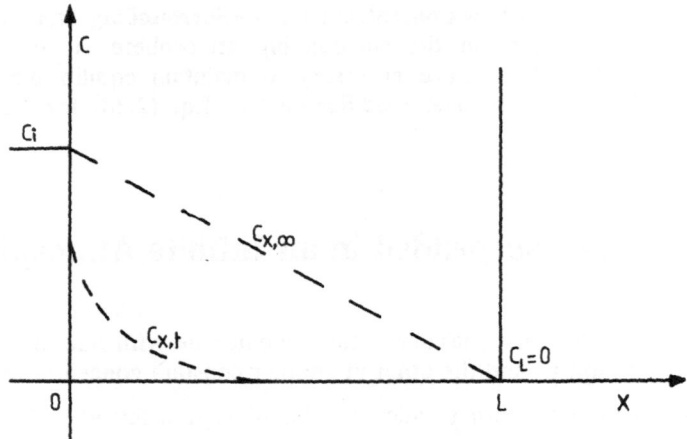

Fig. 2.11. Diffusion through a membrane and evaporation of the substance at an infinite rate.

The equation for diffusion in one dimension through the membrane with constant diffusivity is

$$\frac{\partial C}{\partial t} = D\frac{\partial^2 C}{\partial x^2} \tag{2.1}$$

The initial and boundary conditions are obviously

$$t = 0, \quad 0 < x < L, \quad C = 0 \quad \text{(membrane)} \tag{2.44}$$

$$t > 0, \quad x = 0, \quad C = C_i \quad \text{(surface in contact with a liquid)}$$

$$x = L, \quad C = 0 \quad \text{(evaporating surface)} \tag{2.45}$$

A general solution is considered:

$$C_{x,t} = C_i\left(1 - \frac{x}{L}\right) + \sum_{n=0}^{\infty} (A_n \sin\lambda_n x + B_n \cos\lambda_n x) \exp(-\lambda_n^2 Dt) \quad (2.46)$$

The boundary conditions (2.45) demand that

$$B_n = 0 \quad \lambda_n = \frac{n\pi}{L} \quad (2.47)$$

and the initial conditions become

$$C_{x,t} - C_i\left(1 - \frac{x}{L}\right) = \sum_{n=1}^{\infty} A_n \sin\frac{n\pi x}{L} \exp\left(-\frac{n^2\pi^2}{L^2}Dt\right) \quad (2.48)$$

By multiplying both sides of Eq. (2.48) by $\sin(n\pi x/L)$ and integrating from 0 to L, it is found that

$$C_{x,t} = C_i\left(1 - \frac{x}{L}\right) - \frac{2C_i}{\pi} \sum_{n=1}^{\infty} \frac{1}{n} \sin\frac{n\pi x}{L} \exp\left(-\frac{n^2\pi^2}{L^2}Dt\right) \quad (2.49)$$

The amount of liquid which evaporates after time t per unit area is given by Fick's first law:

$$Q_{L,t} = -D \int_0^t \left(\frac{\partial C}{\partial x}\right)_L dt \quad (2.50)$$

where the gradient of concentration is expressed at the position L of the membrane.

The amount of liquid which has evaporated after time t is then obtained by the series (App. 2.G)

$$Q_{L,t} = \frac{DC_i}{L}t + \frac{2LC_i}{\pi^2} \sum_{n=1}^{\infty} \frac{(-1)^n}{n^2}\left[1 - \exp\left(-\frac{n^2\pi^2 Dt}{L^2}\right)\right] \quad (2.51)$$

For long times, the exponential tends to zero, and Eq. (2.51) reduces to

$$Q_t = \frac{DC_i}{L}\left(t - \frac{L^2}{6D}\right) \quad (2.52)$$

because of the value of the series

$$\sum_{n=1}^{\infty} \frac{(-1)^n}{n^2} = -\frac{\pi^2}{12}$$

The amount of liquid which has evaporated from the membrane as a function of time is shown in Fig. 2.12.

2.3.2 Steady-State Conditions with Infinite Rate of Evaporation

As shown in the previous section, after a given time, a steady state can be considered to be reached, in which the concentration remains constant at all points of the sheet. The diffusion equation in one dimension with constant diffusivity then reduces to

$$\frac{\partial^2 C}{\partial x^2} = 0 \quad (2.53)$$

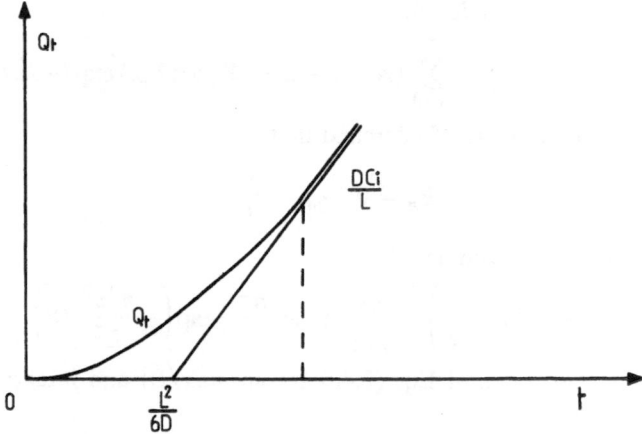

Fig. 2.12. Amount of matter which evaporates from the membrane versus time. Approach to the steady state.

On integrating twice with respect to x, and on introducing the conditions at $x = 0$ and $x = L$,

$$x = 0, \quad C = C_i \qquad \text{(surface in contact with the liquid)}$$
$$x = L, \quad C = 0 \qquad \text{(evaporating surface)} \tag{2.45}$$

we have

$$\frac{C_i - C_{x,t}}{C_i} = \frac{x}{L} \tag{2.54}$$

Of course, Eqs (2.53) and (2.54) show that the gradient of concentration is constant through the sheet. Therefore, the rate of transfer of the liquid is the same across all sections of the membrane and is given by

$$F = \frac{DC_i}{L} \tag{2.55}$$

Sometimes, the surface concentrations C_i and C_0 are not known, but only the vapour or gas pressures P_i and P_0 on the two sides of the membrane. The rate of transfer is then given by

$$F = \text{Per} \frac{P_i - P_0}{L} \tag{2.56}$$

where P_i and P_0 are the inlet and outlet pressure, and Per is the permeability constant.

When the diffusivity is constant, and if the sorption isotherm is linear, Eqs (2.55) and (2.56) are equivalent. The linear isotherm can be written as follows:

$$C = SP \tag{2.57}$$

where C is the concentration in the membrane in equilibrium with the external pressure P and S is the solubility. It is then easy to find with due regard to units that

$$\text{Per} = DS \tag{2.58}$$

2.3.3 Composite Membrane Under Steady-State Conditions with Infinite Rate of Evaporation

A composite membrane composed of sheets of thicknesses L_1, L_2 and L_3, and diffusivities D_1, D_2 and D_3 is shown in Fig. 2.13. Since the rate of transfer is the same across each section within the composite membrane, the total drop in concentration through the whole membrane is the sum of the falls in concentration through the component sheets.

$$F \sum \frac{L_i}{D_i} \qquad (2.59)$$

where L_i/D_i represents the resistance to diffusion of the sheet i.

Two facts are worth noting:

- The resistance to diffusion of the whole membrane is the sum of the resistances of the separate components, by assuming that there are no barriers to diffusion between them.
- The gradient of concentration is inversely proportional to the value of the diffusivity, in each layer. So the higher the diffusivity of a layer, the lower the gradient of vapour through this layer.

2.3.4 Steady-State Conditions With Constant Concentration on One Surface and Finite Rate of Evaporation on the Other

The surface $x = 0$ is maintained at a constant concentration C_i and at $x = L$, the liquid evaporates in the surrounding atmosphere with a finite rate of evaporation. The concentration of liquid on this surface is C_0, and the concentration at equilibrium with the surrounding atmosphere is C_{ext} (Fig. 2.14).

The conditions at the surface $x = L$ are

$$-D\frac{\partial C}{\partial x} = F_0(C_0 - C_{ext}) \qquad (2.60)$$

As the gradient of concentration is constant within the membrane, integration of Eq. (2.60) gives

$$D\left(\frac{C_i - C_0}{L}\right) = F_0(C_0 - C_{ext}) \qquad (2.61)$$

The concentration of liquid on the surface $x = L$ is then

$$C_0 = \frac{C_i D + F_0 L C_{ext}}{D + F_0 L} \qquad (2.62)$$

Then it is found that

$$C_0 - C_i = (C_{ext} - C_i)\frac{F_0 L}{D + F_0 L} \qquad (2.63)$$

and the rate of transfer of liquid through the membrane is

$$Ra = (C_i - C_{ext})\frac{D F_0}{D + F_0 L} \qquad (2.64)$$

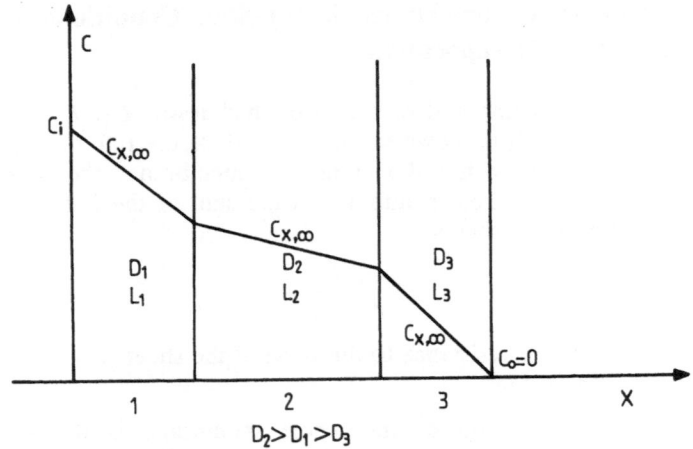

Fig. 2.13. Composite membrane in steady state, with infinite rate of evaporation.

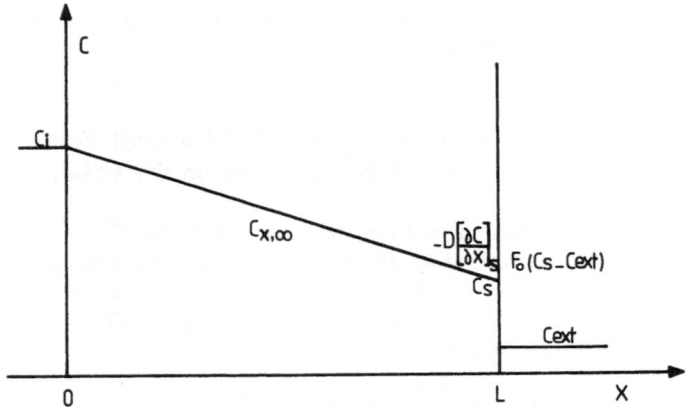

Fig. 2.14. Steady-state transfer of substance through a membrane with constant concentration on one surface and finite rate of evaporation on the other.

2.4   Conclusions

The following conclusions on the diffusion–evaporation process for a plane sheet are worth noting.

2.4.1   Edge Effects

Various cases of one-dimensional diffusion in a solid bounded by two parallel planes have been examined where the analytical solutions obtained apply to sheets so thin that a negligible amount of matter evaporates through the edges. The question arises on how to determine when the effect of the edges can be neglected. Two parameters are of interest: the area of the plane sheet, and its thickness.

The answer can be found from the following two facts:

- The amount of matter transferred after time t is expressed in terms of the square of the thickness through which the diffusion proceeds, by Eq. (2.7):

$$\frac{Q_\infty - Q_t}{Q_\infty} = \frac{8}{\pi^2} \sum \frac{1}{(2n+1)^2} \exp\left(-\frac{(2n+1)^2\pi^2}{L^2}Dt\right) \qquad (2.7)$$

- The rate of evaporation is proportional to the area of the surface from which the liquid has evaporated.

2.4.2 Constant Diffusivity

When the diffusivity is not constant, being concentration dependent for instance, numerical models must be used (Chap. 5).

2.4.3 Effect of the Volume of the Surrounding Atmosphere

The effect of the ratio of the volumes of the surrounding atmosphere and sheet, α, is shown in Sect. 2.1.4 for an infinite rate of evaporation. Of course, Eq. (2.31) expressing the amount of matter leaving the sheet after time t, Q_t, in terms of α

$$\frac{Q_\infty - Q_t}{Q_\infty} = \sum_{n=1}^{\infty} \frac{2\alpha(1+\alpha)}{1 + \alpha + \alpha^2 q_n^2} \exp\left(-\frac{q_n^2}{L^2}Dt\right) \qquad (2.31)$$

can also be used for an atmosphere of very large volume.

When $\alpha \to \infty$, it is obvious that Eq. (2.65)

$$\frac{Q_\infty - Q_t}{Q_\infty} = \sum_{n=1}^{\infty} \frac{2}{q_n^2} \exp\left(-\frac{q_n^2}{L^2}Dt\right) \qquad (2.65)$$

with the q_n given in Table 2.1 for $\alpha \to \infty$ is the same as Eq. (2.7), with the thickness of the sheet $2L$, because in this case

$$q_n = (2n+1)\frac{\pi}{2} \qquad (2.66)$$

2.4.4 Effect of the Rate of Evaporation

Eq. (2.41) established for a finite rate of evaporation in Sect. 2.2

$$\frac{Q_\infty - Q_t}{Q_\infty} = \sum_{n=1}^{\infty} \frac{2S^2}{\beta_n^2(\beta_n^2 + S^2 + S)} \exp\left(-\frac{\beta_n^2}{L^2}Dt\right) \qquad (2.41)$$

becomes, when the rate of evaporation is very high ($F_0 \to \infty$, $S \to \infty$):

$$\frac{Q_\infty - Q_t}{Q_\infty} = \sum_{n=1}^{\infty} \frac{2}{\beta_n^2} \exp\left(-\frac{\beta_n^2}{L^2}Dt\right) \qquad (2.67)$$

As the β_ns given in Table 2.2 for $S = \infty$ are

$$\beta_n = (2n + 1)\frac{\pi}{2} \tag{2.68}$$

it is clear that Eq. (2.7) and Eq. (2.67) are the same when the rate of evaporation is very high, for a sheet of thickness $2L$.

Appendixes. Methods of Solution of Fick's Law When the Diffusivity is Constant

2.A Separation of Variables

A solution of a partial differential equation can be obtained by assuming that the variables are separated. An attempt to find a solution is made by putting

$$C_{x,t} = C_x C_t \tag{2A.1}$$

where C_x and C_t are functions of space x and time t, respectively.
    Substitution in Fick's equation

$$\frac{\partial C}{\partial t} = D\frac{\partial^2 C}{\partial x^2} \tag{2A.2}$$

yields

$$\frac{1}{C_t}\frac{\partial C_t}{\partial t} = \frac{D}{C_x}\frac{\partial^2 C_x}{\partial x^2} \tag{2A.3}$$

The expression on the left-hand side depends on t only, while the right-hand side depends on x only. Both sides must be equal whatever x and t, and they are thus equal to the same constant, which is conveniently taken as $-\lambda^2 D$.
    Therefore, we have two differential equations:

$$\frac{1}{C_t}\frac{\partial C_t}{\partial t} = -\lambda^2 D \tag{2A.4}$$

and

$$\frac{D}{C_x}\frac{\partial^2 C_x}{\partial x^2} = -\lambda^2 D \tag{2A.5}$$

of which solutions are, respectively:

$$C_t = \exp(-\lambda^2 Dt) + \text{Cte} \tag{2A.6}$$

and

$$C_x = A\sin\lambda x + B\cos\lambda x \tag{2A.7}$$

which leads to a solution

$$C_{x,t} = (A\sin\lambda x + B\cos\lambda x)\exp(-\lambda^2 Dt) \tag{2A.8}$$

where A and B are constants of integration.

The most general solution is obtained by summing up Eq. (2A.8):

$$C_{x,t} = \sum_{n=1}^{\infty} (A_n \sin \lambda_n x + B_n \cos \lambda_n x) \exp(-\lambda_n^2 Dt) \tag{2A.9}$$

2.B Diffusion–Evaporation From a Plane Sheet of Thickness L, with Infinite Rate of Evaporation

The diffusing substance is initially uniformly distributed (C_{in}) and the surfaces are kept at zero concentration. The initial and boundary conditions are

$$t = 0, \quad 0 < x < L, \qquad\qquad C = C_{in} \tag{2B.1}$$

$$t > 0, \quad x = 0$$

$$x = L, \qquad\qquad C = 0 \tag{2B.2}$$

The general equation is

$$C_{x,t} = \sum_{n=1}^{\infty} (A_n \sin \lambda_n x + B_n \cos \lambda_n x) \exp(-\lambda_n^2 Dt) \tag{2B.3}$$

where the constants A_n, B_n and λ_n are determined by the boundary and initial conditions of the problem.

The boundary condition $C = 0$ for $x = 0$ necessitates

$$B_n = 0 \tag{2B.4}$$

and the other condition $C = 0$ for $x = L$

$$\lambda_n = \frac{n\pi}{L} \tag{2B.5}$$

The initial conditions (2B.1) become

$$C_{in} = \sum_{1}^{\infty} A_n \sin \frac{n\pi x}{L} \text{ for } 0 < x < L \tag{2B.6}$$

By multiplying both sides of Eq. (2.6) by $\sin(n\pi x/L)$ and integrating from 0 to L,

$$C_{in} \int_0^L \sin \frac{n\pi x}{L} dx = A_p \int_0^L \sin \frac{p\pi x}{L} \sin \frac{n\pi x}{L} dx + A_n \int_0^L \sin^2 \frac{n\pi x}{L} dx \tag{2B.7}$$

By using the relationship

$$2 \sin \frac{p\pi x}{L} \sin \frac{n\pi x}{L} = \cos \frac{(n-p)\pi x}{L} - \cos \frac{(n+p)\pi x}{L} \tag{2B.8}$$

it is obvious that

$$\int_0^L \sin\frac{p\pi x}{L}\sin\frac{n\pi x}{L}\,dx = \begin{cases} 0 & \text{when } n \neq p \\ L/2 & \text{when } n = p \end{cases} \tag{2B.9}$$

Eq. (2B.7) becomes

$$\left[-\frac{C_{in}L}{n\pi}\cos\frac{n\pi x}{L}\right]_0^L = \frac{C_{in}L}{n\pi}[1-(-1)^n] = \frac{A_n L}{2} \tag{2B.10}$$

All the terms for which n is even are equal to zero, and n must be odd (written $2n + 1$)

$$A_n = \frac{4C_{in}}{(2n+1)\pi} \tag{2B.11}$$

The final solution is therefore

$$C_{x,t} = \frac{4C_{in}}{\pi}\sum_{n=0}^{\infty}\frac{1}{2n+1}\sin\frac{(2n+1)\pi x}{L}\exp\left(-\frac{(2n+1)^2\pi^2}{L^2}Dt\right) \tag{2B.12}$$

The amount of liquid which has left one side of the sheet of area A at time t is

$$Q_{0,t} = Q_{L,t} = A\int_0^t D\left|\frac{\partial C}{\partial x}\right|dt \tag{2B.13}$$

$\partial C/\partial x$ being the gradient of concentration at $x = 0$ or $x = L$.

For $x = 0$, we have

$$\left|\frac{\partial C}{\partial x}\right|_{x=0} = \frac{4C_{in}}{L}\sum_{n=0}^{\infty}\exp\left(-\frac{(2n+1)^2\pi^2}{L^2}Dt\right) \tag{2B.14}$$

and Eq. (2B.13) becomes

$$Q_{0,t} = \frac{4C_{in}LA}{\pi^2}\sum_{n=0}^{\infty}\frac{1}{(2n+1)^2}\left\{1-\exp\left(-\frac{(2n+1)^2\pi^2}{L^2}Dt\right)\right\} \tag{2B.15}$$

As the series in Eq. (2B.15) is

$$\sum_{n=0}^{\infty}\frac{1}{(2n+1)^2} = \frac{\pi^2}{8} \tag{2B.16}$$

Eq. (2B.15) can be rewritten:

$$Q_{0,t} = \frac{C_{in}LA}{2}\sum_{n=0}^{\infty}\left\{1-\frac{8}{\pi^2}\frac{1}{(2n+1)^2}\exp\left(-\frac{(2n+1)^2\pi^2}{L^2}Dt\right)\right\} \tag{2B.17}$$

The total amount of the diffusing substance which has left the sheet at time t, is of course

$$Q_t = 2Q_{0,t} = 2Q_{L,t} \tag{2B.18}$$

For infinite time, all the diffusing substance has left the sheet:

$$Q_\infty = C_{in}LA \tag{2B.19}$$

The amount of diffusing substance leaving the sheet after time t, can thus be written in terms of the total amount of substance Q_∞:

$$\frac{Q_\infty - Q_t}{Q_\infty} = \frac{8}{\pi^2}\sum_{n=0}^{\infty}\frac{1}{(2n+1)^2}\exp\left(-\frac{(2n+1)^2\pi^2}{L^2}Dt\right) \tag{2B.20}$$

2.C Half-Life of Desorption Process

When the rate of evaporation is infinite, the volume of surrounding atmosphere very high and the diffusivity is constant, the amount of matter which evaporates after time t is expressed in terms of the corresponding amount after infinite time by the well-known series

$$\frac{Q_\infty - Q_t}{Q_\infty} = \frac{8}{\pi^2} \sum_{n=0}^{\infty} \frac{1}{(2n+1)^2} \exp\left(-\frac{(2n+1)^2\pi^2}{L^2}Dt\right) \tag{2C.1}$$

For long times, the series can be reduced to the first term:

$$\frac{Q_\infty - Q_t}{Q_\infty} = \frac{8}{\pi^2} \exp\left(-\frac{\pi^2}{L^2}Dt\right) \tag{2C.2}$$

The half-life of the desorption process is obtained when the ratio is

$$\frac{Q_t}{Q_\infty} = \tfrac{1}{2} \tag{2C.3}$$

In this case, Eq. (2C.2) becomes

$$\ln\tfrac{1}{2} = \ln\frac{8}{\pi^2} - \pi^2\left(\frac{Dt}{L^2}\right) \quad \text{for} \quad \frac{Q_t}{Q_\infty} = \tfrac{1}{2} \tag{2C.4}$$

and the diffusivity can be easily determined from the relation

$$\left(D\frac{t}{L^2}\right) = 0.0489 \quad \text{for} \quad \frac{Q_t}{Q_\infty} = \tfrac{1}{2} \tag{2C.5}$$

2.D Reflection and Superposition

The equation for diffusion in one dimension through a unit cross-section when D is constant

$$\frac{\partial C}{\partial t} = D\frac{\partial^2 C}{\partial x^2} \tag{2D.1}$$

has a solution,

$$C = \frac{B}{\sqrt{t}} \exp\left(-\frac{x^2}{4Dt}\right) \tag{2D.2}$$

Eq. (2D.2) is symmetrical with respect to $x = 0$; it tends to zero as $x \to \pm\infty$ for $t > 0$, and it tends to zero for $t = 0$ except at $x = 0$ where it is infinite.

Thin Plane Source

The initial boundary conditions are

$$t = 0, \quad x = 0, \quad Q \tag{2D.3}$$

$$t > 0, \qquad \int_{-\infty}^{+\infty} C\,dx = Q \qquad\qquad (2D.4)$$

Q being the total amount of diffusing substance, and C its concentration.
By combining Eqs (2D.2) and (2D.4), the amount of the substance is

$$Q = \frac{B}{\sqrt{t}} \int_{-\infty}^{+\infty} \exp\left(-\frac{x^2}{4Dt}\right) dx \qquad\qquad (2D.5)$$

As

$$\int_{-\infty}^{+\infty} \exp(-v^2)\,dv = \sqrt{\pi}$$

Eq. (2D.5) becomes

$$Q = 2B\sqrt{\pi D} \qquad\qquad (2D.6)$$

On substituting for B from Eq. (2D.6) in Eq. (2D.2):

$$C = \frac{Q}{2\sqrt{\pi Dt}} \exp\left(-\frac{x^2}{4Dt}\right) \qquad\qquad (2D.7)$$

Plane Source of Finite Thickness

The initial conditions are

$$t = 0, \quad -L < x < L, \quad C = C_{in} \qquad \text{(plane source)}$$
$$x < -L,$$
$$x > L \qquad\qquad C = 0 \qquad\qquad (2D.8)$$

The solution to such a problem is obtained by considering the extended distribution to be composed of an infinite number of line sources and by superposing the corresponding number of the elementary solution.

Considering the diffusing substance in an element of width $d\varepsilon$ to be a line source of volume $C_{in}d\varepsilon$ (Fig. 2D.1), from Eq. (2D.7) the concentration at point p, a distance ε from the element $d\varepsilon$ is at time t

$$C_{p,t} = \frac{C_{in}d\varepsilon}{2\sqrt{\pi Dt}} \exp\left(-\frac{\varepsilon^2}{4Dt}\right) \qquad\qquad (2D.9)$$

The complete solution due to the initial distribution described in Eq. (2D.8) is given by summing up over successive elements $d\varepsilon$:

$$C_{x,t} = \frac{C_{in}}{2\sqrt{\pi Dt}} \int_{x-L}^{x+L} \exp\left(-\frac{\varepsilon^2}{4Dt}\right) d\varepsilon \qquad\qquad (2D.10)$$

or

$$C_{x,t} = \frac{C_{in}}{\sqrt{\pi}} \int_{(x-L)/2\sqrt{Dt}}^{(x+L)/2\sqrt{Dt}} \exp(-v^2)\,dv \qquad\qquad (2D.10')$$

where

$$v^2 = \frac{\varepsilon^2}{4Dt} \qquad\qquad (2D.11)$$

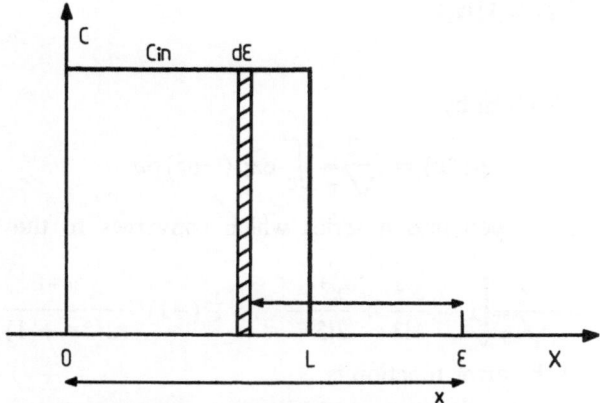

Fig. 2D.1. Extended initial distribution of diffusing substance.

Eq. (2D.10′) can be rewritten

$$C_{x,t} = \frac{C_{in}}{\sqrt{\pi}} \int_0^{(x+L)/2\sqrt{Dt}} \exp(-v^2)\,dv + \frac{C_{in}}{\sqrt{\pi}} \int_{(x-L)/2\sqrt{Dt}}^0 \exp(-v^2)\,dv$$

$$(2D.10'')$$

The error function, a standard mathematical function, is

$$\text{erf}(x) = \frac{2}{\sqrt{\pi}} \int_0^x \exp(-v^2)\,dv \qquad (2D.12)$$

with the following properties:

$$\text{erf}(x) = -\text{erf}(-x) \qquad (2D.13)$$

$$\text{erf}(0) = 0 \qquad (2D.13')$$

$$\text{erf}(\infty) = 1 \qquad (2D.13'')$$

$$1 - \text{erf}(x) = \text{erfc}(x) \qquad (2D.13''')$$

where erfc is the error function complement.

Eq. (2D.10″) becomes by using erfc:

$$\frac{C_{in} - C_{x,t}}{C_{in}} = \sum_{n=0}^{\infty} (-1)^n \text{erfc}\frac{(2n+1)L - x}{2\sqrt{Dt}} + \sum_{n=0}^{\infty} (-1)^n \text{erfc}\frac{(2n+1)L + x}{2\sqrt{Dt}}$$

$$(2D.14)$$

The amount of matter which has left the sheet at time t, Q_t, is then expressed in terms of the corresponding quantity after infinite time Q_∞ by using the following equations:

$$Q_t = 2A \int_0^t D\left|\frac{\partial C}{\partial x}\right| dt \qquad (2D.15)$$

and

$$\frac{Q_t}{Q_\infty} = \frac{2}{L}\sqrt{Dt}\left\{\frac{1}{\sqrt{\pi}} + 2\sum_{n=1}^{\infty} (-1)^n \text{ierfc}\frac{nL}{\sqrt{Dt}}\right\} \qquad (2D.16)$$

with the thickness of the sheet $2L$.

2.E   Error Function

The error function is given by

$$\operatorname{erf}(x) = \frac{2}{\sqrt{\pi}} \int_0^x \exp(-v^2)\,dv \tag{2E.1}$$

This function is developed into a series which converges to the limit 1 when $x \to \infty$.

$$\operatorname{erfc}(x) = \frac{2}{\sqrt{\pi}} \left[x - \frac{x^3}{1!3} + \frac{x^5}{2!5} - \cdots + (-1)^n \frac{x^{2n+1}}{n!(2n+1)} \cdots \right] \tag{2E.2}$$

The derivative of the error function is

$$\frac{d\{\operatorname{erf}(x)\}}{dx} = \frac{2}{\sqrt{\pi}} \exp(-x^2) \tag{2E.3}$$

The error function complement is

$$\operatorname{erfc}(x) = 1 - \operatorname{erf}(x) \tag{2E.4}$$

The integral of the error function complement is

$$\operatorname{ierfc}(x) = \int_x^\infty \operatorname{erfc}(v)\,dv \tag{2E.5}$$

and after integration it becomes:

$$\operatorname{ierfc}(x) = \frac{1}{\sqrt{\pi}} \exp(-x^2) - x\operatorname{erfc}(x) \tag{2E.6}$$

Tables of the error function, error function complement, and their integrals are available in books devoted to mathematics.

2.F   Layer of Diffusing Substance on a Permeable Sheet, With Infinite Rate of Evaporation

The solution for diffusion in one dimension with constant diffusivity and infinite rate of evaporation

$$\frac{\partial C}{\partial t} = D \frac{\partial^2 C}{\partial x^2} \tag{2F.1}$$

is in the general form

$$C_{x,t} = \sum_{n=1}^\infty (A_n \sin \lambda_n x + B_n \cos \lambda_n x) \exp(-\lambda_n^2 D t) \tag{2F.2}$$

Because of the boundary conditions, we have

$$x = 0, \qquad C_{0,t} = 0, \qquad\qquad B_n = 0$$

$$x = L, \qquad C_{L,t} = 0, \qquad \sin \lambda_n L = 0, \qquad \lambda_n = \frac{n\pi}{L} \tag{2F.3}$$

The general solution becomes

$$C_{x,t} = \sum_{n=1}^{\infty} A_n \sin\left(\frac{n\pi x}{L}\right) \exp\left(-\frac{n^2\pi^2}{L^2}Dt\right)$$

(2F.4)

With the initial conditions, this solution is written as follows:

$$C_{x,0} = A_1 \sin\frac{\pi x}{L} + A_2 \sin\frac{2\pi x}{L} \dots A_n \sin\frac{n\pi x}{L} \dots$$

(2F.5)

By multiplying both sides by $\sin(n\pi x/L)dx$ and integrating from 0 to L,

$$\int_0^L C_{x,0}\sin\frac{n\pi x}{L}dx = \int_0^L A_1 \sin\frac{\pi x}{L}\sin\frac{n\pi x}{L}dx \dots + \int_0^L A_n \sin^2\frac{n\pi x}{L}dx \dots$$

(2F.6)

By considering the initial conditions, it is obvious that the left-hand side of this equation is also

$$\int_0^L C_{x,0}\sin\frac{n\pi x}{L}dx = \int_0^h C_{\text{in}}\sin\frac{\pi\pi x}{L}dx + \int_h^L 0\sin\frac{n\pi x}{L}dx$$

(2F.7)

After calculation, A_n is found as

$$A_n = \frac{4C_{\text{in}}}{n\pi}\sin^2\left(\frac{n\pi h}{2L}\right)$$

(2F.8)

and the concentration is thus given by the relation

$$C_{x,t} = \frac{4C_{\text{in}}}{\pi}\sum_{n=1}^{\infty}\frac{1}{n}\sin^2\left(\frac{n\pi h}{2L}\right)\sin\left(\frac{n\pi x}{L}\right)\exp\left(-\frac{n^2\pi^2}{L^2}Dt\right)$$

(2F.9)

2.G Membrane with Constant Concentration on Each Surface and Infinite Rate of Evaporation

In the case of flow through a membrane, one surface $x = 0$ of the membrane is kept at a constant concentration C_i and the other $x = L$ at 0, and the membrane is initially at uniform zero concentration (Fig. 2G.1).

The initial and boundary conditions are

$t = 0,$	$0 < x < L,$	$C = 0$	(membrane)	(2G.1)
$t > 0,$	$x = 0,$	$C = C_i$	(liquid)	
	$x = L,$	$C = 0$	(evaporation)	(2G.2)

The equation for diffusion in one dimension with a constant diffusivity

$$\frac{\partial C}{\partial t} = D\frac{\partial^2 C}{\partial x^2}$$

(2G.3)

has the general solution

$$C_{x,t} = \sum_{n=0}^{\infty}(A_n \sin\lambda_n x + B_n \cos\lambda_n x)\exp(-\lambda_n^2 Dt)$$

(2G.4)

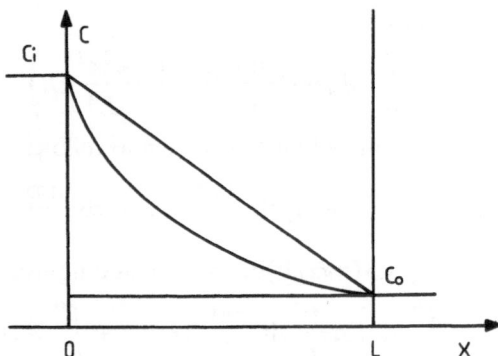

Fig. 2G.1. Membrane with constant concentration of diffusing substance on each surface initially at uniform concentration C_0.

which can be written as follows:

$$C_{x,t} = C_i\left(1 - \frac{x}{L}\right) + \sum_{n=0}^{\infty} (A_n \sin \lambda_n x + B_n \cos \lambda_n x) \exp(-\lambda_n^2 Dt) \quad (2G.5)$$

From the boundary conditions, it is obvious that

$$x = 0, \qquad C_{0,t} = C_i, \qquad B_n = 0 \quad\quad\quad\quad\quad (2G.6)$$

$$x = L, \qquad C_{L,t} = 0, \qquad \sin \lambda_n L = 0, \quad \lambda_n = \frac{n\pi}{L} \quad (2G.7)$$

The general solution then becomes, by using the initial conditions

$$C_{x,t} = C_i\left(1 - \frac{x}{L}\right) + \sum_{n=0}^{\infty} A_n \sin \frac{n\pi x}{L} \exp\left(-\frac{n^2\pi^2}{L^2}Dt\right) \quad (2G.8)$$

The general term A_n is obtained by multiplying both sides of the equation by $\sin(n\pi x/L)$ and integrating from 0 to L:

$$A_n = \frac{2C_i}{n\pi} \quad\quad\quad\quad\quad (2G.9)$$

The concentration in the membrane during the transient conditions is given by the series

$$C_{x,t} = C_i\left(1 - \frac{x}{L}\right) - \frac{2C_i}{\pi} \sum_{n=1}^{\infty} \frac{1}{n} \sin \frac{n\pi x}{L} \exp\left(-\frac{n^2\pi^2}{L^2}Dt\right) \quad (2G.10)$$

The amount of matter which has left the membrane after time t is per unit area

$$Q_t = -\int_0^t D\left(\frac{\partial C}{\partial x}\right)_{x=L} dt \quad\quad\quad\quad (2G.11)$$

and after calculation

$$Q_t = \frac{DC_i t}{L} + \frac{2C_i L}{\pi^2} \sum_{n=1}^{\infty} \frac{(-1)^n}{n^2}\left[1 - \exp\left(-\frac{n^2\pi^2}{L^2}Dt\right)\right] \quad (2G.12)$$

For long time, the exponential tends to zero and Q_t becomes

$$Q_t = \frac{DC_i}{L}\left[t + \frac{2L^2}{\pi^2 D} \sum_{n=1}^{\infty} \frac{(-1)^n}{n^2}\right] \quad\quad\quad (2G.13)$$

By considering that

$$\sum_{n=1}^{\infty} \frac{(-1)^n}{n^2} = -\frac{\pi^2}{12} \qquad (2G.14)$$

the amount of matter leaving the membrane per unit area after a long time is

$$Q_t = \frac{DC_i}{L}\left(t - \frac{L^2}{6D}\right) \qquad (2G.15)$$

The above equations show that after a time for which the exponential in Eq. (2G.12) is very small, the rate of matter evaporated from the membrane is constant, the matter being transported through the membrane under steady-state conditions.

Symbols

A	Area of each surface of the plane sheet
$2a$	Dimension characterising the thickness of air
$C_{x,t}$	Concentration of liquid at position x and t
C_{in}	Initial concentration in the solid
C_i, C_0	Inlet and outlet concentration in a membrane
C_{ext}	Concentration on the surface necessary to maintain equilibrium with the surrounding atmosphere
D	Diffusivity ($cm^2\ s^{-1}$)
F_0	Rate of evaporation of the pure liquid ($cm\ s^{-1}$)
h	Thickness of the layer of diffusing substance (Sect. 2.1.2)
$2h$	Thickness of the layer of diffusing substance in the sandwich (Sect. 2.1.3)
K	Partition coefficient between the sheet and air
L	Thickness of the plane sheet
n	Integer
P_i, P_0	Inlet and outlet pressure, respectively
Per	Permeability of the membrane (volume of gas per second through 1 cm^2 of the surface of a membrane 1 cm thick when $P_i - P_0 = 10$ mmHg)
Q'_t	Amount of substance remaining in the sheet at time t
Q_t, Q_∞	Amount of substance evaporated after time t and after infinite time, respectively
q_n	Non-zero positive roots of $\tan q_n = -\alpha q_n$
R	Constant for ideal gas
Ra	Rate of transfer of substance per unit area
S	Dimensionless number in Eq. (2.40)
t	time
T	Temperature (K)
x	Abscissa
α	Ratio of a and L
β_n	Positive roots of $\beta \tan \beta = S$
λ_n	Constant used in series

Reference

1. Crank J. The mathematics of diffusion, 2nd edn. Clarendon Press, Oxford, 1975, pp. 44–68

Chapter 3

Cylinder

Various cases of diffusion with constant diffusivity in cylinders are considered, when:

- The cylinder is sufficiently long to be considered infinite, so that the diffusion is essentially radial through the circular cross-section.
- The cylinder is of finite length, and longitudinal and radial diffusion take place simultaneously, with the same diffusivity.
- The rate of evaporation is either infinite, so that the concentration of the substance on the evaporating surface falls to zero as soon as the process starts, or finite.
- The transfer is carried out under transient conditions in the case of a solid cylinder.
- The transfer under stationary conditions is studied in the case of a tube where these conditions can be attained.

3.1 Solid Cylinder of Infinite Length, Non-steady State with Constant Diffusivity

Three cases are of interest for the solid cylinder of infinite length:

- When the volume of the surrounding atmosphere is so great that it can be considered infinite, and the rate of evaporation is sufficiently high to be considered infinite.
- When the surrounding atmosphere is of finite volume, and the rate of evaporation can be considered infinite.
- When the volume of the surrounding atmosphere can be considered as infinite, and the rate of evaporation has a finite value.

3.1.1   Infinite Atmosphere with Infinite Rate of Evaporation

The circular cross-section of the cylinder only is considered and the diffusion is radial with a constant diffusivity. The initial concentration of liquid in the cylinder is uniform, and the rate of evaporation on the solid surface is so high that the concentration of liquid on the surface falls to zero as soon as the process starts.

The equation of diffusion with constant diffusivity is

$$\frac{\partial C}{\partial t} = \frac{D}{r}\frac{\partial}{\partial r}\left(r\frac{\partial C}{\partial r}\right) \tag{3.1}$$

and the initial and boundary conditions can be written as

$$t = 0, \quad 0 < r < R, \quad C = C_{in} \quad \text{(cylinder)} \tag{3.2}$$

$$t > 0, \quad r = R, \quad C = C_{ext} \quad \text{(surface)} \tag{3.3}$$

The solution is given [1, 2] (App. 3.A) by

$$\frac{C_{r,t} - C_{ext}}{C_{in} - C_{ext}} = \frac{2}{R}\sum_{n=1}^{\infty}\frac{J_0(r\alpha_n)}{\alpha_n J_1(R\alpha_n)}\exp(-\alpha_n^2 Dt) \tag{3.4}$$

where $J_0(r\alpha_n)$ is the zero-order Bessel function of the first kind and $J_1(R\alpha_n)$ is the first-order Bessel function, while the α_ns are roots of

$$J_0(R\alpha_n) = 0 \tag{3.5}$$

The roots $R\alpha_n$ of Eq. (3.5) are shown in Table 3.1. These values can be approximated by

$$R\alpha_n = \pi(n - 0.25) \tag{3.5'}$$

The quantity of diffusing substance which has evaporated in time t, Q_t, is then expressed in terms of the corresponding quantity after infinite time Q_∞:

$$\frac{Q_\infty - Q_t}{Q_\infty} = 4\sum_{n=1}^{\infty}\frac{1}{(R\alpha_n)^2}\exp(-\alpha_n^2 Dt) \tag{3.6}$$

Another equation is of interest for short times:

$$\frac{Q_t}{Q_\infty} = \frac{4}{\sqrt{\pi}}\left(\frac{Dt}{R^2}\right)^{0.5} - \frac{Dt}{R^2} - \frac{1}{3\sqrt{\pi}}\left(\frac{Dt}{R^2}\right)^{1.5} \tag{3.7}$$

which can be reduced to a simple relation for very short times:

$$\frac{Q_t}{Q_\infty} = \frac{4}{R}\sqrt{\frac{Dt}{\pi}} \tag{3.8}$$

expressing the square root time relationship of the amount of liquid which evaporates.

The value of diffusivity can be easily obtained from the slope of the straight line obtained by plotting the amount of liquid which evaporates as a function of the square root of time, as shown in the above equation.

The value of diffusivity can also be determined by iteration; for long times when $Q_t/Q_\infty > 0.7$, the first term in the series in Eq. (3.6) dominates, and expressed in logarithmic form becomes

$$\ln\frac{Q_\infty - Q_t}{Q_\infty} = -\alpha_1^2 Dt + \ln\frac{4}{(R\alpha_1)^2} \tag{3.6'}$$

Table 3.1. Roots of $J_0(R\alpha_n) = 0$

α_1	α_2	α_3	α_4	α_5	α_6
2.4048	5.5201	8.6537	11.7915	14.9309	18.0711
α_7	α_8	α_9	α_{10}	α_{11}	α_{12}
21.2125	24.3540	27.4957	30.6375	33.7790	36.9205

The slope of the straight line drawn by plotting the left-hand term as a function of time gives the diffusivity if the radius is known.

The kinetics of evaporation from cylinders of infinite length with infinite rate of evaporation are shown in Fig. 3.1, when the volume of atmosphere is infinite. A general curve is obtained by plotting the ratio Q_t/Q_∞ as a function of the dimensionless number Dt/R^2.

3.1.2   Finite Atmosphere with Infinite Rate of Evaporation

The problem is about the same as that shown for the plane sheet immersed in an atmosphere of finite volume (Sect. 2.1.4), when the atmosphere is strongly stirred, so that the concentration is uniform. The atmosphere is initially free of vapour, and the concentration of liquid in the cylinder is initially uniform. The rate of evaporation of the liquid is infinite, so that the concentration of liquid on the cylinder surface falls to zero as soon as the process starts.

The initial and boundary conditions are

$$t = 0, \quad r < R, \qquad\qquad C = C_{\text{in}} \quad \text{(cylinder)} \qquad\qquad (3.9)$$

$$a > r > R, \qquad C = 0 \quad \text{(atmosphere)}$$

$$t > 0 \quad a\,\frac{\partial C}{\partial t} = -D\,\frac{\partial C}{\partial r} \qquad\qquad (3.10)$$

The total amount of vapour in the atmosphere after time t, Q_t, is expressed as a fraction of the corresponding amount Q_∞ after infinite time by the relation [1]:

$$\frac{Q_\infty - Q_t}{Q_\infty} = \sum_{n=1}^{\infty} \frac{4\alpha(1 + \alpha)}{4 + 4\alpha + \alpha^2 q_n^2} \exp\left(-\frac{q_n^2}{R^2}Dt\right) \qquad\qquad (3.11)$$

where the q_n s are the positive, non-zero roots of

$$\alpha q_n J_0(q_n) + 2J_1(q_n) = 0 \qquad\qquad (3.12)$$

and α is the ratio of the volumes of atmosphere and cylinder.

When the external atmosphere is cylindrical, with a radius a, the value of α is obviously given by

$$\alpha = \left(\frac{a}{R}\right)^2 \qquad\qquad (3.13)$$

If there is a partition factor K between the evaporating substance in equilibrium in the cylinder and in the atmosphere, Eq. (3.13) becomes

$$\alpha = \frac{1}{K}\left(\frac{a}{R}\right)^2 \qquad\qquad (3.13')$$

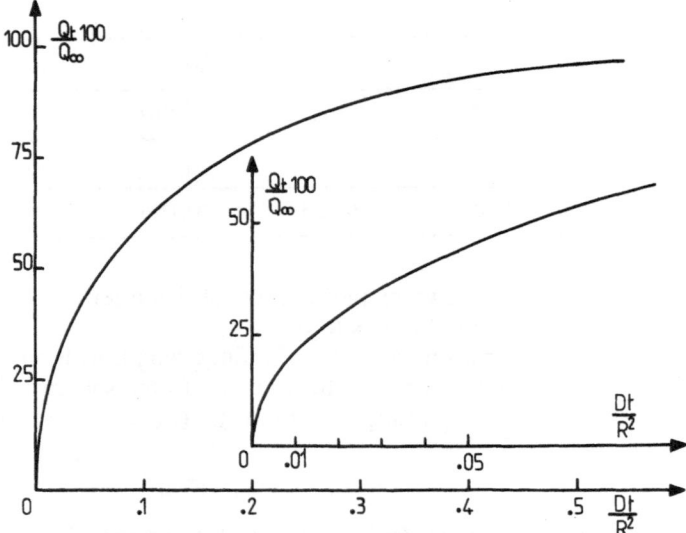

Fig. 3.1. Kinetics of evaporation from cylinders of infinite length with infinite rate of evaporation in a surrounding atmosphere of infinite volume.

The dimensionless parameter α is expressed in terms of the final fractional uptake of evaporating substance by the atmosphere, by the equation

$$\frac{Q_\infty}{\pi R^2 C_{in}} = \frac{\alpha}{1 + \alpha} \qquad (3.14)$$

The concentration of diffusing substance $C_{r,t}$ within the cylinder is given by

$$\frac{C_\infty - C_{r,t}}{C_{in} - C_\infty} = \sum_{n=1}^{\infty} \frac{4(\alpha + 1)}{(4 + 4\alpha + \alpha^2 q_n^2)} \frac{J_0\left(\frac{q_n r}{R}\right)}{J_0(q_n)} \exp\left(-\frac{q_n^2}{R^2} Dt\right) \qquad (3.15)$$

where C_{in} and C_∞ are the uniform cencentration in the cylinder, at the beginning and at the end of the process of drying, respectively.

The values of q_n are given in Table 3.2 for several values of the ratio α of the volumes of the surrounding atmosphere and cylinder.

The value of diffusivity can be determined from the kinetics under these conditions, by using long-time experiments.

When $Q_t/Q_\infty > 0.75$, the series in Eq. (3.11) reduces to the first term, and expressed in logarithmic form becomes

$$\ln \frac{Q_\infty - Q_t}{Q_\infty} = \ln \frac{4\alpha(1 + \alpha)}{4 + 4\alpha + \alpha^2 q_1^2} - \frac{q_1^2}{R^2} Dt \qquad (3.11')$$

The diffusivity can be easily calculated from the value of the slope of the straight line drawn by plotting the left-hand term as a function of time.

The kinetics of evaporation with a finite volume of the surrounding atmosphere are shown in Fig. 3.2, with various values of the ratio of the volume α. General curves are obtained by plotting the ratio Q_t/Q_∞ as a function of the dimensionless number Dt/R^2.

Table 3.2.   Roots of $\alpha q_n J_0(q_n) + 2J_1(q_n) = 0$

Fractional uptake	α	q_1	q_2	q_3	q_4	q_5	q_6
0	∞	2.4048	5.5201	8.6537	11.7915	14.9309	18.0711
0.1	9.0000	2.4922	5.5599	8.6793	11.8103	14.9458	18.0833
0.2	4.0000	2.5888	5.6083	8.7109	11.8337	14.9643	18.0986
0.3	2.3333	2.6962	5.6682	8.7508	11.8631	14.9879	18.1183
0.4	1.5000	2.8159	5.7438	8.8028	11.9026	15.0192	18.1443
0.5	1.0000	2.9496	5.8411	8.8727	11.9561	15.0623	18.1803
0.6	0.6667	3.0989	5.9692	8.9709	12.0334	15.1255	18.2334
0.7	0.4286	3.2645	6.1407	9.1156	12.1529	15.2255	18.3188
0.8	0.2500	3.4455	6.3710	9.3397	12.3543	15.4031	18.4754
0.9	0.1111	3.6374	6.6694	9.6907	12.7210	15.7646	18.8215
1.0	0	3.8317	7.0156	10.173	13.3237	16.4706	19.6159

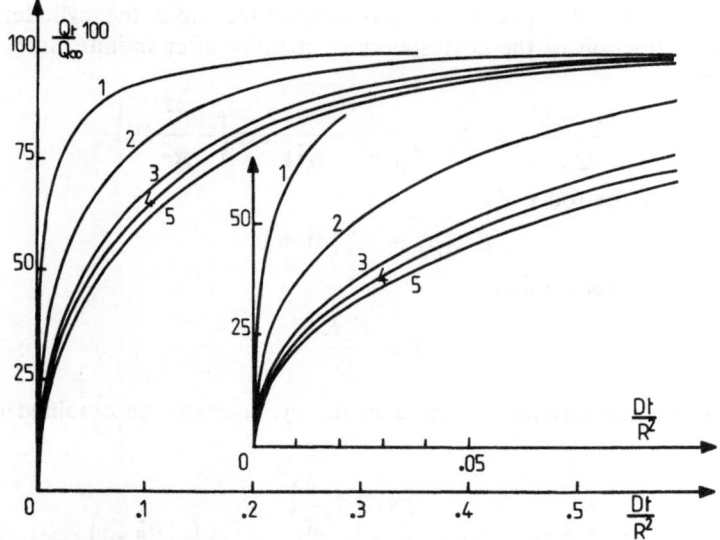

Fig. 3.2. Kinetics of evaporation from cylinders of infinite length in atmosphere of finite volume, with infinite rate of evaporation. Effect of the ratio of the volume of cylinder and surrounding atmosphere α.
1: $\alpha = 0.25$; 2: $\alpha = 1$; 3: $\alpha = 4$; 4: $\alpha = 9$; 5: $\alpha = \infty$.

3.1.3   Infinite Atmosphere with Finite Rate of Evaporation

The cylinder is initially at the uniform concentration C_{in}, and the diffusivity is constant. The initial and boundary conditions are

$$t = 0, \quad r < R, \quad C = C_{in} \qquad \text{(cylinder)} \qquad (3.16)$$

$$t > 0, \quad -D\frac{\partial C}{\partial r} = F_0(C_s - C_{ext}) \qquad \text{(surface)} \qquad (3.17)$$

where C_s is the concentration of liquid on the surface, C_{ext} the concentration on the surface which is at equilibrium with the surrounding atmosphere and F_0 the rate of evaporation of the pure liquid under the same conditions (cm s^{-1}).

Table 3.3. Roots of $\beta J_1(\beta) - S J_0(\beta) = 0$

S	β_1	β_2	β_3	β_4	β_5	β_6
0	0	3.8137	7.0156	10.1735	13.3237	16.4706
0.010	0.1412	3.8343	7.0170	10.1745	13.3244	16.4712
0.100	0.4417	3.8577	7.0298	10.1833	13.3312	16.4767
0.200	0.6170	3.8835	7.0440	10.1931	13.3387	16.4828
0.500	0.9408	3.9594	7.0864	10.2225	13.3611	16.5010
1.000	1.2558	4.0795	7.1558	10.2710	13.3984	16.5312
2.000	1.5994	4.2910	7.2884	10.3658	13.4719	16.5910
5.000	1.9898	4.7131	7.6177	10.6223	13.6786	16.7630
10.00	2.1795	5.0332	7.9569	10.9363	13.9580	17.0099
100.0	2.3809	5.4652	8.5678	11.6747	14.7834	17.8931
∞	2.4048	5.5201	8.6537	11.7915	14.9309	18.0711

The total amount of liquid which has evaporated from the cylinder, Q_t, is expressed as a fraction of the corresponding quantity after infinite time, Q_∞, by the equation [1]

$$\frac{Q_\infty - Q_t}{Q_\infty} = \sum_{n=1}^{\infty} \frac{4S^2}{\beta_n^2(\beta_n^2 + S^2)} \exp\left(-\frac{\beta_n^2}{R^2} Dt\right) \tag{3.18}$$

where the β_ns are the roots of

$$\beta J_1(\beta) - S J_0(\beta) = 0 \tag{3.19}$$

with the dimensionless number S

$$S = \frac{F_0 R}{D} \tag{3.20}$$

The profiles of concentration of liquid in the cylinder can be obtained by using the equation

$$\frac{C_{ext} - C_{r,t}}{C_{ext} - C_{in}} = \sum_{n=1}^{\infty} \frac{2S J_0\left(\beta_n \frac{r}{R}\right)}{(\beta_n^2 + S^2) J_0(\beta_n)} \exp\left(-\frac{\beta_n^2}{R^2} Dt\right) \tag{3.21}$$

Roots of Eq. (3.19) are shown in Table 3.3 for several values of the dimensionless number S characterising the volatility of the liquid.

It is difficult to obtain the value of diffusivity when the amount of liquid which evaporates is low. But when $Q_t/Q_\infty > 0.75$, Eq. (3.18) is simplified as the first term of the series dominates, and expressed in logarithmic form becomes

$$\ln \frac{Q_\infty - Q_t}{Q_\infty} = -\frac{\beta_1^2}{R^2} Dt + \ln \frac{4S^2}{\beta_1^2(\beta_1^2 + S^2)} \tag{3.18'}$$

By plotting the left-hand term as a function of time, the value of the slope can yield the diffusivity by iteration in spite of the fact that β_1 also depends on the value of diffusivity through Eq. (3.20).

The kinetics of evaporation with a finite rate of evaporation are shown in Fig. 3.3 for various values of the parameter S. General curves are obtained by plotting the ratio Q_t/Q_∞ as a function of the dimensionless number Dt/R^2.

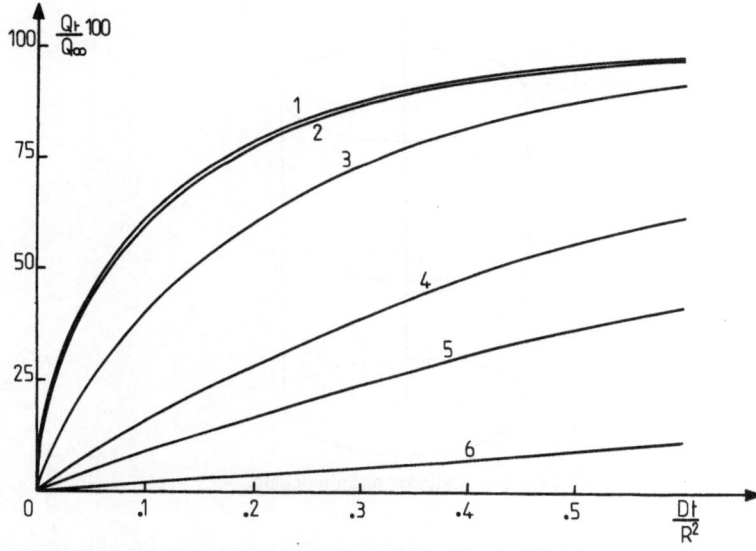

Fig. 3.3. Kinetics of evaporation from cylinders of infinite length in atmosphere of infinite volume, with finite rate of evaporation. Effect of the rate of evaporation S. 1: $S = 9000$; 2: $S = 100$; 3: $S = 5$; 4: $S = 1$; 5: $S = 0.5$; 6: $S = 0.1$.

3.2 Cylinder of Finite Length, Non-steady State

3.2.1 Infinite Atmosphere with Infinite Rate of Evaporation

The equation of diffusion with constant diffusivity is written as follows:

$$\frac{\partial C}{\partial t} = \left[\frac{\partial^2 C}{\partial r^2} + \frac{1}{r} \frac{\partial C}{\partial r} + \frac{\partial^2 C}{\partial z^2} \right] \tag{3.22}$$

with the coordinates r and z (Fig. 3.4).

The cylinder, initially full of liquid, has a radius R and a length $2l$. The surrounding atmosphere of infinite volume is free of vapour. The initial and boundary conditions are then:

$$
\left. \begin{array}{lll}
t = 0, & 0 \leqslant r \leqslant R, & C = C_{in} \\
& -l \leqslant z \leqslant +l, & C = C_{in}
\end{array} \right\} \quad \text{(cylinder)}
\tag{3.23}
$$

$$
\left. \begin{array}{lll}
t > 0, & r = R, & C = 0 \\
& z = \pm l, & C = 0
\end{array} \right\} \quad \text{(surface)}
\tag{3.24}
$$

A solution is obtained by assuming that the variables are separable. Thus we may attempt to find a solution of Eq. (3.22) by putting [2]

$$C_{r,z,t} = R_r Z_z T_t \tag{3.25}$$

Eq. (3.22) can be written in the form

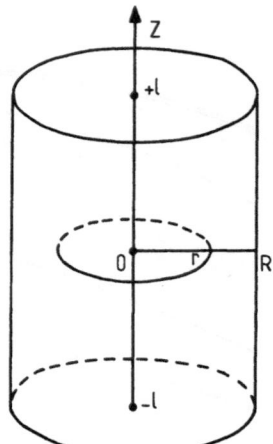

Fig. 3.4. Cylinder of finite length.

$$D\left[\frac{R''}{R} + \frac{1}{r}\frac{R'}{R} + \frac{Z''}{Z}\right] = \frac{T'}{T} \tag{3.26}$$

which has a solution

$$R_r = KJ_0(\alpha r) \tag{3.27}$$

$$Z_z = A \sin \lambda z + B \cos \lambda z \tag{3.27'}$$

$$T_t = K' \exp\left(-(\alpha^2 + \lambda^2)Dt\right) \tag{3.27''}$$

The general solution is then

$$C_{r,z,t} = \sum_{n=1}^{\infty} \sum_{m=1}^{\infty} K_m J_0(\alpha_m r)[A_n \sin \lambda_n z + B_n \cos \lambda_n z] \exp\left[-(\alpha_m^2 + \lambda_n^2)Dt\right] \tag{3.28}$$

where α_ms are roots of

$$J_0(\alpha_m R) = 0 \tag{3.5}$$

because of the boundary condition $C = 0$ for $r = R$.

By introducing the secondary condition, $C = 0$ for $z = \pm l$, the coefficient λ_n is determined. The general solution becomes

$$C_{r,z,t} = \sum_{n=1}^{\infty} \sum_{m=1}^{\infty} A_m J_0(\alpha_m r) \sin\left(\frac{n\pi(l+z)}{2l}\right) \exp\left[-\left(\alpha_m^2 + \frac{n^2\pi^2}{4l^2}\right)Dt\right] \tag{3.29}$$

The coefficient A_m is obtained by considering the initial conditions, and the final equation for the concentration is then

$$C_{r,z,t} = \frac{8C_{in}}{\pi R} \sum_{n=0}^{\infty} \sum_{m=1}^{\infty} \frac{J_0(\alpha_m r)}{(2n+1)\alpha_m J_1(\alpha_m R)} \sin\left(\frac{(2n+1)\pi(z+l)}{2l}\right)$$

$$\times \exp\left[-\left(\alpha_m^2 + \frac{(2n+1)^2\pi^2}{4l^2}\right)Dt\right] \tag{3.30}$$

The amount of diffusing substance remaining in the cylinder after time t is given by

$$Q'_t = \int_0^R 2\pi r \, dr \int_{-l}^{+l} C \, dz \tag{3.31}$$

and becomes

$$Q'_t = \frac{64 C_{in} l R^2}{\pi} \sum_{m=1}^{\infty} \frac{1}{(\alpha_m R)^2} \exp\left[-(\alpha_m R)^2 \frac{Dt}{R^2}\right]$$

$$\times \sum_{n=0}^{\infty} \frac{1}{(2n+1)^2} \exp\left[-\frac{(2n+1)^2 \pi^2}{4l^2} Dt\right] \tag{3.32}$$

As the initial amount of diffusing substance is

$$Q_\infty = 2\pi R^2 l C_{in} \tag{3.33}$$

the amount of diffusing substance which has left the cylinder after time t, Q_t, is expressed in terms of the corresponding value after infinite time Q_∞ by

$$\frac{Q_\infty - Q_t}{Q_\infty} = \frac{32}{\pi^2} \sum_{m=1}^{\infty} \frac{1}{(\alpha_m R)^2} \exp\left[-(\alpha_m R)^2 \frac{Dt}{R^2}\right] \sum_{n=0}^{\infty} \frac{1}{(2n+1)^2}$$

$$\times \exp\left[-\frac{(2n+1)^2 \pi^2}{4l^2} Dt\right] \tag{3.34}$$

This equation can be simplified by making the following assumptions. The roots $\alpha_m R$ of the equation $J_0(x) = 0$ can be approximated to

$$\alpha_m R = \pi(m - 0.25) \tag{3.35}$$

Then, when $\sqrt{Dt}/R$ is small (less than 0.05),

$$\sum_{m=1}^{\infty} \frac{1}{(\alpha_m R)^2} \exp\left[-(a_m R)^2 \frac{Dt}{R^2}\right] = \frac{1}{4} - \frac{1}{R} \sqrt{\frac{Dt}{\pi}}$$

When $\sqrt{Dt}/l$ is small,

$$\sum_{n=0}^{\infty} \frac{1}{(2n+1)^2} \exp\left[-\frac{(2n+1)^2 \pi^2}{4l^2} Dt\right] = \frac{\pi^2}{8} \left(1 - \frac{4}{l} \sqrt{\frac{Dt}{\pi}}\right)$$

Eq. (3.34) is thus reduced to

$$\frac{Q_t}{Q_\infty} = 4\sqrt{\frac{Dt}{\pi}} \left(\frac{1}{R} + \frac{1}{l}\right) \tag{3.36}$$

In fact, a more general equation is also established for the cylinder, sphere and parallelepiped:

$$\frac{Q_t}{Q_\infty} = 2\sqrt{\frac{Dt}{\pi}} \frac{A}{V}$$

where A and V are the external area and the volume of solid, respectively.

The kinetics of evaporation are shown in Fig. 3.5 for various values of the ratio of the radius and length. General curves are obtained by plotting Q_t/Q_∞ as a function of the dimensionless number Dt/R^2.

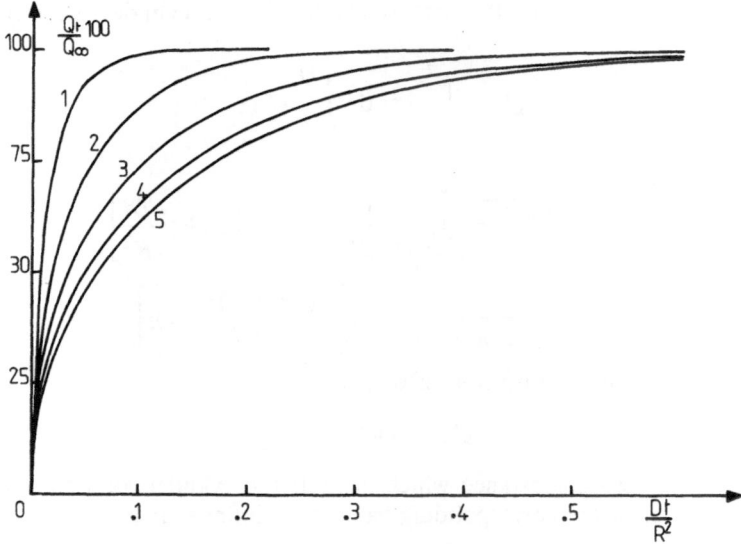

Fig. 3.5. Kinetics of evaporation from cylinders of finite length in atmosphere of infinite volume, with infinite rate of evaporation. Effect of the ratio L of the height and radius. 1: $L = 0.5$; 2: $L = 1$; 3: $L = 2$; 4: $L = 5$; 5: $L = 20$.

3.2.2 Infinite Atmosphere with Finite Rate of Evaporation

The problem can be resolved by using numerical methods with finite differences (see Chap. 6).

3.3 Liquid-Filled Hollow Cylinder of Infinite Length, Steady State

3.3.1 Infinite Atmosphere with Infinite Rate of Evaporation

The medium is a hollow cylinder whose inner and outer radii are R_i and R_0 respectively, and the diffusivity is constant. The internal surface in contact with a liquid is kept at constant concentration C_i and external surface at constant concentration C_0 because of the high rate of evaporation.

The initial and boundary conditions are

$$t = 0 \quad R_i < r < R_0 \quad C_{in} \qquad \text{(solid)} \qquad (3.37)$$

$$t > 0 \quad r = R_i \qquad\quad C = C_i \qquad \text{(internal surface)}$$

$$\qquad\quad r = R_0 \qquad\quad C = C_0 \qquad \text{(external surface)} \quad (3.38)$$

In the infinitely long cylinder the diffusion is everywhere radial. The diffusion equation is

$$\frac{\partial C}{\partial t} = \frac{1}{r}\frac{\partial}{\partial r}\left(Dr\,\frac{\partial C}{\partial r} \right) \qquad (3.1)$$

where concentration is a function of radius and time only.

When the diffusivity is constant, this equation becomes

$$\frac{\partial}{\partial r}\left(r\,\frac{\partial C}{\partial r}\right) = 0 \text{ for } R_i < r < R_0 \tag{3.39}$$

The general solution is

$$C = K_1 + K_2 \ln r \tag{3.40}$$

where K_1 and K_2 are constants to be determined from the boundary conditions:

$$C = \frac{C_i \ln \dfrac{R_0}{r} - C_0 \ln \dfrac{R_i}{r}}{\ln \dfrac{R_0}{R_i}} \tag{3.41}$$

The quantity of diffusing substance which evaporates through a unit length of the cylinder in time t is then, by using Fick's first law,

$$Q_t = 2\pi r D \int_0^t \left|\frac{\partial C}{\partial r}\right|_{R_0} dt \tag{3.42}$$

$$Q_t = \frac{2\pi(C_i - C_0)}{\ln \dfrac{R_0}{R_i}} Dt \tag{3.43}$$

3.3.2 Infinite Atmosphere with Finite Rate of Evaporation

The internal surface of the hollow cylinder R_i is kept at the constant concentration C_i, while there is evaporation with a finite rate from the external surface R_0. The boundary conditions are

$$-\left(\frac{dC}{dr}\right)_{R_0} = \frac{F_0}{D}(C_0 - C_{ext}), \; r = R_0 \tag{3.44}$$

where C_{ext} is the concentration on the external surface which is at equilibrium with the surrounding atmosphere.

The concentration C_r under steady-state conditions is:

$$C_r = \frac{C_i\left(1 + \dfrac{F_0 R_0}{D}\ln\dfrac{R_0}{r}\right) + \dfrac{F_0 R_0}{D}C_0 \ln\dfrac{r}{R_i}}{1 + \dfrac{F_0 R_0}{D}\ln\dfrac{r}{R_i}} \tag{3.45}$$

and the rate of evaporation per unit length of the hollow cylinder is

$$\frac{dQ}{dt} = 2\pi D(C_i - C_0)\;\frac{\dfrac{F_0 R_0}{D}}{1 + \dfrac{F_0 R_0}{D}\ln\dfrac{R_0}{R_i}} \tag{3.46}$$

3.4   Conclusions

The following conclusions are drawn on the drying process of cylinders.

Non-steady and Steady States
In the case of a solid cylinder, the process of drying always occurs in a non-steady state.

In the interesting case of hollow cylinders, evaporation takes place under non-steady-state conditions at the beginning of the process. After a period of time, steady-state conditions are attained, and the concentration of substance depends only on space. This steady-state transfer needs constant concentrations on both surfaces of the hollow cylinder: this is the case, for instance, when a liquid is in contact with the internal surface.

Cylinder of Finite and Infinite Length
The problem of transfer with a cylinder of infinite length is simple in the sense that the transfer is only radial, the effect of the edges being neglected.

Cylinders of finite length offer a two-dimensional transfer (radial and long-itudinal). This case remains simple, as generally the diffusivity is the same for these two transfers. The effect of the ratio of the length and radius on the process is of interest. Only an infinite rate of evaporation is considered.

Solid Cylinder of Infinite Length in a Finite Atmosphere with Infinite Rate of Evaporation
The effect of the ratio of the volumes of external atmosphere and cylinder on the process of drying is of interest. When this ratio is very large ($\alpha \to \infty$), Eq. (3.11) simplifies to:

$$\frac{Q_\infty - Q_t}{Q_\infty} = \frac{4}{q_1^2} \exp\left(-\frac{q_1^2}{R^2} Dt\right) \tag{3.47}$$

which is the same as Eq. (3.6), as

$$q_n = (\alpha_n R) \quad \text{when} \quad \alpha \to \infty \tag{3.48}$$

Solid Cylinder of Infinite Length in an Infinite Atmosphere With Finite Rate of Evaporation
The effect of the rate of evaporation on the rate of drying is of concern, in spite of the fact that the process is also controlled by diffusion of the liquid through the solid.

Eq. (3.18) obtained for the kinetics of drying in terms of the parameters S and β_n can also be used for an infinite rate of evaporation. In this case, $F_0 \to \infty$, and so $S \to \infty$, and

$$\beta_n = (\alpha_n R) \quad \text{for} \quad S \to \infty$$

Eq. (3.18) reduces to

$$\frac{Q_\infty - Q_t}{Q_\infty} = \sum_{n=1}^{\infty} \frac{4}{\beta_n^2} \exp\left(-\frac{\beta_n^2}{R^2} Dt\right)$$

which is the same as Eq. (3.6) obtained for an infinite rate of evaporation.

Appendix. Radial Diffusion in a Solid Cylinder of Infinite Length with Infinite Rate of Evaporation

The equation of the radial diffusion in a solid cylinder

$$\frac{\partial C}{\partial t} = \frac{D}{r} \frac{\partial}{\partial r}\left(r \frac{\partial C}{\partial r}\right) \tag{3A.1}$$

is solved by separation of the variables:

$$C_{r,t} = C_r C_t = C_r \exp\left(-\alpha^2 D t\right) \tag{3A.2}$$

C_r being a function of space r, and C_t a function of time, and α a constant.

Eq. (3A.1) can thus be written in the form

$$\frac{d^2 C_r}{dr^2} + \frac{1}{r}\frac{dC_r}{dr} + \alpha^2 C_r = 0 \tag{3A.3}$$

Eq. (3A.3) is Bessel's equation of order zero, with the general solution [2]

$$C_{r,t} = \sum_{n=1}^{\infty} [A_n J_0(\alpha_n r) + B_n Y_0(\alpha_n r)] \exp\left(-\alpha_n^2 D t\right) \tag{3A.4}$$

The initial and boundary conditions can be written as follows:

$$t = 0, \quad 0 < r < R, \qquad C_{\text{in}} \qquad \text{(cylinder)} \tag{3A.5}$$

$$t > 0, \quad r = R, \qquad C_R = 0 \qquad \text{(surface)} \tag{3A.6}$$

The boundary conditions are satisfied by

$$0 = \sum_{n=1}^{\infty} A_n J_0(\alpha_n R) \exp\left(-\alpha_n^2 D t\right) \tag{3A.7}$$

where the α_ns are roots of

$$J_0(\alpha_n R) = 0 \tag{3A.8}$$

$J_0(x)$ being the zero-order Bessel function of the first kind, which tends to 1 when $r = 0$. The term B_n is zero, as for $r = 0$, $Y_0(\alpha r) \to \infty$. These roots are given in tables of Bessel functions.

The initial conditions become

$$C_{\text{in}} = \sum_{n=1}^{\infty} A_n J_0(\alpha_n r) \tag{3A.9}$$

The A_ns are determined by multiplying both sides of Eq. (3A.9) by $r J_0(\alpha_n r)$ and integrating from 0 to R, using the following results:

$$\int_0^R r J_0(\alpha r) J_0(\beta r)\, dr = 0 \quad \text{with} \quad \alpha \neq \beta \tag{3A.10}$$

$$\int_0^R r J_0^2(\alpha r)\, dr = \frac{R^2}{2} J_1^2(\alpha R) \tag{3A.11}$$

where $J_1(x)$ is the first-order Bessel function and α and β are roots of Eq. (3A.8).

The concentration of liquid at position r and time t is then

$$\frac{C_{r,t}}{C_{\text{in}}} = \frac{2}{R} \sum_{n=1}^{\infty} \frac{J_0(\alpha_n r)}{\alpha_n J_1(\alpha_n R)} \exp\left(-\alpha_n^2 D t\right) \tag{3A.12}$$

When the concentration of liquid on the surface is C_{ext} instead of zero, the left-hand term in the above equation becomes

$$\frac{C_{r,t} - C_{ext}}{C_{in} - C_{ext}} \text{ instead of } \frac{C_{r,t}}{C_{in}} \tag{3A.12'}$$

The amount of liquid remaining in the cylinder at time t, Q'_t, is given by

$$Q'_t = 2\pi \int_0^R C_{r,t} r \, dr \tag{3A.13}$$

By substituting in this equation the value of $C_{r,t}$ expressed by Eq. (3A.13), and by considering that

$$Q_{in} = \pi R^2 C_{in} \text{ per unit length} \tag{3A.14}$$

and

$$\int_0^R r J_0(\alpha_n r) \, dr = \frac{R J_1(\alpha_n R)}{\alpha_n} \tag{3A.15}$$

Eq. (3A.13) becomes

$$\frac{Q'_t}{Q_{in}} = \sum_{n=1}^{\infty} \frac{4}{(R\alpha_n)^2} \exp(-\alpha_n^2 Dt) \tag{3A.16}$$

If the amount of liquid which has left the cylinder at time t is Q_t, with

$$Q_t + Q'_t = Q_{in} \tag{3A.17}$$

the final expression of the amount of liquid evaporated after time t is then

$$\frac{Q_{in} - Q_t}{Q_{in}} = \sum_{n=1}^{\infty} \frac{4}{(R\alpha_n)^2} \exp(-\alpha_n^2 Dt) = \frac{4}{R^2} \sum_{n=1}^{\infty} \frac{1}{(\alpha_n)^2} \exp(-\alpha_n^2 Dt) \tag{3A.18}$$

Symbols

a	Radius of the external atmosphere (cylindrical in shape)
C_i, C_0	Concentration at internal and external surface (hollow cylinder)
C_{in}, C_∞	Uniform concentration of the substance in the cylinder, at the beginning and at the end of the process
$C_{r,t}$	Concentration of the substance in the cylinder, at position r and time t
D	Diffusivity ($cm^2 s^{-1}$)
F_0	Rate of evaporation of the pure liquid ($cm \, s^{-1}$)
K	Partition factor
$2l$	Height of the cylinder of finite length
q_n	Positive, non-zero roots of Eq. (3.12)
Q_t, Q_∞	Amount of substance evaporated after time t, after infinite time, respectively
r	Position in the cylinder
R	Radius of the cylinder

R_i, R_0	Internal and external radii of the hollow cylinder
S	Dimensionless number $\left(\dfrac{F_0 R}{D}\right)$
t	Time
z	Coordinate for the cylinder of finite length
α	Ratio of the volumes of the atmosphere and solid
$\alpha_n R$	Roots of the Bessel function of the first kind of order zero
β_n	Roots of $\beta J_1(\beta) = S J_0(\beta)$

References

1. Crank J. The mathematics of diffusion, 2nd edn, Clarendon Press, Oxford, 1975, pp 69–88
2. Adda Y, Philibert J. La diffusion dans les solides, vol. 1. Presses Universitaires de France, Paris, 1966

Chapter 4

Sphere

The following cases of diffusion–evaporation for a solid sphere with radial transfer and constant diffusivity, are of interest. The initial concentration within the sphere is considered uniform:

- The surrounding atmosphere is sufficiently large to be considered infinite and the rate of evaporation is also considered infinite.
- The surrounding atmosphere is of finite volume and the rate of evaporation is infinite.
- The surrounding atmosphere is infinite and the rate of evaporation is finite.

Of course, in the above three cases, transfer takes place under transient conditions.

The hollow sphere can also be of interest, and three cases are considered when the diffusivity is constant:

- The substance diffuses through the wall of the sphere under transient conditions with constant concentrations on both surfaces (the rate of evaporation on the external surface is infinite).
- The substance diffuses under steady-state conditions, and the rate of evaporation on the external surface is infinite.
- The substance diffuses under steady-state conditions, with a finite rate of evaporation on the external surface.

4.1 Solid Sphere, Non-steady State with Infinite Rate of Evaporation

The following two cases are considered, where the surrounding atmosphere is of either an infinite or finite volume.

4.1.1   Infinite Atmosphere

The diffusion is radial, with constant diffusivity. The diffusion equation is

$$\frac{\partial C}{\partial t} = D\left[\frac{\partial^2 C}{\partial r^2} + \frac{2}{r}\frac{\partial C}{\partial r}\right] = \frac{D}{r^2}\frac{\partial}{\partial r}\left[r^2\frac{\partial C}{\partial r}\right] \tag{4.1}$$

On putting

$$u = Cr \tag{4.2}$$

the equation for radial diffusion becomes

$$\frac{\partial U}{\partial t} = D\frac{\partial^2 U}{\partial r^2} \tag{4.3}$$

The matter in the sphere is initially at a uniform concentration, and the concentration in the surrounding atmosphere is constant.

Because the rate of evaporation is considered infinite, the concentration of substance on the surface is constant as soon as the process starts.

Initial and boundary conditions are

$t = 0$	$0 < r < R,$	$C = C_{in},$	$u = rC_{in}$	(sphere)	(4.4)
$t > 0$	$r = R,$	$C = C_0,$	$u = RC_0$	(surface)	
	$r = 0,$		$u = 0$	(centre)	(4.5)

where C_{in} is the initial concentration in the sphere and C_0 the constant concentration on the surface.

The solution for the concentration within the sphere is [1, 2], with the method of separation of variables

$$\frac{C_{r,t} - C_{in}}{C_0 - C_{in}} = 1 + \frac{2R}{\pi r}\sum_{n=1}^{\infty}\frac{(-1)^n}{n}\sin\frac{n\pi r}{R}\exp\left(-\frac{n^2\pi^2}{R^2}Dt\right) \tag{4.6}$$

At the centre the concentration is reduced as $\sin r/r \to 1$ when $r \to 0$:

$$\frac{C_{0,t} - C_{in}}{C_0 - C_{in}} = 1 + 2\sum_{n=1}^{\infty}(-1)^n\exp\left(-\frac{n^2\pi^2}{R^2}Dt\right) \tag{4.7}$$

with $C_{r,t}(C_{0,t})$ the concentration at position r (and at the centre) at time t.

The total amount of substance leaving the sphere after time t, Q_t, is expressed in terms of the corresponding quantity after infinite time, Q_∞, by the equation

$$\frac{Q_\infty - Q_t}{Q_\infty} = \frac{6}{\pi^2}\sum_{n=1}^{\infty}\frac{1}{n^2}\exp\left(-\frac{n^2\pi^2}{R^2}Dt\right) \tag{4.8}$$

By using the method of reflection and superposition, with the general solution

$$C = \frac{Q}{\sqrt{t}}\exp\left(-\frac{r^2}{4Dt}\right) \tag{4.9}$$

the other solutions are obtained [1]:

$$\frac{C_{r,t} - C_{in}}{C_0 - C_{in}} = \frac{R}{r}\sum_{n=0}^{\infty}\left\{\operatorname{erfc}\frac{(2n+1)R - r}{2\sqrt{Dt}} - \operatorname{erfc}\frac{(2n+1)R + r}{2\sqrt{Dt}}\right\} \tag{4.10}$$

and

$$\frac{Q_t}{Q_\infty} = \frac{6\sqrt{Dt}}{R} \left\{ \frac{1}{\sqrt{\pi}} + 2 \sum_{n=1}^{\infty} \text{ierfc}\frac{nR}{\sqrt{Dt}} \right\} - 3\frac{Dt}{R^2} \tag{4.11}$$

By using dimensionless numbers, the ratio Q_t/Q_∞ is expressed in terms of time (Fig. 4.1). The following conclusions can be drawn:

1. The series converges slowly for short times, and 40 terms are needed when $Dt/R^2 < 0.2$.
2. For longer times, when $0.2 < Dt/R^2 < 0.9$, 10 terms are enough.
3. For very long times, when $Dt/R^2 > 0.9$, the first term dominates.
4. The time necessary for the amount of evaporating substance to reach a given value is proportional to the square of the radius of the sphere.
5. This same time is inversely proportional to the diffusivity of the substance.

Remark 4.1: Long Times. For long times, when $Q_t/Q_\infty > 0.8$, the first term of the series in Eq. (4.8) dominates, and the useful equation is obtained in the logarithmic form

$$\ln\left(\frac{Q_\infty - Q_t}{Q_\infty}\right) = -\frac{\pi^2}{R^2} Dt + \ln\frac{6}{\pi^2} \tag{4.12}$$

The diffusivity can thus be obtained from the straight line obtained by plotting the quantity on the left-hand side of Eq. (4.12) as a function of time.

Remark 4.2: Short Times. Eq. (4.11) is very useful for short times, when $Q_t/Q_\infty < 0.3$, because in this case it reduces to

$$\frac{Q_t}{Q_\infty} = \frac{6}{R} \sqrt{\frac{Dt}{\pi}} \tag{4.13}$$

showing a linear relationship.

4.1.2 Finite Atmosphere

The initial concentration within the sphere C_{in} is uniform. The sphere is surrounded by a well-stirred atmosphere of volume V, initially free from vapour. The sphere occupies the space $r < R$.

The initial and boundary conditions are

$$t = 0, \quad r < R, \quad C_{in} \text{ (sphere)} \tag{4.14}$$

$$t > 0, \quad r = R, \qquad \text{(boundary conditions)} \tag{4.15}$$

The boundary conditions express the fact that the rate at which the substance is evaporated from the surface is always equal to that at which the substance is brought to the surface by internal diffusion:

$$t > 0, \quad V\frac{\partial C}{\partial t} = \pm D\left(\frac{\partial C}{\partial r}\right)_R \pi R^2, \quad r = R \text{ (surface)} \tag{4.15}$$

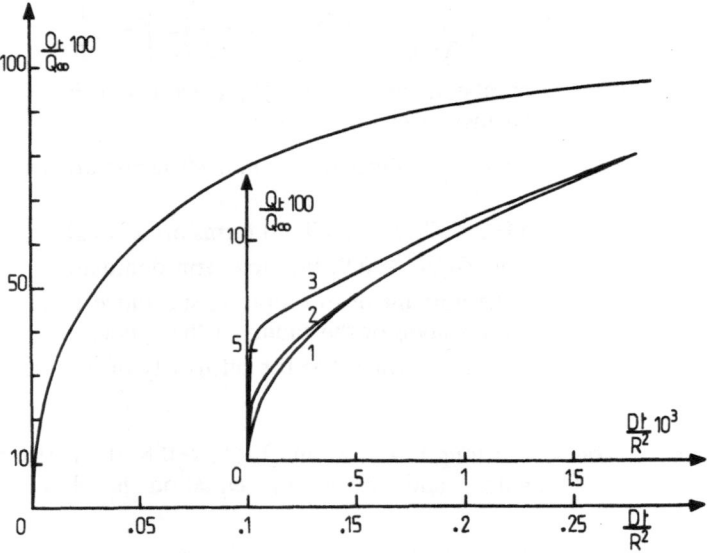

Fig. 4.1. Kinetics of evaporation for infinite rate of evaporation in an infinite surrounding atmosphere, for various values of the integer n. 1: $n = 40$; 2: $n = 20$; 3: $n = 10$.

The concentration of diffusing substance in the sphere, $C_{r,t}$ at position r and time t is given by

$$\frac{C_{in} - C_{r,t}}{C_{in} - C_\infty} = 1 + \frac{R}{r} \sum_{n=1}^{\infty} \frac{6(1 + \alpha)}{9 + 9\alpha + q_n^2\alpha^2} \frac{\sin\dfrac{q_n r}{R}}{\sin q_n} \exp\left(-\frac{q_n^2}{R^2} Dt\right) \quad (4.16)$$

where C_∞ is the concentration of substance in the sphere after infinite time. The q_ns are the non-zero roots of

$$\tan q_n = \frac{3q_n}{3 + \alpha q_n^2} \quad (4.17)$$

and α is the ratio of the volumes of the surrounding atmosphere and the sphere:

$$\alpha = \frac{3V}{4\pi R^3} \quad (4.18)$$

The ratio α can be expressed in terms of the final uptake of substance by the surrounding atmosphere by the relation

$$\frac{3M_\infty}{4\pi R^3 C_{in}} = \frac{\alpha}{1 + \alpha} \quad (4.19)$$

When there is a partition factor K between the substance in equilibrium in the sphere and in the atmosphere, α becomes:

$$\alpha = \frac{3V}{4\pi R^3 K} \quad (4.18')$$

The total amount of substance which evaporates after time t, Q_t, is expressed as a fraction of the corresponding quantity after infinite time by the equation

$$\frac{Q_\infty - Q_t}{Q_\infty} = \sum_{n=1}^{\infty} \frac{6\alpha(\alpha + 1)}{9 + 9\alpha + q_n^2\alpha^2} \exp\left(-\frac{q_n^2}{R^2} Dt\right) \quad (4.20)$$

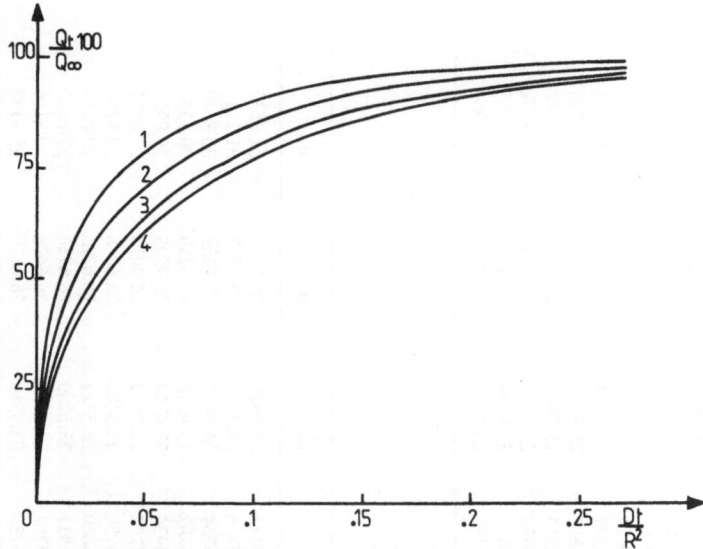

Fig. 4.2. Kinetics of evaporation for an infinite rate of evaporation in a finite atmosphere, for various values of the ratio α. 1: $\alpha = 1$; 2: $\alpha = 2.33$; 3: $\alpha = 9$; 4: $\alpha = 50$.

The effect of the ratio of volumes α on the kinetics of evaporation is illustrated in Fig. 4.2 where the partial amount of substance which has evaporated from the sphere, Q_t/Q_∞ is expressed in terms of the dimensionless number Dt/R^2 for various values of α.

The roots of Eq. (4.17) are given in Table 4.1, for various values of α.

The diffusivity can be easily obtained from the kinetics of drying determined after long experiments. Eq. (4.20) can then be reduced to the first term of the series, when $Q_t/Q_\infty > 0.8$. By expressing the simplified equation in a logarithmic form,

$$\ln\left(\frac{Q_\infty - Q_t}{Q_\infty}\right) = -\frac{q_1^2}{R^2} Dt + \ln\left(\frac{6\alpha(\alpha + 1)}{9 + 9\alpha + q_1^2\alpha^2}\right) \tag{4.20'}$$

A linear relationship is shown between the left-hand term and time. The diffusivity is determined from the slope of this straight line, when the other parameters such as the radius and the ratio α are known.

4.2   Solid Sphere, Non-steady State with Finite Rate of Evaporation

The sphere, initially at a uniform concentration C_{in} is placed in a surrounding atmosphere of infinite volume, so that the concentration of vapour in this atmosphere remains constant as evaporation proceeds.

On the surface of the sphere, the rate of evaporation of the substance is always equal to the rate at which this substance arrives at the surface by internal diffusion.

Table 4.1. Roots of $\tan q_n = 3q_n/(3 + \alpha q_n^2)$

α	q_1	q_2	q_3	q_4	q_5	q_6	q_7	q_8	q_9	q_{10}
9.0000	3.2410	6.3353	9.4599	12.5928	15.7291	18.8672	22.0063	25.1460	28.2861	31.4265
4.0000	3.3485	6.3978	9.5029	12.6254	15.7554	18.8892	22.0251	25.1625	28.3008	31.4398
2.3333	3.4650	6.4736	9.5567	12.6667	15.7888	18.9172	22.0492	25.1836	28.3196	31.4567
1.5000	3.5909	6.5664	9.6254	12.7204	15.8326	18.9541	22.0811	25.2117	28.3446	31.4792
1.0000	3.7264	6.6814	9.7156	12.7927	15.8924	19.0048	22.1251	25.2504	28.3793	31.5106
0.6666	3.8711	6.8246	9.8369	12.8940	15.9780	19.0785	22.1895	25.3075	28.4305	31.5570
0.4286	4.0236	7.0019	10.0039	13.0423	16.1082	19.1932	22.2914	25.3989	28.5131	31.6322
0.2500	4.1811	7.2169	10.2355	13.2689	16.3211	19.3898	22.4719	25.5647	28.6656	31.7732
0.1111	4.3395	7.4645	10.5437	13.6134	16.6831	19.7565	22.8350	25.9189	29.0082	32.1025
0.0000	4.4934	7.7253	10.9041	14.0662	17.2208	20.3713	23.5195	26.6661	29.8116	32.9564

Table 4.2. Roots of $\beta_n \cot \beta_n + S - 1 = 0$

S	β_1	β_2	β_3	β_4	β_5	β_6	β_7	β_8	β_9	β_{10}
500.00	3.1353	6.2706	9.4059	12.5412	15.6766	18.8119	21.9472	25.0825	28.2178	31.3532
100.00	3.1102	6.2204	9.3308	12.4414	15.5521	18.8632	21.7746	24.8865	27.9987	31.1114
50.000	3.0788	6.1582	9.2384	12.3200	15.4034	18.4888	21.5764	24.6664	27.7589	30.8540
10.000	2.8363	5.7172	8.6587	11.6532	14.6869	17.7481	20.8282	23.9218	27.0250	30.1354
2.5000	2.1746	5.0036	8.0385	11.1295	14.2421	17.3649	20.4934	23.6254	26.7595	29.8953
1.0000	1.5708	4.7124	7.8540	10.9956	14.1372	17.2788	20.4203	23.5619	26.7035	29.8451
0.5000	1.1656	4.6042	7.7899	10.9499	14.1017	17.2498	20.3958	23.5407	26.6848	29.8284
0.3000	0.9208	4.5601	7.7641	10.9316	14.0875	17.2382	20.3860	23.5322	26.6773	29.8217
0.2000	0.7593	4.5379	7.7511	10.9225	14.0804	17.2324	20.3811	23.5280	26.6736	29.8183
0.1000	0.5423	4.5157	7.7382	10.9133	14.0733	17.2266	20.3762	23.5237	26.6698	29.8150
0.0500	0.3854	4.5045	7.7317	10.9087	14.0697	17.2237	20.3738	23.5216	26.6679	29.8133
0.0100	0.1730	4.4956	7.7265	10.9050	14.0669	17.2213	20.3718	23.5199	26.6664	29.8119
0.0000	0.0000	4.4934	7.7253	10.9041	14.0662	17.2208	20.3713	23.5195	26.6661	29.8116

The initial and boundary conditions are

$$t = 0, \quad r < R, \quad C_{in} \text{ (sphere)} \tag{4.21}$$

$$t > 0, \quad -D\left(\frac{\partial C}{\partial r}\right)_R = F_0(C_{R,t} - C_{eq}) \tag{4.22}$$

where $C_{R,t}$ is the concentration of substance on the surface at time t, and C_{eq} is the concentration on the surface required to maintain equilibrium with the constant vapour pressure in the surrounding atmosphere.

The profiles of concentration of the substance within the sphere are given by [1]

$$\frac{C_{r,t} - C_{eq}}{C_{in} - C_{eq}} = \frac{2SR}{r} \sum_{n=1}^{\infty} \frac{1}{(\beta_n^2 + S^2 - S)} \frac{\sin\frac{\beta_n r}{R}}{\sin\beta_n} \exp\left(-\frac{\beta_n^2}{R^2} Dt\right) \tag{4.23}$$

where the β_ns are roots (Table 4.2) of the following equation:

$$\beta_n \cot \beta_n + S - 1 = 0 \tag{4.24}$$

and the dimensionless number S is

$$S = F_0 \frac{R}{D} \tag{4.25}$$

The amount of diffusing substance which has evaporated after time t, Q_t, as a fraction of the corresponding quantity which has evaporated after infinite time, is expressed in terms of time and β_n and S:

$$\frac{Q_\infty - Q_t}{Q_\infty} = \sum_{n=1}^{\infty} \frac{6S^2}{\beta_n^2(\beta_n^2 + S^2 - S)} \exp\left(-\frac{\beta_n^2}{R^2} Dt\right) \tag{4.26}$$

The kinetics of evaporation are expressed in terms of the dimensionless number Dt/R^2, for various values of S ranging from 0.2 to 50 (Fig. 4.3).

The effect of S (which is proportional to the rate of evaporation) on the rate of drying is of great importance.

The diffusivity can be determined from the kinetics obtained from long experiments. For long times, when for instance $Q_t/Q_\infty > 0.8$, Eq. (4.26) can be reduced to the first term of the series, becoming in a logarithmic form

$$\ln\left(\frac{Q_\infty - Q_t}{Q_\infty}\right) = -\frac{\beta_1^2}{R^2} Dt + \ln\left(\frac{6S^2}{\beta_1^2(\beta_1^2 + S^2 - S)}\right) \tag{4.26'}$$

From the slope of the straight line derived from this equation, the diffusivity can be calculated by iteration as β_1 is also a function of diffusivity through the number S.

4.3 Liquid-Filled Hollow Sphere

4.3.1 Non-steady State with Infinite Rate of Evaporation

The hollow sphere is between the region $R_i < r < R_0$ shown in Fig. 4.4. The internal surface in contact with a liquid is maintained at constant concentration C_i, and the external surface is maintained at zero concentration, because the rate

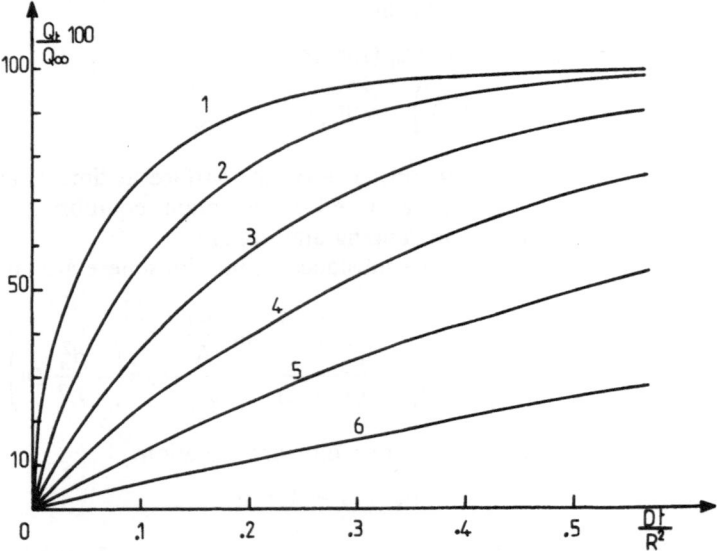

Fig. 4.3. Kinetics of evaporation for a finite rate of evaporation in an infinite surrounding atmosphere, for various values of S. 1: $S = 50$; 2: $S = 5$; 3: $S = 2$; 4: $S = 1$; 5: $S = 0.5$; 6: $S = 0.2$.

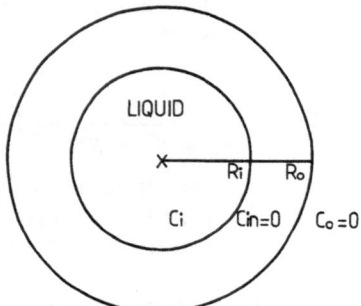

Fig. 4.4. Spherical wall with constant concentration on each surface. Internal surface, C_i; external surface, $C_0 = 0$.

of evaporation is considered infinite. The region $R_i < r < R_0$ is initially at zero concentration.

The initial and boundary conditions are

$$t = 0, \quad R_i < r < R_0, \qquad C_{in} = 0 \text{ (solid)} \tag{4.27}$$

$$
\left.
\begin{aligned}
t > 0, \quad r = R_i, \qquad & C_i \qquad \text{(liquid–surface)} \\
r = R_0, \qquad & C_0 = 0 \text{ (external surface)} \\
R_i < r < R_0, \qquad & C_{r,t} \qquad \text{(solid)}
\end{aligned}
\right\}
\tag{4.28}
$$

The concentration $C_{r,t}$ in the solid is given by [1]

$$C_{r,t} = \frac{R_i(R_0 - r)}{r(R_0 - R_i)} \, C_i - \frac{2R_i C_i}{\pi r} \sum_{n=1}^{\infty} \frac{1}{n} \sin \frac{n\pi(r - R_i)}{R_0 - R_i} \exp\left(-\frac{n^2\pi^2}{(R_0 - R_i)^2} Dt\right)$$

$$\tag{4.29}$$

The amount of matter evaporating from the outer surface R_0 is given by

$$\frac{Q_t}{4\pi R_i R_0 \Delta R C_i} = \frac{Dt}{\Delta R^2} - \frac{1}{6} - \frac{2}{\pi^2} \sum_{n=1}^{\infty} \frac{(-1)^n}{n^2} \exp\left(-\frac{n^2\pi^2}{\Delta R^2} Dt\right) \qquad (4.30)$$

where $\Delta R = R_0 - R_i$ is the thickness of the spherical wall.

For long times, the series in Eq. (4.30) becomes negligible, and the rate of evaporation is constant:

$$Q_t = \frac{4\pi R_i R_0 C_i}{\Delta R} Dt - \frac{4\pi R_i R_0 \Delta R C_i}{6} \qquad (4.31)$$

The straight line obtained by plotting the amount of substance which has evaporated against time has an intercept on the time axis given by

$$\frac{Dt}{\Delta R^2} - \frac{1}{6} = 0 \qquad (4.32)$$

This intercept as well as the slope of the straight line can be used to determine the diffusivity under steady-state conditions (Fig. 4.5).

4.3.2 Steady State with Infinite Rate of Evaporation

The internal surface is in contact with a liquid, and there is an infinite rate of evaporation from the external surface.

The boundary conditions are

$$t > 0, \qquad r = R_0, \qquad C_0 \text{ (external surface)}$$

$$r = R_i, \qquad C_i \text{ (internal surface)} \qquad (4.33)$$

The concentration in the hollow sphere as well as the quantity of substance which evaporates from the external surface can easily be determined from the Eqs (4.29) and (4.30) obtained under non-steady-state conditions. When the time is long enough as shown in Fig. 4.5, the series in these equations become negligible.

Another way of calculating the concentration in the hollow sphere and the amount of substance which evaporates is by considering the equation of diffusion in a steady state. The equation of radial diffusion with constant diffusivity (Eq. (4.1)) can also be written

$$\frac{\partial C}{\partial t} = \frac{D}{r^2} \frac{\partial}{\partial r}\left(r^2 \frac{\partial C}{\partial r}\right) \qquad (4.34)$$

This equation becomes for a steady state

$$\frac{d}{dr}\left(r^2 \frac{dC}{dr}\right) = 0 \qquad (4.35)$$

The general solution is

$$C = K_1 + \frac{K_2}{r} \qquad (4.36)$$

where K_1 and K_2 are constants to be determined from the boundary conditions.

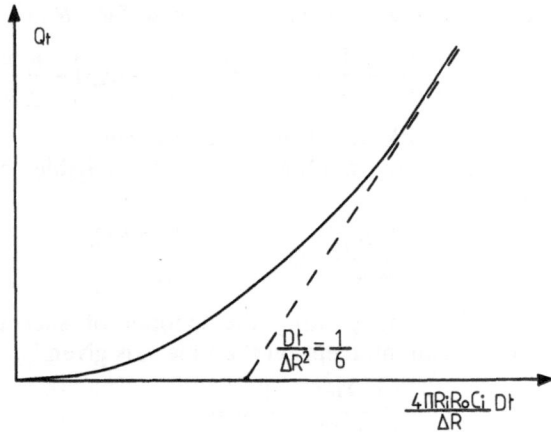

Fig. 4.5. Spherical wall with constant concentration on each surface. Approach to steady-state conditions.

The concentration at position r within the hollow sphere is given by

$$C_r = \frac{R_i(R_0 - r)C_i + R_0(r - R_i)C_0}{r(R_0 - R_i)} \tag{4.37}$$

The amount of diffusing substance which evaporates from the external surface in time t is expressed in terms of the constant concentrations on each surface:

$$Q_t = \frac{4\pi R_i R_0}{R_0 - R_i} Dt \tag{4.38}$$

4.3.3 Steady State with Finite Rate of Evaporation

The internal surface of the hollow sphere is in contact with a liquid which maintains a constant concentration C_i on this surface and there is a finite rate of evaporation from the external surface.

The boundary conditions are

$$t > 0, \quad r = R_i, \qquad\qquad C_i \qquad\qquad \text{(internal surface)} \left.\vphantom{\frac{dC}{dr}}\right\}$$

$$r = R_0, \quad -D\frac{dC}{dr} = F_0(C_{R_0} - C_{eq}) \text{ (external surface)} \tag{4.39}$$

where C_{R_0} is the concentration on the external surface and C_{eq} is the concentration on this surface which is at equilibrium with the surrounding atmosphere.

The concentration at position r is

$$C_r = \frac{R_i[F_0R_0^2 + r(D - F_0R_0)]C_i + F_0R_0^2(r - R_i)C_0}{r[F_0R_0^2 + R_i(D - F_0R_0)]} \tag{4.40}$$

and the amount of liquid evaporating from the external surface in time t is expressed in terms of the difference of concentrations:

$$Q_t = \frac{4\pi F_0 R_i R_0^2(C_i - C_0)}{F_0R_0^2 + R_i(D - F_0R)} Dt \tag{4.41}$$

4.4 Conclusions

The following conclusions can be drawn from the mathematical treatment of the process of diffusion–evaporation for a solid sphere.

Transient and Steady-State Conditions
For a solid sphere, the process is performed under non-steady-state conditions.

In the particular case of the hollow sphere with a liquid in contact with the internal surface, and evaporation from the external surface, the steady conditions are attained after a period of time in non-steady state.

Kinetics
The expression for the kinetics in the case of a sphere is about the same as that obtained for the plane sheet. However, the series for the sphere does not converge as fast as the series for the plane sheet.

Diffusivity
The diffusivity can be determined from the kinetics in three ways:

1. With short tests, where F_0 is very high:

$$\frac{Q_t}{Q_\infty} = \frac{6}{R} \sqrt{\frac{Dt}{\pi}} \quad \text{with} \quad \frac{Q_t}{Q_\infty} < 0.3 \tag{4.13}$$

2. With long tests either when F_0 is infinite or finite. When F_0 is infinite, we have Eq. (4.12):

$$\ln\left(\frac{Q_\infty - Q_t}{Q_\infty}\right) = -\frac{\pi^2}{R^2} Dt + \ln\frac{6}{\pi^2} \tag{4.12}$$

When F_0 is finite, Eq. (4.26) can be reduced to the first term when $Q_t/Q_\infty > 0.8$, and by plotting in a logarithmic form, we obtain

$$\ln\left(\frac{Q_\infty - Q_t}{Q_\infty}\right) = -\frac{\beta_1^2}{R^2} Dt + \ln\left(\frac{6S^2}{\beta_1^2(\beta_1^2 + S^2 - S)}\right)$$

From the slope of the straight line, the diffusivity D can be determined by iteration in spite of the dependence of β_1 on the diffusivity.

3. With the half-life test, when F_0 is infinite, when

$$\frac{Q_t}{Q_\infty} = \tfrac{1}{2}, \quad \frac{Dt}{R^2} = 0.0308$$

Volume of Surrounding Atmosphere
When the rate of evaporation is very high, the Eq. (4.20) is simplified in the case of $\alpha \to \infty$, and reduces to Eq. (4.8), as the terms q_n are related to the value of π.

$$q_n = n\pi \quad \text{for } \alpha \to \infty$$

Rate of Evaporation
When the rate of evaporation is very high, the dimensionless number shown in Eq. (4.25), $S \to \infty$, and Eq. (4.26) then reduces to Eq. (4.8):

$$\beta_n = n\pi \quad \text{when } S \to \infty$$

Symbols

C_{in}, C_∞ Concentration in the sphere at time 0 and at infinite time, respectively

$C_{r,t}$ Concentration at position r and time t

D Diffusivity ($cm^2\,s^{-1}$)

F_0 Rate of evaporation ($cm\,s^{-1}$)

K Partition factor

q_n Non-zero roots of $\tan q_n = \dfrac{3q_n}{3 + \alpha q_n^2}$

Q_t, Q_∞ Amount of substance evaporated after time t and after infinite time, respectively

R Radius of the sphere

R_i, R_0 Internal and external radius of the hollow sphere

S Dimensionless number $\left(\dfrac{F_0 R}{D}\right)$

t Time

α Ratio of the volumes of the atmosphere and sphere

β_n Roots of $\beta_n \cot \beta_n + S - 1 = 0$

References

1. Crank J. The mathematics of diffusion. 2nd edn. Clarendon Press, Oxford, 1975 pp 89–103
2. Adda Y, Philibert J. La diffusion dans les solides, vol. 1. Presses Universitaires de France, Paris, 1966

Chapter 5

Numerical Analysis for a Plane Sheet

When no analytical solution can be determined for the equation of diffusion and evaporation by a classical mathematical treatment, the numerical analysis solution remains the only possibility to resolve the problems.

The following cases can be resolved by using numerical analysis:

- Concentration-dependent diffusivity, or non-constant diffusivity.
- Special boundary conditions.
- Special initial conditions, and especially when the initial concentration is not uniform.
- The solid is anisotropic, e.g. wood.
- The shape of the solid is complex.

The principle of explicit numerical methods based on finite differences is described, and illustrated in various examples of interest such as

- The plane sheet with an infinite or finite rate of evaporation, and with constant or concentration-dependent diffusivity.
- The cylinder of infinite or finite length, with an infinite or finite rate of evaporation, and with constant or concentration-dependent diffusivity.
- The sphere with an infinite or finite rate of evaporation, with constant or concentration-dependent diffusivity.
- The parallelepiped with three main directions and three main diffusivities, which can be constant or concentration dependent.

Explicit numerical methods are very useful and fairly simple. They can easily accommodate all the known facts, being very flexible. Moreover, these techniques are especially appropriate to microcomputers as well as to more powerful computers.

5.1 Infinite Rate of Evaporation

The following two cases are considered, where the diffusivity is constant or concentration dependent.

5.1.1   Constant Diffusivity

This problem is very easy and it can be resolved by mathematical treatment. However, numerical analysis is performed in order to accustom the reader to this new way of calculation.

The equation for diffusion of a liquid through the thickness of a plane sheet is

$$\frac{\partial C}{\partial t} = D\frac{\partial^2 C}{\partial x^2} \tag{5.1}$$

with constant diffusivity.

Fig. 5.1 shows a cross-section of the plane sheet of thickness L, with a uniform cross-sectional area A. The sheet is divided into a number of equal finite slices Nu of thickness Δx by concentration-reference planes.

Within the Solid
The matter balance is expressed for the slice centred at position n, and located between the planes $(n + \frac{1}{2})$ and $(n - \frac{1}{2})$, by considering the amount of substance which enters and leaves the slices during the increment of time Δt:

$$A\left[-D\left(\frac{\partial C}{\partial x}\right)_{n+0.5} + D\left(\frac{\partial C}{\partial x}\right)_{n-0.5}\right]\Delta t = A\Delta x[CN_n - C_n] \tag{5.2}$$

where $\partial C/\partial x$ is the gradient of concentration at position $(n + 0.5)$ or $(n - 0.5)$, C_n is the concentration at position n at time t and CN_n is the concentration at position n after time Δt.

Another way of writing this is

$$C_{n,t} \text{ for } C_n, \text{ and } C_{n,t+\Delta t} \text{ for } CN_n \tag{5.3}$$

The slopes $\partial C/\partial x$ at plane $(n + 0.5)$ and $(n - 0.5)$ are approximated by the chord slope.

$$-\left(\frac{\partial C}{\partial x}\right)_{n+0.5} = \frac{C_{n+1} - C_n}{\Delta x} \tag{5.4}$$

$$-\left(\frac{\partial C}{\partial x}\right)_{n-0.5} = \frac{C_n - C_{n-1}}{\Delta x} \tag{5.4'}$$

The concentration at plane n approximates the average concentration of slice located between $(n + 0.5)$ and $(n - 0.5)$.

The resulting matter balance is

$$D\frac{\Delta t}{\Delta x}(C_{n-1} - 2C_n + C_{n+1}) = \Delta x(CN_n - C_n) \tag{5.5}$$

Upon replacing the dimensionless number by M:

$$M = \frac{(\Delta x)^2}{\Delta t}\frac{1}{D} \tag{5.6}$$

the new concentration after the finite time increment Δt at position n is expressed in terms of the concentrations obtained at the previous time at position n and adjacent positions $(n - 1)$ and $(n + 1)$:

$$CN_n = \frac{1}{M}[C_{n-1} + (M - 2)C_n + C_{n+1}] \tag{5.7}$$

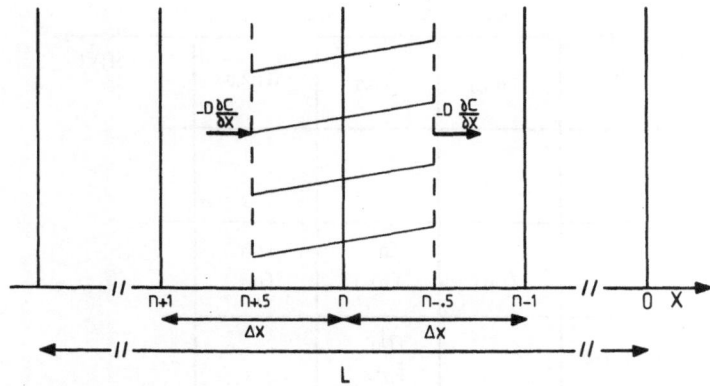

Fig. 5.1. Scheme for one-dimensional transfer of matter by diffusion within the solid.

The scheme for calculation is shown in Fig. 5.2, using the two ways of writing the concentration shown in Eq. (5.3).

Surface of the Sheet
As the rate of evaporation is infinite, the concentration of liquid on the sheet surface falls to zero as soon as the drying process starts, or in the more general case this surface concentration immediately reaches the value which is at equilibrium with the surrounding atmosphere.

$$C_0 = C_{eq} \text{ or } C_0 = 0 \text{ (surface)} \qquad (5.8)$$

Of course, the general equation (5.7) can also be used for the plane at 1 next to the surface at 0 (Fig. 5.3).

$$CN_1 = \frac{1}{M} [C_2 + (M - 2)C_1 + C_{eq}] \qquad (5.7')$$

Amount of Matter Remaining
It is easy to determine the amount of matter remaining in the sheet at time t, by integrating with respect to space the concentrations obtained at this time.

$$Q_t = A \sum_{n=0}^{n} C_{n,t} \Delta x \qquad (5.9)$$

Remark 5.1. The problem is symmetrical, and the midplane of the sheet is a plane of symmetry with a gradient of concentration equal to zero.

5.1.2   Concentration-Dependent Diffusivity

The drying problem with concentration-dependent diffusivity is much more complex than the problem with constant diffusivity. Generally, no analytical solution is obtained, and numerical methods must be used.

Fig. 5.1, with the cross-section of the plane sheet divided into Nu equal finite slices of thickness Δx is again considered.

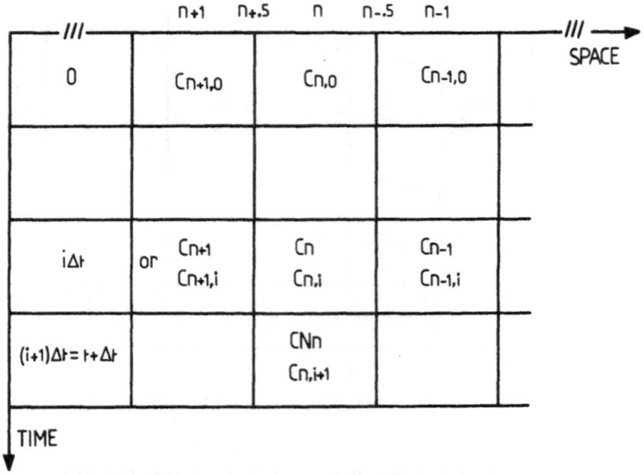

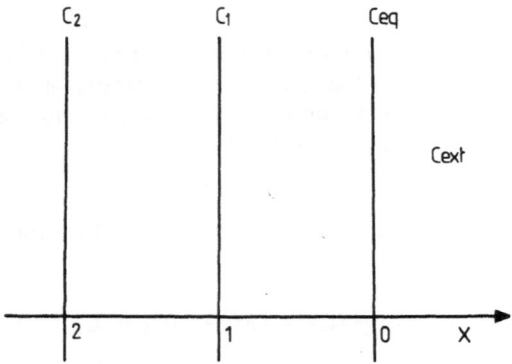

Fig. 5.2. Space–time diagram for calculating the new concentration CN_n after the time increment Δt.

Fig. 5.3. One-dimensional transfer with infinite rate of evaporation on the surface and diffusion within the solid.

The matter balance expressed for the slice of thickness Δx, centred at position n, during the increment of time Δt, with a cross-sectional area A, is again

$$A\left[-D\left(\frac{\partial C}{\partial x}\right)_{n+0.5} + D\left(\frac{\partial C}{\partial x}\right)_{n-0.5}\right]\Delta t = A\Delta x[CN_n - C_n] \qquad (5.2)$$

As the diffusivity is concentration dependent, the approximation for the chord slope made in Eq. (5.4) cannot be made.

It is of interest to introduce a function G defined by

$$G_{n-0.5} = D_{n-0.5}\left(\frac{\partial C}{\partial x}\right)_{n-0.5} = D_{n-0.5}\frac{C_{n-1} - C_n}{\Delta x} \qquad (5.10)$$

The new concentration after time Δt at position n is thus expressed in terms of the concentrations obtained at the previous time at the same position and adjacent positions:

$$CN_n = C_n + (G_{n-0.5} - G_{n+0.5})\frac{\Delta t}{\Delta x} \qquad (5.11)$$

Of course, Eq. (5.11) can be used for all positions, and especially for the plane at 1 next to the surface at 0:

$$CN_1 = C_1 + (G_{0.5} - G_{1.5})\frac{\Delta t}{\Delta x} \qquad (5.11')$$

with

$$G_{0.5} = D_{0.5}\frac{C_0 - C_1}{\Delta x} \qquad (5.10')$$

Amount of Matter Transferred

In the same way as for constant diffusivity, the amount of matter remaining in the sheet is calculated by integrating with respect to space the concentrations within the sheet.

Remark 5.2: Expression of the Diffusivity. The diffusivity can be expressed as a function of the concentration of liquid by various expressions, such as

$$D_{n+0.5} = D_0 \exp(Cte - C_{n+0.5}) \qquad (5.12)$$

where D_0 and Cte are constants and the concentration at position $(n + 0.5)$ is easily obtained by the mean value:

$$C_{n+0.5} = \tfrac{1}{2}(C_n + C_{n+1}) \qquad (5.13)$$

Other expressions of the diffusivity can also be found.

When the profiles of concentration are very flat, and the diffusivity varies little with the liquid concentration, the mean value of the diffusivity can be used:

$$D_{n+0.5} = \tfrac{1}{2}(D_n + D_{n+1}) \qquad (5.14)$$

5.2 Finite Rate of Evaporation

Two cases are of interest, when the diffusivity is constant and when it is concentration dependent.

5.2.1 Constant Diffusivity

The same numerical treatment is followed as in the previous cases, either within the sheet or on the sheet face.

Within the Solid

Eqs (5.6) and (5.7) obtained within the solid with an infinite rate of evaporation can also be used with a finite rate of evaporation:

$$CN_n = \frac{1}{M}(C_{n-1} + (M - 2)C_n + C_{n+1}) \qquad (5.7)$$

$$M = \frac{(\Delta x)^2}{\Delta t}\frac{1}{D} \qquad (5.6)$$

Sheet Surfaces

A difficulty arises with the determination of the concentration of the evaporating substance on each surface of the sheet.

The matter balance is determined during the time increment Δt within the slice located between the parallel planes of abscissa 1 and 0 (Fig. 5.4), by considering the rate of matter which passes through the plane at 1 by diffusion and the rate of matter which evaporates from the plane at 0:

$$A\left[-D\frac{\partial C}{\partial x}\Big|_{\text{at }1} - \frac{F_0}{\rho}(C_0 - C_{\text{ext}})\right]\Delta t = A\Delta x(CN_{0.5} - C_{0.5}) \qquad (5.15)$$

C_0 is the concentration of liquid on the surface at time t and C_{ext} the concentration on the surface which is at equilibrium with the surrounding atmosphere, F_0 is the rate of evaporation of the pure liquid (g cm^{-2} s^{-1}) and ρ the density of the liquid (g cm^{-3}).

Some assumptions must be made for the differences in the concentrations $CN_{0.5} - C_{0.5}$, on the right-hand side of Eq. (5.15).

First Assumption. Consider a slice of thickness Δx next to the surface with

$$CN_{0.5} - C_{0.5} = CN_0 - C_0 \qquad (5.16)$$

As the diffusivity is constant, another assumption is made for the gradient of concentration at the plane at 1:

$$-D\left(\frac{\partial C}{\partial x}\right)_1 = D\frac{C_2 - C_0}{2\Delta x} \qquad (5.17)$$

On defining the dimensionless number N,

$$N = \frac{F_0}{\rho}\frac{\Delta x}{D} \qquad (5.18)$$

the new concentration on the sheet surface after time Δt, CN_0, is expressed in terms of the concentrations at the previous time at various planes, by the equation

$$CN_0 = \frac{1}{2M}[C_2 + (2M - 1 - 2N)C_0 + 2NC_{\text{ext}}] \qquad (5.19)$$

where C_{ext} is the surface concentration which is at equilibrium with the surrounding atmosphere and M and N are the dimensionless numbers given in Eqs (5.18) and (5.6):

$$M = \frac{(\Delta x)^2}{\Delta t}\frac{1}{D} \qquad (5.6)$$

Second Assumption. Consider a slice of thickness $\Delta x/2$ next to the surface (Fig. 5.4). The matter balance is calculated for this slice during the increment of time Δt by taking into account the rate of diffusion at the plane at 0.5 and the rate of evaporation on the surface:

$$A\left[-D\frac{\partial C}{\partial x}\Big|_{\text{at }0.5} - \frac{F_0}{\rho}\Big|_{\text{at }0}(C_0 - C_{\text{ext}})\right]\Delta t = \frac{A\Delta x}{2}(CN_{0.25} - C_{0.25}) \qquad (5.20)$$

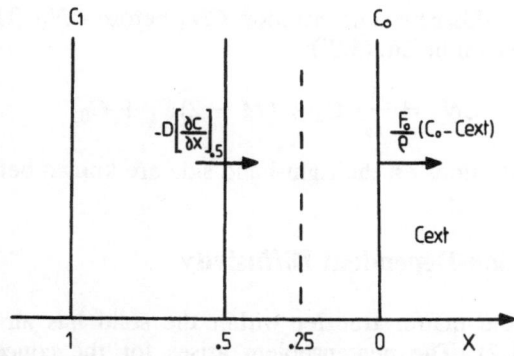

Fig. 5.4. One-dimensional transfer with finite rate of evaporation on the surface and constant diffusivity. Slice next to the surface.

$C_{0.25}$, $CN_{0.25}$ are the concentrations at position 0.25 at time t, and after time Δt respectively, and $\partial C/\partial x$ is the gradient of concentration at the plane at 0.5.

By making the assumptions

$$CN_{0.25} - C_{0.25} = CN_0 - C_0$$

and

$$-D\left(\frac{\partial C}{\partial x}\right)_{0.5} = D\frac{C_1 - C_0}{\Delta x}$$

Eq. (5.20) can be rewritten as follows:

$$D\frac{(C_1 - C_0)}{\Delta x} - \frac{F_0}{\rho}(C_0 - C_{\text{ext}}) = \frac{\Delta x}{2\Delta t}(CN_0 - C_0) \qquad (5.21)$$

and leads to the expression of the concentration of liquid on the surface:

$$CN_0 = \frac{1}{M}\left[2C_1 + (M - 2 - 2N)C_0 + 2NC_{\text{ext}}\right] \qquad (5.22)$$

with the dimensionless numbers M and N given by Eqs (5.6) and (5.18).

Third Assumption. Consider a slice of thickness $\Delta x/2$ next to the surface, and the equation:

$$CN_{0.25} - C_{0.25} = \tfrac{3}{4}(CN_0 - C_0) + \tfrac{1}{4}(CN_1 - C_1) \qquad (5.23)$$

With the assumption of Eq. (5.23), the matter balance calculated within the slice of thickness $\Delta x/2$ next to the surface leads to the new concentration CN_0:

$$CN_0 = C_0 - \tfrac{1}{3}(CN_1 - C_1) + \frac{8}{3M}\left[C - (N + 1)C_0 + NC_{\text{ext}}\right] \qquad (5.24)$$

where CN_1 is the new concentration at the plane at 1 after time Δt and M and N are the dimensionless numbers given by Eqs (5.6) and (5.18).

Eq. (5.24) results from the best assumption. However, attention must be drawn to the method of calculating the concentration: as the new concentration on the surface CN_0 is expressed in terms of the new concentration at plane 1, CN_1, it is necessary to compute this new concentration CN_1 before CN_0. This calculation of CN_1 is possible as shown in Eq. (5.7'):

$$CN_1 = \frac{1}{M}[C_2 + (M - 2)C_1 + C_0] \qquad (5.7')$$

where all the concentrations on the right-hand side are known before calculation.

5.2.2 Concentration-Dependent Diffusivity

The calculation of the matter transfer within the solid has already been made previously (Sect. 5.1.2). The new problem arises for the concentration on the sheet surface.

Within the Solid
The new concentration at plane n, CN_n, is expressed in terms of the previous concentration C_n and of the function G by

$$G_{n-0.5} = D_{n-0.5}\left(\frac{\partial C}{\partial x}\right)_{n-0.5} \qquad (5.10)$$

$$CN_n = C_n + \frac{\Delta t}{\Delta x}(G_{n-0.5} - G_{n+0.5}) \qquad (5.11)$$

Sheet Surface
The slice of thickness $\Delta x/2$ next to the surface is considered (Fig. 5.5), with concentration-dependent diffusivity. The matter balance is obtained during the increment of time by taking into account the rate of matter diffusing through the plane at 0.5 and the rate of evaporation at the plane at 0:

$$A\left[-D\left(\frac{\partial C}{\partial x}\right)_{\text{at } 0.5} - \frac{F_0}{\rho}(C_0 - C_{\text{ext}})\right]\Delta t = \frac{A\Delta x}{2}(CN_{0.25} - C_{0.25}) \qquad (5.20')$$

As shown in the previous section, various assumptions can be made with the right-hand side.

First Assumption. With the simple assumption

$$CN_{0.25} - C_{0.25} = CN_0 - C_0 \qquad (5.16)$$

and by using the function G already given,

$$G_{0.5} = D_{0.5}\left(\frac{\partial C}{\partial x}\right)_{0.5} = D_{0.5}\frac{C_0 - C_1}{\Delta x} \qquad (5.10')$$

the new concentration at the surface is obtained by the following expression:

$$CN_0 = C_0 - \frac{2\Delta t}{\Delta x}\left[G_{0.25} + \frac{F_0}{\rho}(C_0 - C_{\text{ext}})\right] \qquad (5.25)$$

Second Assumption. A better assumption is made with

$$CN_{0.25} - C_{0.25} = \tfrac{3}{4}(CN_0 - C_0) + \tfrac{1}{4}(CN_1 - C_1) \qquad (5.23)$$

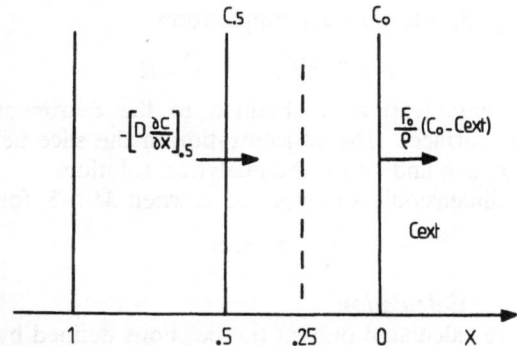

Fig. 5.5. One-dimensional transfer with finite rate of evaporation on the surface and concentration-dependent diffusivity. Slice next to the surface.

The new concentration on the surface after time Δt, CN_0, is thus expressed in terms of the function G and of the new concentration CN_1 and other concentrations obtained at the previous time:

$$CN_0 = C_0 - \frac{1}{3}(CN_1 - C_1) - \frac{8\Delta t}{3\Delta x}\left[G_{0.5} + \frac{F_0}{\rho}(C_0 - C_{ext})\right] \quad (5.26)$$

Of course, it is necessary to compute the value of the concentration CN_1 before that of CN_0. The concentration CN_1 is given by the simple relation

$$CN_1 = C_1 + \frac{\Delta t}{\Delta x}(G_{0.5} - G_{1.5}) \quad (5.11')$$

Amount of Matter Remaining
The amount of matter remaining in the sheet is defined by integrating the concentrations with respect to space.

5.3 Conclusions

The following three conclusions are worth noting:

Wide Use of the Numerical Models
The numerical models can be used whatever the initial profile of concentration of substance. This is an advantage over the analytical solutions which can be obtained only for simple shapes of the initial profiles of concentration.

Of course, the numerical models are also of help when there is no analytical solution, for instance when the diffusivity is concentration-dependent.

Conditions of Stability for Calculation
In many cases, the dimensionless number M is found, as for instance Eq.(5.7) when the diffusivity is constant.

Some authors have proposed [1] to determine the value of the increments of space and of time Δx and Δt, so as to obtain

$$M = 2 \quad (5.27)$$

in order to reduce Eq. (5.7) to the very simple form

$$CN_n = \tfrac{1}{2}[C_{n-1} + C_{n+1}]$$ (5.28)

Unfortunately, this simplification is obtained to the detriment of calculation, especially next to the surfaces. The concentration in the slice next to the surface oscillates alternately above and below the analytical solution.

A better value for dimensionless number M is when $M > 3$, for instance

$$4 < M < 8$$ (5.29)

Assumptions Made for Calculation
The concentrations are calculated only at the positions defined by an integer. The concentration at a position $(n - 0.5)$ located between the slices $(n - 1)$ and n must then be calculated in the following way:

$$C_{n-0.5} = \tfrac{1}{2}(C_{n-1} + C_n)$$ (5.30)

and the diffusivity at the place $(n - 0.5)$ can thus be calculated in either way.

When the diffusivity varies little with concentration, and when the profile of concentration is fairly flat, the simple equation can be used:

$$D_{n+0.5} = \tfrac{1}{2}(D_{n+1} + D_n)$$ (5.14)

A better expression is obtained by calculating the diffusivity in terms of the new concentration at position $(n + 0.5)$:

$$D_{n+0.5} = D_0 \exp\left(\text{Cte} - \frac{C_n + C_{n+1}}{2}\right)$$ (5.31)

and then the function G becomes

$$G_{n+0.5} = D_{n+0.5} \frac{C_n - C_{n+1}}{\Delta x}$$ (5.10)

Expression of the Rate of Evaporation
The rate of evaporation is expressed in two ways:

- By F_0, when F_0 is expressed by the volume of liquid (cm^3) evaporated per second through an area of 1 cm^2 (cm s^{-1}).
- By F_0/ρ, when F_0 is expressed by the mass of liquid (g) evaporated per second through an area of 1 cm^2. As the density of liquid ρ is expressed in g cm^{-3}, the rate of evaporation F_0/ρ is thus expressed in the same way (cm s^{-1}).

Symbols

A	Area of the surface through which the matter is transferred
C_n, $C_{n,t}$	Concentration at position n at time t
CN_n, $C_{n,t+\Delta t}$	Concentration at position n after time Δt
C_0	Concentration on the surface
D	Diffusivity (cm^2 s^{-1})

F_0	Rate of evaporation of the pure liquid ($\mathrm{g\,cm^{-2}\,s^{-1}}$)
G	Function (Eq. (5.10))
M	Dimensionless number (Eq. (5.6))
n	Position
N	Number of slices ($L = N\Delta x$)
N	Dimensionless number (Eq. (5.18))
Q'_t	Amount of substance remaining in the solid
$\Delta t, \Delta x$	Increments of time and of space, respectively
ρ	Density of the substance ($\mathrm{g\,cm^{-3}}$)

Reference

1. Wuithier P. Raffinage et genie chimique, vol. 1. Editions Technip, Paris, 1972, p 295

Chapter 6

Numerical Analysis for a Cylinder

Many problems appear with cylinders because various kinds of cylinders are encountered. Moreover, interesting applications are drawn either from solid cylinders or from hollow cylinders, with infinite or finite length in both cases.

The rate of evaporation can be finite or infinite, and diffusivity can be constant or concentration dependent. The following cases are examined, for constant and concentration-dependent diffusivity:

1. Solid cylinder of infinite length with infinite rate of evaporation.
2. Solid cylinder of infinite length with finite rate of evaporation.
3. Solid cylinder of finite length with infinite rate of evaporation.
4. Solid cylinder of finite length with finite rate of evaporation.
5. Liquid-filled hollow cylinder of infinite length with infinite rate of evaporation.
6. Liquid-filled hollow cylinder of infinite length with finite rate of evaporation.
7. Liquid-filled hollow cylinder of finite length with infinite rate of evaporation.
8. Liquid-filled hollow cylinder of finite length with finite rate of evaporation.

The problem of diffusion–evaporation of a substance with constant diffusivity can generally be resolved by a mathematical treatment. Nevertheless, these cases are considered in this chapter using numerical analysis in order to help the reader to get accustomed to this way of working.

The advantage of the numerical model over the analytical solution is that the model can be used whatever the initial profile of concentration of the diffusing substance within the solid. An analytical solution can be found only for simple initial profiles.

Moreover, the results obtained for both constant and concentration-dependent diffusivity can play the role of basic numerical models, and more complex difficulties (concerning for example the behaviour of the surrounding atmosphere) can be added to them.

6.1 Solid Cylinder of Infinite Length with Infinite Rate of Evaporation

6.1.1 Constant Diffusivity

In this case, the transfer of substance is radial only. The circular cross-section of the solid cylinder of radius R is considered (Fig. 6.1). It is divided into N_r concentric circular elements of constant thickness Δr, with

$$R = N_r \Delta r \tag{6.1}$$

A circular element of height U, radius r and thickness Δr, is considered, as well as the integer j defined by

$$r = j\Delta r \quad \text{with} \quad 0 \leqslant j \leqslant N_r \tag{6.2}$$

Within the Cylinder $1 \leqslant j \leqslant N_r - 1$
The matter balance is calculated within this circular element during the increment of time Δt as follows:

$$U\left[- AD \underset{\text{at } (j-0.5)}{\frac{\partial C}{\partial r}} + AD \underset{\text{at } (j+0.5)}{\frac{\partial C}{\partial r}}\right] \Delta t = \underset{\text{at } j}{UA\Delta r(CN_j - C_j)} \tag{6.3}$$

where A is the length of the circumference through which the substance diffuses, i.e. at $(j + 0.5)$ and $(j - 0.5)$, respectively.
The gradient of concentration is

$$\left(\frac{\partial C}{\partial r}\right)_{j+0.5} = \frac{C_{j+1} - C_j}{\Delta r} \tag{6.4}$$

and the matter balance can be determined:

$$2\pi\Delta r(j - 0.5)D \frac{C_{j-1} - C_j}{\Delta r} + 2\pi\Delta r(j + 0.5)D \frac{C_{j+1} - C_j}{\Delta r}$$
$$= 2\pi\frac{(\Delta r)^2}{\Delta t} j(CN_j - C_j) \tag{6.5}$$

where C_j and CN_j represent the mean concentration within this circular element at time t, and after the time increment Δt, respectively.
The new concentration CN_j can thus be expressed in terms of the previous concentrations obtained in the same element and in the adjacent elements.

$$CN_j = \frac{1}{M}[C_{j+1} + (M - 2)C_j + C_{j-1}] + \frac{1}{2jM}[C_{j+1} - C_{j-1}] \tag{6.6}$$

with the dimensionless number M

$$M = \frac{(\Delta r)^2}{D\Delta t} \tag{6.7}$$

Surface of the Cylinder, $j = N_r$ or $r = R$
As the rate of evaporation is infinite, the concentration on the surface falls to zero as soon as the process starts.

$$C_N = 0 \tag{6.8}$$

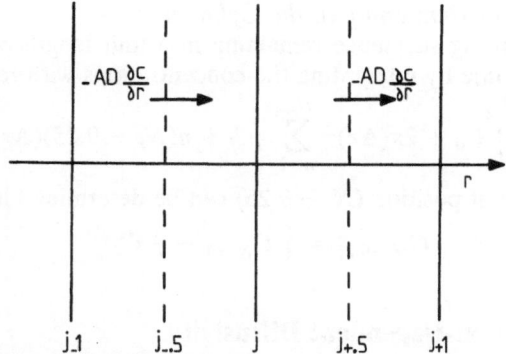

Fig. 6.1. Scheme for diffusion within the cylinder annulus of radius $j\Delta r$ and thickness Δr.

Centre of the Cylinder, $j = 0$ or $r = 0$
The above equation obtained for the cylinder cannot be used at the centre. A cylindrical element of radius $\Delta r/2$ and height U, is considered (Fig. 6.2). The matter balance for the element during the increment of time Δt gives

$$UA \frac{\partial C}{\partial r} \Delta t = \pi U \left(\frac{\Delta r}{2}\right)^2 (CN_0 - C_0) \qquad (6.9)$$

$$\text{at } j = 0.5$$

The gradient of concentration being

$$-\frac{\partial C}{\partial r} = \frac{C_0 - C_1}{\Delta r} \qquad (6.4)$$

the matter balance becomes

$$-2\pi \left(\frac{\Delta r}{2}\right) D \frac{C_0 - C_1}{\Delta r} \Delta t = \pi \left(\frac{\Delta r}{2}\right)^2 (CN_0 - C_0) \qquad (6.10)$$

The new concentration at the centre of the cylinder after time Δt is thus easily obtained:

$$CN_0 = C_0 - \frac{4}{M} (C_0 - C_1) \qquad (6.11)$$

Three equations are necessary for calculating the concentration of substance within the cylinder, at the centre and on the surface.

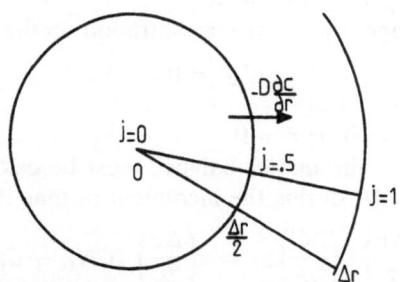

Fig. 6.2. Scheme for diffusion at the centre of the cylinder. Cylinder of radius $\Delta r/2$.

Amount of Substance Remaining in the Cylinder

The amount of diffusing substance remaining in a unit length of the cylinder can be obtained at any time by integrating the concentrations with respect to space:

$$Q_t = \pi \left(\frac{\Delta r}{2}\right)^2 C_0 + 2\pi(\Delta r)^2 \sum_{n=1}^{N_r-1} jC_j + \pi(N_r - 0.25)(\Delta r)^2 C_{N_r-0.25} \quad (6.12)$$

The concentration at position $(N_r - 0.25)$ can be determined in many ways, e.g.

$$C_{N_r-0.25} = \tfrac{1}{4} C_{N_r-1} + \tfrac{3}{4} C_{N_r} \quad (6.13)$$

6.1.2 Concentration-Dependent Diffusivity

As this case of concentration-dependent diffusivity cannot be resolved by a mathematical treatment, a numerical model is necessary. The circular cross-section of the solid cylinder is considered (Fig. 6.1). It is divided into N_r concentric circular elements of thickness Δr.

Three cases are studied: within the cylinder, on the surface and at the centre.

Within the Cylinder, $1 \leqslant j \leqslant N_r - 1$

The matter balance is written in the same way as for constant diffusivity:

$$U\left[- AD \frac{\partial C}{\partial r} + AD \frac{\partial C}{\partial r}\right] \Delta t = UA\Delta r(CN_j - C_j) \quad (6.3)$$
$$\text{at } (j-0.5) \quad \text{at } (j+0.5) \quad\quad\quad \text{at } j$$

The diffusivity, being concentration dependent, varies with position and time, and so will never have the same value at positions $(j - 0.5)$ and $(j + 0.5)$.

On defining the function H,

$$H_{j+0.5} = (j + 0.5)D \frac{\partial C}{\partial r} \text{ at position } (j + 0.5) \quad (6.14)$$

where D is the value of diffusivity at position $(j + 0.5)$ and $\partial C/\partial r$ is the gradient of concentration at the same position, the new concentration at position j is thus expressed in terms of H and of the various concentrations obtained at the same place:

$$CN_j = C_j + \frac{\Delta t}{j\Delta r} [H_{j+0.5} - H_{j-0.5}] \quad (6.15)$$

The function H has dimensions of concentration $\times$ cm s^{-1}.

Surface of the Cylinder, $j = N_r$ or $r = R$

For an infinite rate of evaporation, the concentration on the surface is

$$C_{N_r} = 0 \quad (6.16)$$

Centre of the Cylinder, $j = 0$ or $r = 0$

As for constant diffusivity, the matter balance must be calculated in the cylinder of radius $\Delta r/2$ and height U, during the increment of time Δt (Fig. 6.2):

$$2\pi\left(\frac{\Delta r}{2}\right)D\left(\frac{\partial C}{\partial h}\right)\Delta t = \pi\left(\frac{\Delta r}{2}\right)^2(CN_0 - C_0) \quad (6.17)$$
$$\quad\quad\quad\quad\quad j=0.5 \quad\quad\quad\quad at\ j=0$$

The new concentration CN_0 is easily obtained:

$$CN_0 = C_0 + \frac{8\Delta t}{\Delta r}H_{0.5} \tag{6.18}$$

with the function $H_{0.5}$

$$H_{0.5} = 0.5D\frac{\partial C}{\partial r} \quad \text{at position 0.5} \tag{6.14}$$

Amount of Substance Remaining in the Cylinder
The amount of substance remaining in the cylinder at any time can be calculated by integrating the concentration with respect to space, as shown in the case of constant diffusivity.

6.2   Solid Cylinder of Infinite Length with Finite Rate of Evaporation

6.2.1   Constant Diffusivity

The following three parts of the cylinder are of interest: within the cylinder, at the centre and on the surface.

Within the Cylinder, $1 \leqslant j \leqslant N_r - 1$
The problem is the same as that already studied for the cylinder with infinite rate of evaporation.

The new concentration at position j after the time increment Δt is thus given as a function of the concentrations found at the previous time at the same place and at the adjacent two places:

$$CN_j = C_j + \frac{1}{M}[C_{j+1} - 2C_j + C_{j-1}] + \frac{1}{2jM}[C_{j+1} - C_{j-1}] \tag{6.6}$$

with the dimensionless number M

$$M = \frac{(\Delta r)^2}{D\Delta t} \tag{6.7}$$

Centre of the Cylinder, $j = 0$ or $r = 0$
For the same reason as that given for the cylinder with infinite rate of evaporation, Eq. (6.6) is not suitable for the centre of the cylinder.
The following equation must be used:

$$CN_0 = C_0 - \frac{4}{M}(C_0 - C_1) \tag{6.11}$$

Surface of the Cylinder, $j = N_r$ or $r = R$
An annulus of radius $(N_r - 0.25)$ and thickness $\Delta r/2$, next to the surface is considered (Fig. 6.3). The matter balance is calculated within this annulus during the increment of time Δt by taking into account the matter transferred by

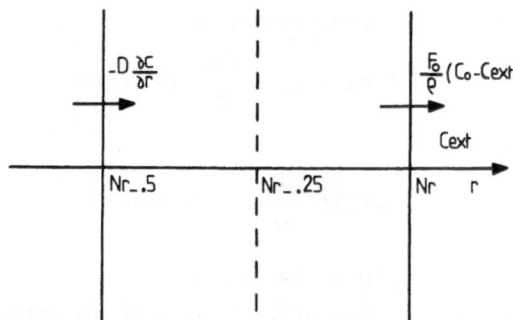

Fig. 6.3. Diffusion–evaporation through the surface of a cylinder of infinite length. Annulus of thickness $\Delta r/2$.

diffusion through the surface at position $(N_r - 0.5)$ and by evaporation from the external surface:

$$\left[- AD \left.\frac{\partial C}{\partial r}\right|_{\text{at } N-0.5} - A \left.\frac{F_0}{\rho}(C_{N_r} - C_{\text{ext}})\right|_{\text{at } N} \right] \Delta t = A \left(\frac{\Delta r}{2}\right)(CN_{N-0.25} - C_{N-0.25}) \quad \text{at } N-0.25 \qquad (6.19)$$

By making the assumption

$$CN_{N-0.25} - C_{N-0.25} = CN_N - C_N \qquad (6.20)$$

and introducing the dimensionless numbers

$$M = \frac{(\Delta r)^2}{D \Delta t} \qquad (6.7)$$

and

$$\text{Na} = \frac{F_0 \Delta r}{\rho D} \qquad (6.21)$$

the new concentration on the surface after time Δt is thus given by

$$CN_N = C_N + \left(\frac{2N-1}{N-0.25}\right)\frac{1}{M}(C_{N-1} - C_N) - \left(\frac{2N}{N-0.25}\right)\frac{\text{Na}}{M}(C_N - C_{\text{ext}})$$

$$(6.22)$$

Amount of Substance Remaining in the Cylinder
The amount of substance remaining in the cylinder at time t is obtained by integrating the concentrations with respect to space, by using Eqs (6.12) and (6.13).

6.2.2   Concentration-Dependent Diffusivity

The following three parts of the cylinder are considered: within the cylinder, at the centre and on the surface.

Within the Cylinder, $1 \leqslant j \leqslant N - 1$ *(Fig. 6.1)*
The problem is the same as that for the infinite rate of evaporation, and the new concentration at position j after the time increment Δt is given by

$$CN_j = C_j + \frac{\Delta t}{j\Delta r}\,[H_{j+0.5} - H_{j-0.5}] \qquad (6.15)$$

with the function H:

$$H_{j+0.5} = (j + 0.5)D\,\frac{\partial C}{\partial r} \quad \text{at } (j + 0.5) \qquad (6.14)$$

Centre of the Cylinder, $j = 0$ *(Fig. 6.2)*
The new concentration at the centre of the cylinder after time Δt is expressed in terms of the previous concentration at the same place and of the function $H_{0.5}$:

$$CN_0 = C_0 + 8\,\frac{\Delta t}{\Delta r}H_{0.5} \qquad (6.18)$$

with

$$H_{0.5} = 0.5D\,\frac{\partial C}{\partial r} \quad \text{at position } 0.5 \qquad (6.14)$$

Surface of the Cylinder, $j = N$, *or* $r = R$
The concentration of the substance can be calculated from the matter balance determined to an annulus next to the external surface. Two cases are worth studying, according to the thickness given to this annulus.

First Case: Annulus With a Thickness of Δr. As shown in Fig. 6.4, the radius of this annulus is $(N - 0.5)\Delta r$ and the thickness Δr. The matter balance is written by considering the matter transferred by diffusion through the surface $(N - 1)$ and by evaporation from the external surface.

$$\left[-\,AD\,\frac{\partial C}{\partial r}_{\text{at } N-1} -\, A\,\frac{F_0}{\rho}\,(C_N - C_{\text{ext}})_{\text{at } N}\right]\Delta t = A(\Delta r)(CN_{N-0.5} - C_{N-0.5})_{\text{at } N-0.5} \qquad (6.19')$$

With the assumption

$$CN_{N-0.5} - C_{N-0.5} = CN_N - C_N \qquad (6.20)$$

the above equation can be written as follows:

$$\left[-\,2\pi(N - 1)\Delta rD\,\frac{\partial C}{\partial r}_{\text{at } N-1} -\, 2\pi N\Delta r\,\frac{F_0}{\rho}\,(C_N - C_{\text{ext}})_{\text{at } N}\right]\Delta t$$

$$= 2\pi(N - 0.5)(\Delta r)^2(CN_N - C_N)_{\text{at } N-0.5} \qquad (6.23)$$

and the new concentration on the surface after time Δt, CN_N, is obtained:

$$CN_N = C_N - \frac{\Delta t}{(N - 0.5)\Delta r}\left[H_{N-1} + N\,\frac{F_0}{\rho}\,(C_N - C_{\text{ext}})\right] \qquad (6.24)$$

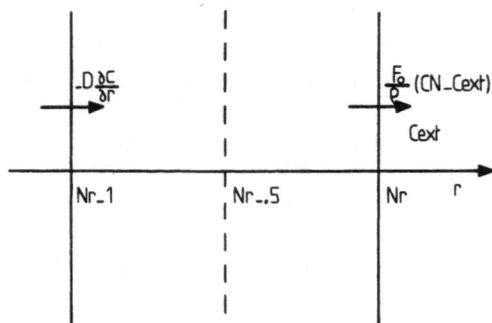

Fig. 6.4. Diffusion–evaporation through the surface of a cylinder of infinite length. Annulus of thickness Δr.

Second Case: Annulus with a Thickness of $\Delta r/2$. The annulus of thickness $\Delta r/2$ next to the surface is considered (Fig. 6.5). The matter balance, by taking into account the diffusion through the surface at position $(N - 0.5)$ and evaporation from the surface, is written as follows:

$$\left[- AD \frac{\partial C}{\partial r} \underset{\text{at } N-0.5}{} - A \frac{F_0}{\rho} (C_N - C_{\text{ext}}) \underset{\text{at } N}{} \right] \Delta t = A \left(\frac{\Delta r}{2} \right) (CN_{N-0.25} - C_{N-0.25}) \underset{\text{at } N-0.25}{} \qquad (6.19'')$$

with the assumption

$$CN_{N-0.25} - C_{N-0.25} = CN_N - C_N \qquad (6.20)$$

and with the function H

$$H_{N-0.5} = (N - 0.5)D \frac{\partial C}{\partial r} \quad \text{at} (N - 0.5) \qquad (6.14)$$

and the new concentration on the surface after time Δt, CN_N is expressed in terms of the previous concentration at the same place by

$$CN_N = C_N - \frac{2\Delta t}{(N - 0.25)\Delta r} \left[H_{N-0.5} + N \frac{F_0}{\rho} (C_N - C_{\text{ext}}) \right] \qquad (6.25)$$

6.3 Solid Cylinder of Finite Length with Infinite Rate of Evaporation

6.3.1 Constant Diffusivity

Various places in the solid cylinder of finite length are examined, as shown in Fig. 6.6.

Within the Cylinder, $0 < j < N_r$ and $0 < n < N_h$
The annulus of radius $j\Delta r$, thickness Δr and height Δh is considered (Fig. 6.7), with the integers j characterising the radial position and n the longitudinal position. The matter balance is calculated within this annulus during the increment of time Δt by taking into account the transfer of substance by diffusion

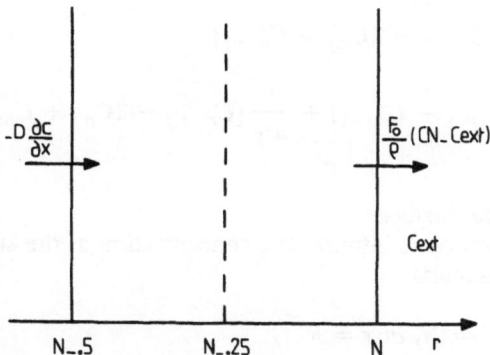

Fig. 6.5. Diffusion–evaporation through the surface of a cylinder of infinite length. Annulus of thickness $\Delta r/2$.

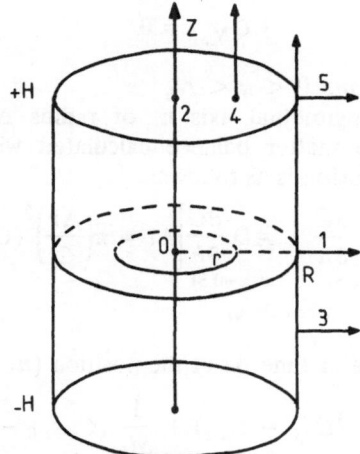

Fig. 6.6. Cylinder of finite length. 1: midheight; 2: longitudinal axis; 3: evaporation from the cylindrical surface; 4: evaporation from the plane surface; 5: evaporation from an edge.

along the radial and longitudinal axes:

$$\left[- AD\,\frac{\partial C}{\partial r} + AD\,\frac{\partial C}{\partial r} - AD\,\frac{\partial C}{\partial h} + AD\,\frac{\partial C}{\partial h} \right]\Delta t = 2\pi j \Delta r \Delta r \Delta h (CN_{n,j} - C_{n,j})$$

at $(j-0.5)$ at $(j+0.5)$ at $(n-0.5)$ at $(n+0.5)$

(radial) (longitudinal)

$$(6.26)$$

With the dimensionless numbers M_r and M_h

$$M_r = \frac{(\Delta r)^2}{D\Delta t} \qquad (6.7)$$

$$M_h = \frac{(\Delta h)^2}{D\Delta t} \qquad (6.7')$$

The new concentration $CN_{n,j}$ is expressed as a function of the concentrations obtained at the previous time at the same place, and at adjacent places.

$$CN_{n,j} = C_{n,j} + \frac{1}{M_r} [C_{n,j-1} - 2C_{n,j} + C_{n,j+1}]$$

$$+ \frac{1}{2jM_r} (C_{n,j+1} - C_{n,j-1}) + \frac{1}{M_h} [C_{n-1,j} - 2C_{n,j} + C_{n+1,j}] \qquad (6.27)$$

Cylindrical and Plane Surfaces
As the rate of evaporation is infinite, the concentration at the surface falls to zero as soon as the process starts.

Cylindrical surface, j = N_r or r = R

$$C_{n,N_r} = 0 \qquad (6.28)$$

Plane surface, n = N_h or h = H

$$C_{N_h,j} = 0 \qquad (6.28')$$

Longitudinal Axis, j = 0 and 0 < n < N_h
The cylinder with the longitudinal axis h, of radius $\Delta r/2$ and height Δh is considered (Fig. 6.8). The matter balance calculated within this cylinder with radial and longitudinal diffusion is as follows:

$$\left[AD \underset{\substack{at(j=0.5)\\(radial)}}{\frac{\partial C}{\partial r}} - AD \underset{\substack{at(n-0.5)}}{\frac{\partial C}{\partial h}} + AD \underset{\substack{at(n+0.5)\\(longitudinal)}}{\frac{\partial C}{\partial h}} \right] \Delta t = \pi \left(\frac{\Delta r}{2} \right)^2 (CN_{n,0} - C_{n,0}) \qquad (6.29)$$

The new concentration after time Δt at the position $(n, 0)$ is given by

$$CN_{n,0} = C_{n,0} - \frac{4}{M_r} (C_{n,0} - C_{n,1}) + \frac{1}{M_h} (C_{n+1,0} - 2C_{n,0} + C_{n-1,0}) \quad (6.30)$$

with the dimensionless numbers M_r and M_h.

Centre of the Cylinder, n = 0 and j = 0
The small cylinder of radius $\Delta r/2$ and height Δh is considered at the position (0, 0) (Fig. 6.9). The matter balance with radial and longitudinal diffusion during the increment of time Δt is written

$$\left[AD \underset{\substack{at\ j=0.5\\(radial)}}{\frac{\partial C}{\partial r}} + 2AD \underset{\substack{at\ n=0.5\\(longitudinal)}}{\frac{\partial C}{\partial h}} \right] \Delta t = \pi \left(\frac{\Delta r}{2} \right)^2 \Delta h (CN_{0,0} - C_{0,0}) \qquad (6.31)$$

After calculation, the new concentration at the centre of the cylinder is given by

$$CN_{0,0} = C_{0,0} - \frac{4}{M_r} (C_{0,0} - C_{0,1}) - \frac{2}{M_h} (C_{0,0} - C_{1,0}) \qquad (6.32)$$

In this case, the matter leaves the small cylinder by longitudinal diffusion through two surfaces.

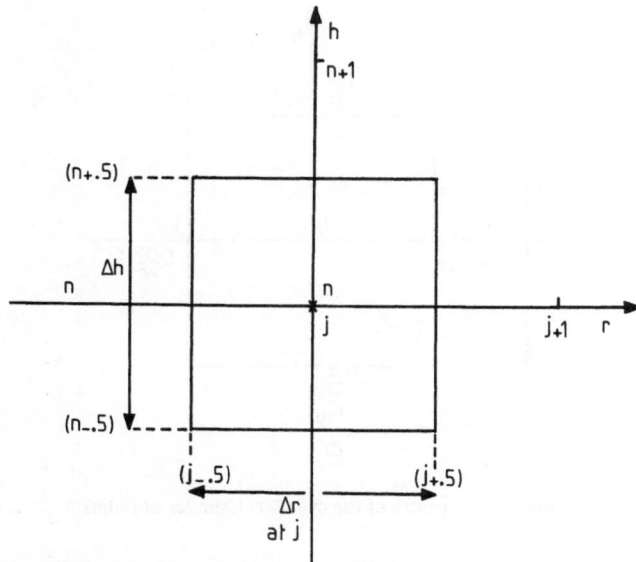

Fig. 6.7. Cylinder of finite length. Transfer within the annulus of radius $j\Delta r$, thickness Δr and height Δh.

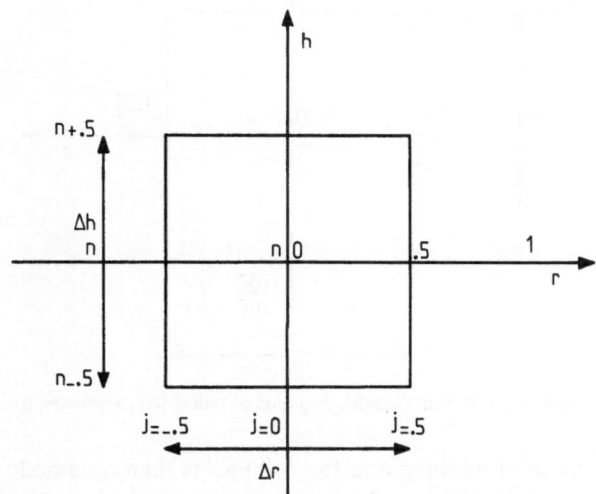

Fig. 6.8. Cylinder of finite length: longitudinal axis. Transfer within the cylinder of radius $\Delta r/2$.

Midheight in the Cylinder, $n = 0$, j

The annulus of thickness Δr and height Δh located at midheight in the cylinder is shown in Fig. 6.10. The matter balance, by considering the radial and longitudinal diffusion during the increment of time Δt, gives

$$\left[-AD \frac{\partial C}{\partial r} + AD \frac{\partial C}{\partial r} + 2AD \frac{\partial C}{\partial h} \right] \Delta t = 2\pi j (\Delta r)^2 \Delta h (CN_{0,j} - C_{0,j}) \quad (6.33)$$

$$\begin{array}{ccc} \text{at } (j-0.5) & \text{at } (j+0.5) & \text{at } (n=0.5) \\ \text{(radial)} & & \text{(longitudinal)} \end{array}$$

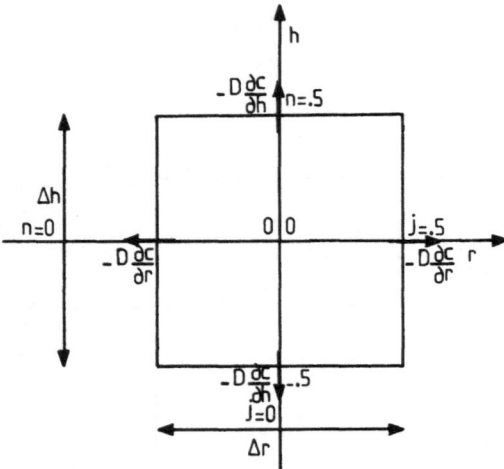

Fig. 6.9. Cylinder of finite length: centre of the cylinder. Cylinder of radius $\Delta r/2$, height Δh.

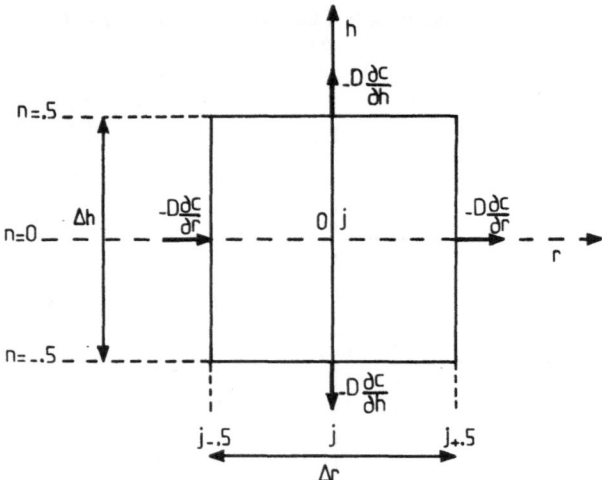

Fig. 6.10. Cylinder of finite length: at midheight. Annulus of radius $j\Delta r$, thickness Δr and height Δh.

The new concentration at midheight in the cylinder is then obtained:

$$CN_{0,j} = C_{0,j} + \frac{1}{M_r} \left(C_{0,j-1} - 2C_{0,j} - C_{0,j+1} \right)$$
$$+ \frac{0.5}{jM_r} \left(C_{0,j+1} - C_{0,j-1} \right) + \frac{2}{M_h} \left(C_{1,j} - C_{0,j} \right) \quad (6.34)$$

The matter leaves this annulus by longitudinal diffusion through two surfaces.

6.3.2 Concentration-Dependent Diffusivity

The positions in the cylinder studied already with constant diffusivity are considered with concentration-dependent diffusivity.

Within the Cylinder, $0 < j < N_r$ and $0 < n < N_h$

The annulus of radius $j\Delta r$, with thickness Δr and height Δh is shown in Fig. 6.7. The matter balance is calculated by taking into account the radial and longitudinal duffusion, during the increment of time Δt:

$$\left[- AD\,\frac{\partial C}{\partial r} + AD\,\frac{\partial C}{\partial r} - AD\,\frac{\partial C}{\partial h} + AD\,\frac{\partial C}{\partial h}\right]\Delta t = 2\pi j(\Delta r)^2 \Delta h(CN_{n,j} - C_{n,j})$$

$$\underset{\text{at }(j-0.5)}{}\quad \underset{\text{at }(j+0.5)}{}\quad \underset{\text{at }(n-0.5)}{}\quad \underset{\text{at }(n+0.5)}{}$$

$$\underset{\text{(radial)}}{}\qquad\qquad \underset{\text{(longitudinal)}}{}$$

$$\tag{6.26}$$

The solution for the new concentration $CN_{n,j}$ after time Δt is thus given by:

$$CN_{n,j} = C_{n,j} + \frac{\Delta t}{j\Delta r}\,[H_{n,j+0.5} - H_{n,j-0.5}] + \frac{\Delta t}{\Delta h}\,[G_{n+0.5,j} - G_{n-0.5,j}] \tag{6.35}$$

with the functions H and G defined as follows:

$$H_{n,j+0.5} = (j+0.5)D\,\frac{\partial C}{\partial r}\ \text{at}\ (n,j+0.5) \tag{6.14}$$

$$G_{n+0.5,j} = D\,\frac{\partial C}{\partial h}\ \text{at}\ (n+0.5,j) \tag{6.36}$$

Cylindrical and Plane Surfaces

As the rate of evaporation is infinite, the concentration at the surface is zero.

Cylindrical surface, $j = N_r$

$$C_{n,N_r} = 0 \tag{6.28}$$

Plane surface, $j = N_h$

$$C_{N_h,j} = 0 \tag{6.28'}$$

Longitudinal Axis, $j = 0$ (Fig. 6.8)

The matter balance in the small cylinder of radius $\Delta r/2$ and height Δh is calculated, by considering the radial and longitudinal diffusion of substance. The matter leaves the cylinder only by radial diffusion

$$\left[AD\,\frac{\partial C}{\partial r} - AD\,\frac{\partial C}{\partial h} + AD\,\frac{\partial C}{\partial h}\right]\Delta t = \pi\left(\frac{\Delta r}{2}\right)^2 \Delta h(CN_{n,0} - C_{n,0}) \tag{6.29}$$

$$\underset{\text{radial}}{}\qquad\qquad \underset{\text{longitudinal}}{}$$

The new concentration on the longitudinal axis after time Δt is thus given by

$$CN_{n,0} = C_{n,0} + \frac{8\Delta t}{\Delta r}\,H_{n,0.5} + \frac{\Delta t}{\Delta h}\,(G_{n+0.5,0} - G_{n-0.5,0}) \tag{6.37}$$

with the functions G (Eq. (6.36)) and

$$H_{n,0.5} = (0.5)D\,\frac{\partial C}{\partial r}\ \text{at}\ (n, 0.5) \tag{6.14}$$

Centre of the Cylinder, with $n = 0$ and $j = 0$ (Fig. 6.9)

A small cylinder of radius $\Delta r/2$ and height Δh is shown in Fig. 6.9. The plane $n = 0$ is a plane of symmetry and the substance leaves the small cylinder by radial and longitudinal diffusion.

The matter balance is calculated, and attention is called to the plane faces through which longitudinal diffusion takes place:

$$\left[+ AD \frac{\partial C}{\partial r} + 2AD \frac{\partial C}{\partial r} \right] \Delta t = \pi \left(\frac{\Delta r}{2} \right)^2 \Delta h (CN_{0,0} - C_{0,0}) \qquad (6.31)$$

at $j=0.5$ at $n=\pm 0.5$
(radial)   (longitudinal)

The new concentration at the centre of the cylinder after time Δt is thus given by

$$CN_{0,0} = C_{0,0} + \frac{8\Delta t}{\Delta r} H_{0,0.5} + \frac{2\Delta t}{\Delta h} G_{0.5,0} \qquad (6.38)$$

where the functions H and G are

$$H_{0,0.5} = (0.5)D \frac{\partial C}{\partial r} \qquad \text{at } (0, 0.5) \qquad (6.14)$$

$$G_{0.5,0} = D \frac{\partial C}{\partial h} \qquad \text{at } (0.5, 0) \qquad (6.36)$$

Midheight in the Cylinder, $n = 0$, j (Fig. 6.10)
An annulus of radius $j\Delta r$, thickness Δr and height Δh is shown in Fig. 6.10. The matter balance is determined by considering the radial and longitudinal diffusion. The matter leaves the annulus through two plane surfaces by longitudinal diffusion:

$$\left[- AD \frac{\partial C}{\partial r} + AD \frac{\partial C}{\partial r} + 2AD \frac{\partial C}{\partial h} \right] \Delta t = 2\pi j (\Delta r)^2 \Delta h (CN_{0,j} - C_{0,j}) \quad (6.33)$$

at $(j-0.5)$ at $(j+0.5)$ at $(n=0.5)$
(radial)   (longitudinal)

The new concentration at midheight after time Δt is thus given by

$$CN_{0,j} = C_{0,j} + \frac{\Delta t}{j\Delta r} (H_{0,j+0.5} - H_{0,j-0.5}) + \frac{2\Delta t}{\Delta h} G_{0.5,j} \qquad (6.39)$$

with the functions H and G previously defined.

Amount of Matter Remaining in the Cylinder at Time t
The amount of matter remaining in the cylinder at time t is determined by integrating the concentration with respect to space, by considering the radial and longitudinal coordinates.

6.4 Solid Cylinder of Finite Length with Finite Rate of Evaporation

6.4.1 Constant Diffusivity

Various places in the cylinder are considered.

Within the Cylinder
The equations determined for the infinite rate of evaporation (Sect. 6.3.1) can be used:

$$CN_{n,j} = C_{n,j} + \frac{1}{M_r}[C_{n,j-1} - 2C_{n,j} + C_{n,j+1}]$$

$$+ \frac{0.5}{jM_r}(C_{n,j+1} - C_{n,j-1}) + \frac{1}{M_h}[C_{n-1,j} - 2C_{n,j} + C_{n+1,j}]$$

$$(6.27)$$

as well as for the longitudinal axis $(j = 0)$

$$CN_{n,0} = C_{n,0} - \frac{4}{M_r}(C_{n,0} - C_{n,1}) + \frac{1}{M_h}[C_{n-1,0} - 2C_{n,0} + C_{n+1,0}] \quad (6.30)$$

and at the centre of the cylinder $(j = 0, \ n = 0)$

$$CN_{0,0} = C_{0,0} - \frac{4}{M_r}(C_{0,0} - C_{0,1}) - \frac{2}{M_h}(C_{0,0} - C_{1,0}) \qquad (6.32)$$

Circular Surface, $j = N_r$ (Fig. 6.11)
An annulus of radius $(N_r - 0.25)\Delta r$ with a thickness of $\Delta r/2$ and height Δh is considered.

The matter balance within this annulus during the increment of time Δt is calculated:

$$\left[-\underset{\text{at } (N_r-0.5)}{AD\frac{\partial C}{\partial r}} - \underset{\substack{\text{at } N_r \\ \text{(radial)}}}{A\frac{F_0}{\rho}(C_{n,N_r} - C_{\text{ext}})} - \underset{\text{at } (n-0.5)}{AD\frac{\partial C}{\partial h}} + \underset{\text{at } (n+0.5)}{AD\frac{\partial C}{\partial h}} \right]\Delta t =$$

$$\underset{\text{(longitudinal)}}{}$$

$$2\pi(N_r - 0.25)\Delta r \frac{\Delta r}{2} \Delta h[CN_{n,N_r-0.25} - C_{n,N_r-0.25}]$$

$$(6.40)$$

With the assumption made at positon $(n, N_r - 0.25)$

$$CN_{n,N_r-0.25} - C_{n,N_r-0.25} = CN_{n,N_r} - C_{n,N_r} \qquad (6.20)$$

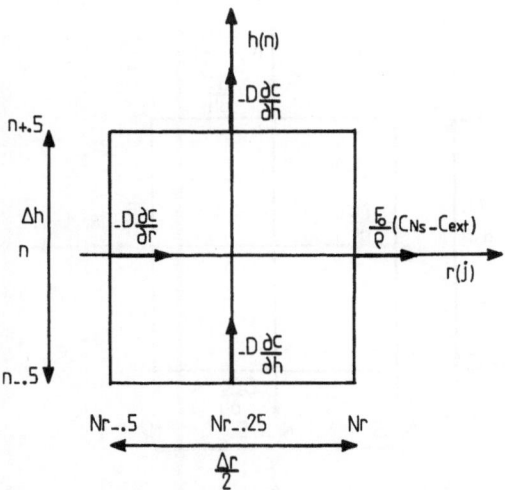

Fig. 6.11. Cylinder of finite length with finite rate of evaporation. Diffusion–evaporation through the cylindrical surface. Annulus of radius $(N_r - 0.25)\Delta r$, thickness $\Delta r/2$ and height Δh.

the new concentration at the surface after time Δt is expressed in terms of the previous concentrations at the surface and in the slice next to the surface:

$$CN_{n,N_r} = C_{n,N_r} + \left(\frac{2N_r - 1}{N_r - 0.25}\right)\frac{1}{M_r}(C_{n,N_r-1} - C_{n,N_r})$$

$$- \left(\frac{2N_r}{N_r - 0.25}\right)\frac{Na}{M_r}(C_{n,N_r} - C_{ext})$$

$$+ \frac{1}{M_h}[C_{n-1,N_r-0.25} - 2C_{n,N_r-0.25} + C_{n+1,N_r-0.25}] \qquad (6.41)$$

Circular Surface at Midheight, $n = 0$, $j = N_r$

An annulus of radius $(N_r - 0.25)\Delta r$, of thickness $\Delta r/2$ and height Δh located next to the circular surface at midheight is shown in Fig. 6.12. The matter balance resulting from radial and longitudinal diffusion as well as from evaporation is determined. The matter leaves the annulus by longitudinal diffusion on both directions:

$$\left[\underset{\substack{\text{at }(N_r-0.5) \\ \text{(radial)}}}{- AD \frac{\partial C}{\partial r}} - \underset{\substack{\text{at }N_r \\ }}{A \frac{F_0}{\rho}(C_{0,N_r} - C_{ext})} + \underset{\substack{\text{at }(n=0.5) \\ \text{(longitudinal)}}}{2AD \frac{\partial C}{\partial h}}\right]\Delta t =$$

$$2\pi(N_r - 0.25)\Delta r\left(\frac{\Delta r}{2}\right)\Delta h[CN_{0,N_r-0.25} - C_{0,N_r-0.25}]$$

$$(6.42)$$

with this assumption

$$CN_{0,N_r-0.25} - C_{0,N_r-0.25} = CN_{0,N_r} - C_{0,N_r} \qquad (6.20)$$

the new concentration at the circular surface at midheight is obtained as a

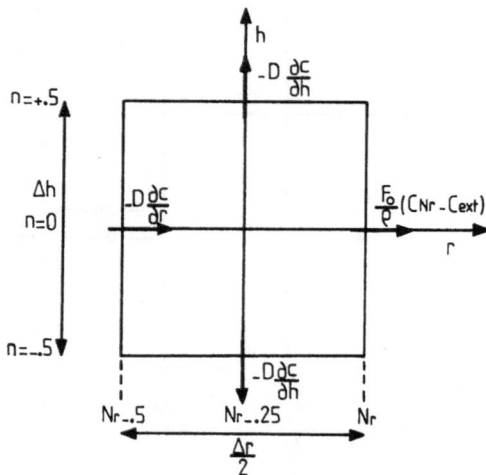

Fig. 6.12. Cylinder of finite length with finite rate of evaporation. Diffusion–evaporation through the cylindrical surface at midheight. Annulus of radius $(N_r - 0.25)\Delta r$, thickness $\Delta r/2$ and height Δh.

function of the previous concentrations at the surface and at the slice next to the surface:

$$CN_{0,N_r} = C_{0,N_r} + \left(\frac{2N_r - 1}{N_r - 0.25}\right) \frac{1}{M_r}(C_{0,N_r-1} - C_{0,N_r})$$

$$- \left(\frac{2N_r}{N_r - 0.25}\right) \frac{Na}{M_r}(C_{0,N_r} - C_{ext}) + \frac{2}{M_h}(C_{1,N_r-0.25} - C_{0,N_r-0.25})$$

$$(6.43)$$

Plane surface, $n = N_h$

An annulus of radius $j\Delta r$, of thickness Δr and height $\Delta h/2$ is considered (Fig. 6.13). The matter balance is determined within this annulus during the increment of time Δt, by taking into account the radial and longitudinal diffusion as well as the evaporation from the plane surface:

$$\left[- AD \frac{\partial C}{\partial h} - A \frac{F_0}{\rho}(C_{N_h,j} - C_{ext}) - AD \frac{\partial C}{\partial r} + AD \frac{\partial C}{\partial r}\right]\Delta t =$$

at $(N_h-0.5)$ at N_h at $(j-0.5)$ at $(j+0.5)$

(longitudinal)   (radial)

$$2\pi j\Delta r\Delta r\frac{\Delta h}{2}(CN_{N_h-0.25,j} - C_{N_h-0.25,j})$$

$$(6.44)$$

By making the assumption

$$CN_{N_h-0.25,j} - C_{N_h-0.25,j} = CN_{N_h,j} - C_{N_h,j} \qquad (6.20)$$

the new concentration at the surface after time Δt is obtained as a function of the previous concentrations on the surface and at various places next to the surface:

$$CN_{N_h,j} = C_{N_h,j} + \frac{2}{M_h}(C_{N_h-1,j} - C_{N_h,j}) - \frac{2F_0\Delta t}{\rho\Delta h}(C_{N_h,j} - C_{ext})$$

$$+ \frac{1}{M_r}[C_{N_h-0.25,j-1} - 2C_{N_h-0.25,j} + C_{N_h-0.25,j+1}]$$

$$+ \frac{0.5}{jM_r}(C_{N_h-0.25,j+1} - C_{N_h-0.25,j-1}) \qquad (6.45)$$

Plane surface, $j = 0$

The cylinder next to the plane surface with the longitudinal axis $j = 0$, of radius $\Delta r/2$ and height $\Delta h/2$ is considered (Fig. 6.14). The matter balance is evaluated within this cylinder by taking into account the radial and longitudinal diffusion as well as the evaporation from the plane surface.

By making the assumption

$$CN_{N_h-0.25,0} - C_{N_h-0.25,0} = CN_{N_h,0} - C_{N_h,0} \qquad (6.20)$$

the new concentration at the plane surface for $j = 0$ can be expressed in terms of the previous concentrations obtained at the plane surface and within the cylinder next to this surface:

$$CN_{N_h,0} = C_{N_h,0} + \frac{2}{M_h}(C_{N_h-1,0} - C_{N_h,0})$$

$$- 2\frac{F_0\Delta t}{\rho\Delta h}(C_{N_h,0} - C_{ext}) + \frac{4}{M_r}(C_{N_h-0.25,1} - C_{N_h-0.25,0}) \qquad (6.46)$$

Edge of the Cylinder, j = N_r and n = N_h
An annulus located at one of the two edges of the cylinder is shown in Fig. 6.15. The radius of this annulus is $(N_r - 0.25)\,\Delta r$, the thickness is $\Delta r/2$ and the height $\Delta h/2$. The matter balance is determined within this annulus during the time increment Δt, by considering the radial and longitudinal diffusion, as well as the evaporation from the circular and plane surfaces:

$$\left[-\underset{\substack{\text{at }(N_r-0.5)}}{AD\frac{\partial C}{\partial r}} - \underset{\substack{\text{at }N_r}}{A\frac{F_0}{\rho}}(C_{N_h-0.25,N_r} - C_{ext}) - \underset{\substack{\text{at }(N_h-0.5)}}{AD\frac{\partial C}{\partial h}} - \underset{\substack{\text{at }N_h}}{A\frac{F_0}{\rho}}(C_{N_h,N_r-0.25} - C_{ext}) \right]\Delta t$$

$$\underset{\text{(radial)}}{\qquad\qquad\qquad\qquad} \underset{\text{(longitudinal)}}{\qquad\qquad\qquad\qquad\qquad}$$

$$= 2\pi(N_r - 0.25)\Delta r\,\frac{\Delta r}{2}\frac{\Delta h}{2}(CN_{N_h-0.25,N_r-0.25} - C_{N_h-0.25,N_r-0.25}) \qquad (6.47)$$

With the assumption

$$CN_{N_h-0.25,0} - C_{N_h-0.25,0} = CN_{N_h,0} - C_{N_h,0} \qquad (6.20)$$

the new concentration at the edge after time Δt is given as a function of the previous concentrations obtained at the edge and within the cylinder next to the edge.

$$CN_{N_h,N_r} = C_{N_h,N_r} + \frac{2N_r - 1}{(N_r - 0.25)M_r}(C_{N_h-0.25,N_r-1} - C_{N_h-0.25,N_r})$$

$$- \left(\frac{2N_r}{N_r - 0.25}\right)\left(\frac{F_0\Delta t}{\rho\Delta r}\right)(C_{N_h-0.25,N_r} - C_{ext})$$

$$+ \frac{2}{M_h}(C_{N_h-1,N_r-0.25} - C_{N_h,N_r-0.25})$$

$$- \frac{2F_0\Delta t}{\rho\Delta h}(C_{N_h,N_r-0.25} - C_{ext}) \qquad (6.48)$$

Amount of Substance Remaining in the Cylinder
The amount of evaporating substance remaining in the cylinder after time t is obtained by integrating the concentrations with respect to space.

6.4.2   Concentration-Dependent Diffusivity

Various places must be considered for the case of concentration-dependent diffusivity.

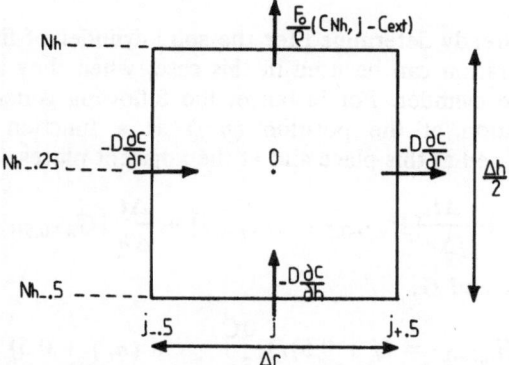

Fig. 6.13. Cylinder of finite length, with finite rate of evaporation. Diffusion–evaporation through the plane surface. Annulus of radius $j\Delta r$, thickness Δr and height $\Delta h/2$.

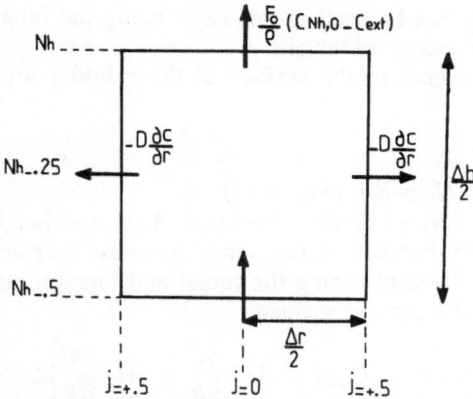

Fig. 6.14. Cylinder of finite length, with finite rate of evaporation. Diffusion–evaporation through the plane surface at the centre. Cylinder of radius $\Delta r/2$, height $\Delta h/2$.

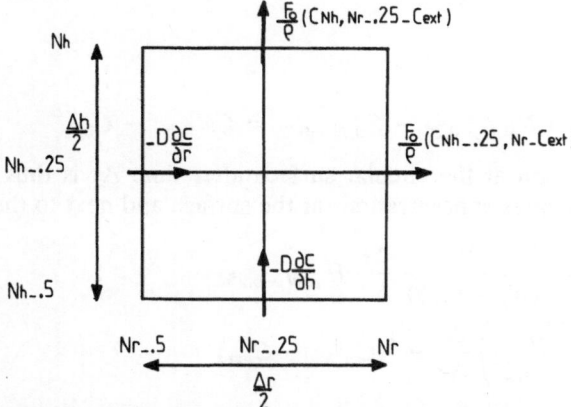

Fig. 6.15. Cylinder of finite length, with finite rate of evaporation. Diffusion–evaporation through the edge. Annulus of radius $(N_r - 0.25)\Delta r$, $\Delta r/2$ thickness and height $\Delta h/2$.

Within the Cylinder

All the equations already determined for the solid cylinder of finite length with a finite rate of evaporation can be used in this case, when they are applied to the internal parts of the cylinder. For instance, the following general equation gives the new concentration at the position (n, j) as a function of the previous concentrations obtained at this place and at the adjacent places.

$$CN_{n,j} = C_{n,j} + \frac{\Delta t}{j \Delta r}[H_{n,j+0.5} - H_{n,j-0.5}] + \frac{\Delta t}{\Delta h}[G_{n+0.5,j} - G_{n-0.5,j}] \quad (6.35)$$

and the functions H and G:

$$H_{n,j+0.5} = (j + 0.5)D \frac{\partial C}{\partial r} \qquad \text{at } (n, j + 0.5) \qquad (6.14)$$

$$G_{n+0.5,j} = D \frac{\partial C}{\partial h} \qquad \text{at } (n + 0.5, j) \qquad (6.36)$$

Other equations have been determined (Sect. 6.3.2) for an infinite rate of evaporation, and they can be used in this case: along the longitudinal axis, at the centre of the cylinder and at midheight.

However, all places next to the surface of the cylinder must be calculated for the present case.

Cylinder Surface, $j = N_r$ and n (Fig. 6.11)

An annulus of radius $(N_r - 0.25)\Delta r$, thickness $\Delta r/2$ and height Δh, is shown in Fig. 6.11. The matter balance within this annulus is calculated during the increment of time Δt, by considering the radial and longitudinal diffusion, as well the evaporation from the circular surface:

$$\left[\underbrace{- AD \frac{\partial C}{\partial r}}_{\substack{\text{at } (N_r-0.5)}} \underbrace{- A \frac{F_0}{\rho}(C_{n,N_r} - C_{\text{ext}})}_{\text{at } N_r} \underbrace{- AD \frac{\partial C}{\partial h}}_{\substack{\text{at } (n-0.5)}} + \underbrace{AD \frac{\partial C}{\partial h}}_{\text{at } (n+0.5)} \right]\Delta t =$$

$$\begin{array}{c} \text{(radial)} \hspace{6em} \text{(longitudinal)} \end{array}$$

$$2\pi(N_r - 0.25)\Delta r \frac{\Delta r}{2} \Delta h[CN_{n,N_r,-0.25} - C_{n,N_r,-0.25}]$$

$$(6.40)$$

With the assumption

$$CN_{0,N_r,-0.25} - C_{0,N_r,-0.25} = CN_{0,N_r} - C_{0,N_r} \qquad (6.20)$$

the new concentration at the circular surface after time Δt is thus obtained as a function of the previous concentrations at the surface and next to the surface:

$$CN_{n,N_r} = C_{n,N_r} - \frac{2}{(N_r - 0.25)} \frac{\Delta t}{\Delta r} H_{n,N_r,-0.25}$$

$$- \left(\frac{2N_r}{N_r - 0.25}\right) \frac{\Delta t}{\Delta r} \frac{F_0}{\rho}(C_{n,N_r} - C_{\text{ext}})$$

$$+ \frac{\Delta t}{\Delta h}[G_{n+0.5,N_r,-0.25} - G_{n-0.5,N_r,-0.25}] \qquad (6.49)$$

Circular Surface at midheight, $n = 0$, $j = N_r$

An annulus of radius $(N_r - 0.25)\Delta r$, with a thickness $\Delta r/2$ and height Δh, which is located in contact with the circular surface at midheight, is considered (Fig. 6.12). The matter balance is calculated for this annulus, by taking into account the radial and longitudinal diffusion as well as the evaporation from the circular surface:

$$\left[\underbrace{- AD \frac{\partial C}{\partial r}}_{\substack{\text{at }(N_r-0.5) \\ \text{(radial)}}} \underbrace{- A \frac{F_0}{\rho}(C_{0,N_r} - C_{\text{ext}})}_{\substack{\text{at }N_r}} + \underbrace{2AD \frac{\partial C}{\partial h}}_{\substack{\text{at }n=0.5 \\ \text{(longitudinal)}}}\right]\Delta t =$$

$$2\pi(N_r - 0.25)\Delta r \frac{\Delta r}{2} \Delta h(CN_{0,Nr-0.25} - C_{0,N_r-0.25}) \tag{6.42}$$

With the assumption

$$CN_{0,N_r-0.25} - C_{0,N_r-0.25} = CN_{0,N_r} - C_{0,N_r} \tag{6.20}$$

the new concentration after time Δt at midheight of the circular surface can be expressed in terms of the previous concentrations obtained at this place and at adjacent places.

$$CN_{0,N_r} = C_{0,N_r} - \left(\frac{2}{N_r - 0.25}\right)\frac{\Delta t}{\Delta r} H_{0,N_r-0.25} + \frac{2\Delta t}{\Delta h} G_{0,N_r-0.25}$$

$$- \left(\frac{2N_r}{N_r - 0.25}\right)\frac{\Delta t}{\Delta r} \frac{F_0}{\rho}(C_{0,N_r} - C_{\text{ext}}) \tag{6.50}$$

Plane Surface, $n = N_h$

An annulus of radius $j\Delta r$, thickness Δr and height $\Delta h/2$, located next to the plane surface is shown in Fig. 6.13. The matter balance is determined by considering the radial and longitudinal diffusion and the evaporation from the plane surface.

$$\left[\underbrace{- AD \frac{\partial C}{\partial h}}_{\substack{\text{at }(N_h-0.5) \\ \text{(longitudinal)}}} \underbrace{- A \frac{F_0}{\rho}(C_{Nh,j} - C_{\text{ext}})}_{\substack{\text{at }N_h}} \underbrace{- AD \frac{\partial C}{\partial r} + AD \frac{\partial C}{\partial r}}_{\substack{\text{at }(j-0.5) \quad \text{at }(j+0.5) \\ \text{(radial)}}}\right]\Delta t =$$

$$2\pi j\Delta r\Delta r \frac{\Delta h}{2}(CN_{N_h-0.25,j} - C_{N_h-0.25,j}) \tag{6.44}$$

With the assumption:

$$CN_{0,N_r-0.25} - C_{0,N_r-0.25} = CN_{0,N_r} - C_{0,N_r} \tag{6.20}$$

the new concentration on the plane surface after time Δt can be given as a function of the previous concentrations at the plane surface and at adjacent positions:

$$CN_{N_h,j} = C_{N_h,j} - \frac{2\Delta t}{\Delta h} G_{N_h-0.5,j} - 2 \frac{\Delta t}{\Delta h} \frac{F_0}{\rho}(C_{N_h,j} - C_{\text{ext}})$$

$$+ \frac{\Delta t}{j\Delta r}[H_{N_h-0.25,j+0.5} - H_{N_h-0.25,j-0.5}] \tag{6.51}$$

Plane Surface, at $j = 0$

The cylinder of radius $\Delta r/2$ and height $\Delta h/2$, located next to the plane surface at the position $j = 0$, must be considered in this case (Fig. 6.14). The matter balance is obtained by taking into account the longitudinal diffusion and evaporation from the plane surface, as well as the radial diffusion:

$$\left[\underset{\substack{\text{at }(N_h-0.5) \\ \text{(longitudinal)}}}{-AD \frac{\partial C}{\partial h}} - \underset{\substack{\text{at }N_h}}{A \frac{F_0}{\rho}(C_{\text{Nh},0} - C_{\text{ext}})} + \underset{\substack{\text{at }n=0.5 \\ \text{(radial)}}}{AD \frac{\partial C}{\partial r}} \right] \Delta t =$$

$$\pi \left(\frac{\Delta r}{2}\right)^2 \frac{\Delta h}{2} (CN_{N_h-0.25,0} - C_{N_h-0.25,0}) \tag{6.52}$$

By making the assumption

$$CN_{N_h-0.25,0} - C_{N_h-0.25,0} = CN_{N_h,0} - C_{N_h,0} \tag{6.20}$$

the new concentration after time Δt is thus expressed by the equation:

$$CN_{N_h,0} = C_{N_h,0} - 2\frac{\Delta t}{\Delta h} G_{N_h-0.5,0} \frac{2\Delta t F_0}{\Delta h \rho}(C_{N_h,0} - C_{\text{ext}}) + 8\frac{\Delta t}{\Delta r} H_{N_h-0.25,0} \tag{6.53}$$

Edge of the Cylinder, $j = N_r$ and $n = N_h$

An annulus located at the edge of the cylinder, of radius $(N_r - 0.25)\Delta r$, thickness $\Delta r/2$ and height $\Delta h/2$ is considered (Fig. 6.15). The matter balance within this annulus during the increment of time Δt can be written as follows:

$$\left[\underset{\substack{\text{at }(N_r-0.5) \\ \text{(radial)}}}{-AD \frac{\partial C}{\partial r}} - \underset{\substack{\text{at }N_r}}{A\frac{F_0}{\rho}(C_{N_h-0.25,N_r} - C_{\text{ext}})} - \underset{\substack{\text{at }(N_h-0.5) \\ \text{(longitudinal)}}}{AD \frac{\partial C}{\partial h}} - \underset{\substack{\text{at }N_h}}{A \frac{F_0}{\rho}(C_{N_h,N_r-0.25} - C_{\text{ext}})} \right] \Delta t$$

$$= 2\pi(N_r - 0.25)\Delta r \frac{\Delta r}{2} \frac{\Delta h}{2} (CN_{N_h-0.25,N_r-0.25} - C_{N_h-0.25,N_r-0.25}) \tag{6.47}$$

with the assumption

$$CN_{N_h-0.25,N_r-0.25} - C_{N_h-0.25,N_r-0.25} = CN_{N_h,N_r} - C_{N_h,N_r} \tag{6.20}$$

the new concentration after time Δt on the edge is expressed in terms of the functions H and G:

$$CN_{N_h,N_r} = C_{N_h,N_r} - \frac{2}{(N_r - 0.25)} \frac{\Delta t}{\Delta h} H_{N_h-0.25,N_r-0.25}$$

$$- \frac{2N_r}{(N_r - 0.25)} \frac{\Delta t F_0}{\Delta r \rho}(C_{N_h-0.25,N_r} - C_{\text{ext}})$$

$$- \frac{2\Delta t}{\Delta h} G_{N_h-0.5,N_r-0.25} - \frac{2\Delta t F_0}{\Delta h \rho}(C_{N_h,N_r-0.25} - C_{\text{ext}}) \tag{6.54}$$

Amount of Matter Remaining in the Cylinder
The amount of diffusing substance remaining in the cylinder at time t is obtained by integrating the concentration with respect to space, as shown in Eq. (6.12).

Remark 6.1. For calculation, it is easier to characterise the position by integers instead of numbers such as $N_h - 0.25$ for instance.
  The problem is resolved by making the following assumptions:

$$C_{N_h-0.25,j} = \tfrac{3}{4} C_{N_h,j} + \tfrac{1}{4} C_{N_h-1,j} \tag{6.55}$$

$$C_{n,N_r-0.25} = \tfrac{3}{4} C_{n,N_r} + \tfrac{1}{4} C_{n,N_r-1} \tag{6.56}$$

in the equations, when necessary. These assumptions must be made when the thickness or height of the annulus of cylinder is $\Delta r/2$ or $\Delta h/2$, located next to the external surfaces and along the axis of the cylinder.

6.5 Liquid-Filled Hollow Cylinder of Infinite Length with Infinite Rate of Evaporation

The concentration of liquid in the case of the hollow cylinder of infinite length is calculated in the same way as for the solid cylinder of infinite length. The only difference arises from the presence of the internal surface.
  The problems created by the internal surface and by the external surface are the same, whatever the rate of evaporation and the diffusivity. The equations found for the internal and external surface have the same shape, and only the indices are modified in order to take into account the position.
  The transfer of substance is radial only.

6.5.1 Constant Diffusivity

The circular cross-section of the hollow cylinder with the internal and external radii R_i and R_e is considered (Fig. 6.16) with the corresponding integers N_i and N_e defined by

$$R_i = N_i \Delta r$$

$$R_e = N_e \Delta r$$

$$r = j\Delta r \tag{6.57}$$

The new concentration after time Δt is thus expressed in terms of the previous concentrations by the following equation.

Within the Cylinder $N_i + 1 \leqslant j \leqslant N_e - 1$

$$CN_j = \frac{1}{M}[C_{j+1} + (M-2)C_j + C_{j-1}] + \frac{1}{2jM}[C_{j+1} - C_{j-1}] \tag{6.58}$$

with the modulus:

$$M = \frac{(\Delta r)^2}{D\Delta t} \tag{6.7}$$

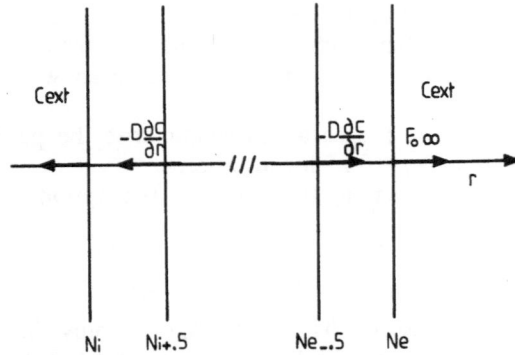

Fig. 6.16. Hollow cylinder of infinite length, with radial transfer by diffusion within the solid and infinite rate of evaporation from the surfaces.

Surfaces of the Cylinder, $j = N_i$ or $j = N_e$
The rate of evaporation being infinite, the concentration of liquid on the surface falls to zero as soon as the drying process starts.

$$C_{N_i} = C_{N_e} = 0 \qquad (6.59)$$

Amount of Matter Remaining in the Cylinder
The amount of matter remaining in the cylinder is determined by integrating the concentration with respect to space as shown in Eq. (6.12), within the boundaries N_i and N_e.

6.5.2 Concentration-Dependent Diffusivity

The problem has been considered for the case of the solid cylinder of infinite length with infinite rate of evaporation. The only difference is due to the presence of the internal surface (Fig. 6.16).

Within the Cylinder, $N_i + 1 \leqslant j \leqslant N_e - 1$
The new concentration CN_j after time Δt is given by

$$CN_j = C_j + \frac{\Delta t}{j\Delta r}[H_{j+0.5} - H_{j-0.5}] \qquad (6.60)$$

with

$$H_{j+0.5} = (j + 0.5)D \frac{\partial C}{\partial r} \text{ at } (j + 0.5) \qquad (6.14)$$

Surfaces of the Cylinder, $j = N_i$ or $j = N_e$
The concentration on each surface is zero:

$$C_{N_i} = C_{N_e} = 0 \qquad (6.59)$$

Amount of Matter Remaining in the Cylinder
The amount of substance remaining in the cylinder at time t is obtained by integrating the concentration with respect to space within the boundaries N_i and N_e.

6.6 Liquid-Filled Hollow Cylinder of Infinite Length With Finite Rate of Evaporation

6.6.1 Constant Diffusivity

The problem is the same as that for the solid cylinder of infinite length with finite rate of evaporation, but special attention must be given to the boundaries defined by internal and external surfaces (Fig. 6.17).

The concentrations can thus be calculated in various places by using the following equations.

Within the Cylinder, $N_i + 1 \leq j \leq N_e - 1$
The new concentration CN_j is expressed in terms of the previous concentrations by

$$CN_j = C_j + \frac{1}{M}[C_{j+1} - 2C_j + C_{j-1}] + \frac{1}{2jM}[C_{j+1} - C_{j-1}] \qquad (6.61)$$

with the dimensionless number M

$$M = \frac{(\Delta r)^2}{D\Delta t} \qquad (6.7)$$

Surfaces of the Cylinder, $j = N_e$ or $j = N_i$
The new concentration CN_N for the external surface is:

$$CN_{N_e} = C_{N_e} + \left(\frac{2N_e - 1}{N_e - 0.25}\right)\frac{1}{M}(C_{N_e-1} - C_{N_e}) - \left(\frac{2N_e}{N_e - 0.25}\right)\frac{Na}{M}(C_{N_e} - C_{ext})$$

$$(6.62)$$

and for the internal surface

$$CN_{N_i} = C_{N_i} + \left(\frac{2N_i - 1}{N_i + 0.25}\right)\frac{1}{M}(C_{N_i+1} - C_{N_i}) - \left(\frac{2N_i}{N_i + 0.25}\right)\frac{Na}{M}(C_{N_i} - C_{ext})$$

$$(6.63)$$

6.6.2 Concentration-Dependent Diffusivity

The solution is the same as for the case of the solid cylinder of infinite length with finite rate of evaporation and concentration-dependent diffusivity (Fig. 6.17).

Within the Cylinder, $N_i + 1 \leq j \leq N_e - 1$
The new concentration after time Δt is given by

$$CN_j = C_j + \frac{\Delta t}{j\Delta r}[H_{j+0.5} - H_{j-0.5}] \qquad (6.64)$$

with the function H

$$H_{j+0.5} = (j + 0.5)D\frac{\partial C}{\partial r} \qquad \text{at } (j + 0.5)$$

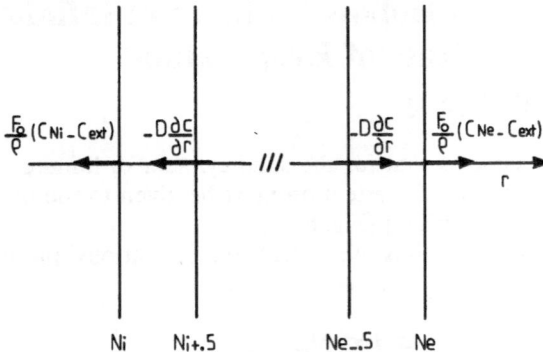

Fig. 6.17. Hollow cylinder of infinite length, with radial diffusion and a finite rate of evaporation.

Surfaces of the Cylinder, $j = N_i$ or $j = N_e$
The new concentration after time Δt is

- For the external surface, $j = N_e$;

$$CN_{N_e} = C_{N_e} - \frac{2\Delta t}{(N_e - 0.25)\Delta r}\left[H_{N_e - 0.5} + N_e \frac{F_0}{\rho}(C_{N_e} - C_{ext})\right] \quad (6.65)$$

- For the internal surface, $j = N_i$:

$$CN_{N_i} = C_{N_i} - \frac{2\Delta t}{(N_i + 0.25)\Delta r}\left[H_{N_i + 0.5} + N_i \frac{F_0}{\rho}(C_{N_i} - C_{ext})\right]$$

6.7 Liquid-Filled Hollow Cylinder of Finite Length with Infinite Rate of Evaporation

The problems which arise from the hollow cylinder of finite length are about the same as those resulting from the solid cylinder of finite length, with attention given to the boundaries and especially to the internal surface.

The transfer within the solid is radial and longitudinal.

6.7.1 Constant Diffusivity

A hollow cylinder of finite length is shown in Fig. 6.18, with the internal and external cylindrical surfaces and the two plane surfaces, as well as the four circular edges at the intersects of the cylindrical and plane surfaces.

The integers previously defined are used:

$$R_i = N_{ri}\Delta r$$
$$R_e = N_{re}\Delta r$$
$$r = j\Delta r$$
$$H = N_h\Delta h$$
$$h = n\Delta h \quad (6.66)$$

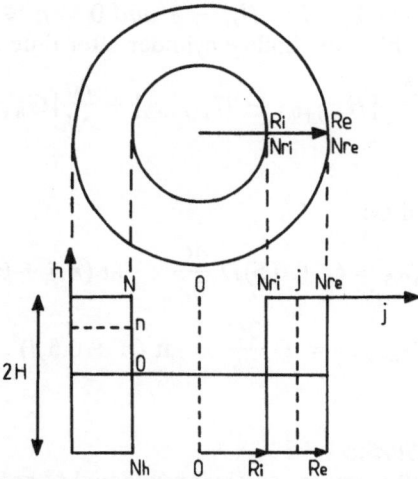

Fig. 6.18. Hollow cylinder of finite length, internal and external radii R_i and R_e, and height $2H$.

with the cylinder of internal and external radii, R_i and R_e, respectively, and of height $2H$.

Within the Cylinder, $N_{ri} + 1 \leqslant j \leqslant N_{re} - 1$ and $0 \leqslant n \leqslant N_h - 1$
The new concentration within the solid after time Δt is expressed in terms of the previous concentrations at the same and adjacent places:

$$CN_{n,j} = C_{n,j} + \frac{1}{M_r}[C_{n,j-1} - 2C_{n,j} + C_{n,j+1}]$$

$$+ \frac{1}{2jM_r}[C_{n,j+1} - C_{n,j-1}] + \frac{1}{M_h}[C_{n-1,j} - 2C_{n,j} + C_{n+1,j}] \qquad (6.67)$$

with the dimensionless numbers M_r and M_h

$$M_r = \frac{(\Delta r)^2}{D\Delta t} \qquad (6.7)$$

$$M_h = \frac{(\Delta h)^2}{D\Delta t} \qquad (6.7')$$

Cylindrical and Plane Surfaces, $j = N_{ri}$, $j = N_{re}$, or $n = N_h$
As the rate of evaporation is infinite, the concentration on the surfaces falls to zero as soon as the drying process starts.

$$C_{n,N_{ri}} = 0 \qquad \text{(internal surface)}$$

$$C_{n,N_{re}} = 0 \qquad \text{(external surface)}$$

$$C_{N_h,j} = 0 \qquad \text{(plane surface)}$$

6.7.2 Concentration-Dependent Diffusivity

This problem was solved for the case of the solid cylinder with infinite rate of evaporation and concentration-dependent diffusivity. The results are given.

Within the Cylinder, $N_{ri} + 1 \leqslant j \leqslant N_{re} - 1$ and $0 \leqslant n \leqslant N_h - 1$
The new concentration within the hollow cylinder after time Δt is given by

$$CN_{n,j} = C_{n,j} + \frac{\Delta t}{j\Delta r}[H_{n,j+0.5} - H_{n,j-0.5}] + \frac{\Delta t}{\Delta h}[G_{n+0.5,j} - G_{n-0.5,j}]$$

(6.69)

with the functions H and G:

$$H_{n,j+0.5} = (j + 0.5)D\frac{\partial C}{\partial r} \quad \text{at } (n,j + 0.5)$$

(6.14)

$$G_{n+0.5,j} = D\frac{\partial C}{\partial h} \quad \text{at } (n + 0.5, j)$$

(6.36)

Cylindrical and Plane Surfaces
The concentrations on all surfaces are obviously equal to zero:

$$\begin{aligned}
C_{N_h,j} &= 0 \qquad \text{(plane surface)} \\
C_{n,N_{ri}} &= 0 \qquad \text{(internal surface)} \\
C_{n,N_{re}} &= 0 \qquad \text{(external surface)} \qquad (6.70)
\end{aligned}$$

Amount of Matter Remaining in the Cylinder
The amount of evaporating substance remaining in the cylinder is obtained by integrating the concentration of this substance with respect to space by taking account of the boundaries.

6.8 Liquid-Filled Hollow Cylinder of Finite Length with Finite Rate of Evaporation

6.8.1 Constant Diffusivity

Some solutions to this problem were given in the case of a solid cylinder of finite length with a diffusing substance exhibiting a finite rate of evaporation and a constant diffusivity. Special attention must be given to the internal surface.

Within the Cylinder, $N_{ri} + 1 \leqslant j \leqslant N_{re} - 1$ and $0 \leqslant n \leqslant N_h - 1$
The new concentration within the hollow cylinder after Δt is expressed in terms of the previous concentration obtained at the same and adjacent places:

$$CN_{n,j} = C_{n,j} + \frac{1}{M_r}[C_{n,j-1} - 2C_{n,j} + C_{n,j+1}]$$

$$+ \frac{0.5}{jM_r}[C_{n,j+1} - C_{n,j-1}] + \frac{1}{M_h}[C_{n-1,j} - 2C_{n,j} + C_{n+1,j}]$$

(6.71)

with the dimensionless numbers M_r and M_h:

$$M_r = \frac{(\Delta r)^2}{D\Delta t} \tag{6.7}$$

$$M_h = \frac{(\Delta h)^2}{D\Delta t} \tag{6.7'}$$

Circular Surface, $j = N_{re}$ or $= N_{ri}$

For the external surface, with $j = N_{re}$, the new concentration $CN_{N_{re}}$ is given by

$$
CN_{N_{re}} = C_{N_{re}} + \left(\frac{2N_{re} - 1}{N_{re} - 0.25}\right)\frac{1}{M_r}(C_{n,N_{re}-1} - C_{ext})
$$
$$
- \left(\frac{2N_{re}}{N_{re} - 0.25}\right)\frac{\text{Na}}{M_r}(C_{n,N_{re}} - C_{ext})
$$
$$
+ \frac{1}{M_h}[C_{n-1,N_{re}-0.25} - 2C_{n,N_{re}-0.25} + C_{n+1,N_{re}-0.25}] \tag{6.72}
$$

For the internal surface, with $j = N_{ri} + 1$, the equation expressing the new concentration after time Δt differs slightly from the equation obtained for the external surface.

$$
CN_{N_{ri}} = C_{N_{ri}} + \left(\frac{2N_{ri} + 1}{N_{ri} + 0.25}\right)\frac{1}{M_r}(C_{n,N_{ri}+1} - C_{ext})
$$
$$
- \left(\frac{2N_{ri}}{N_{ri} + 0.25}\right)\frac{\text{Na}}{M_r}(C_{n,N_{ri}} - C_{ext})
$$
$$
+ \frac{1}{M_h}[C_{n-1,N_{ri}+0.25} - 2C_{n,N_{ri}+0.25} + C_{n+1,N_{ri}+0.25}] \tag{6.72}
$$

with the dimensionless numbers M_r, M_h and Na:

$$\text{Na} = \frac{F_0\Delta r}{\rho D} \tag{6.21}$$

Plane Parallel Surfaces, $n = N_h$

The new concentration CN_{N_h} on each plane parallel surface is given by the same relationship as that shown for the solid cylinder of finite length, with a finite rate of evaporation and constant diffusivity.

$$
CN_{N_h,j} = C_{N_h,j} + \frac{2}{M_h}(C_{N_h-1,j} - C_{N_h,j}) - \frac{2F_0\Delta t}{\rho\Delta h}(C_{N_h,j} - C_{ext})
$$
$$
+ \frac{1}{M_r}[C_{N_h-0.25,j-1} - 2C_{N_h-0.25,j} + C_{N_h-0.25,j+1}]
$$
$$
+ \frac{0.5}{jM_r}(C_{N_h-0.25,j+1} - C_{N_h-0.25,j-1}) \tag{6.73}
$$

Edges of the Cylinder

The new concentration for the edge in contact with the external cylindrical surface has already been found for the case of the solid cylinder of finite length.

Edge in contact with the external circular surface, with $n = N_h$ and $j = N_{re}$:

$$CN_{N_h,N_{re}} = C_{N_h,N_{re}} + \left(\frac{2N_{re} - 1}{N_{re} - 0.25}\right)\frac{1}{M_r}(C_{N_h-0.25,N_{re}-1} - C_{N_h-0.25,N_{re}})$$

$$- \left(\frac{2N_{re}}{N_{re} - 0.25}\right)\frac{F_0\Delta t}{\rho\Delta r}(C_{N_h-0.25,N_{re}} - C_{ext})$$

$$- \frac{2}{M_h}(C_{N_h-1,N_{re}-0.25} - C_{N_h,N_{re}-0.25})$$

$$- \frac{2F_0\Delta t}{\rho\Delta h}(C_{N_h,N_{re}-0.25} - C_{ext}) \tag{6.74}$$

Edge in contact with the internal circular surface $n = N_h$ and $j = N_{ri}$:

$$CN_{N_h,N_{ri}} = C_{N_h,N_{ri}} + \left(\frac{2N_{ri} + 1}{N_{ri} + 0.25}\right)\frac{1}{M_r}(C_{N_h-0.25,N_{ri}+1} - C_{N_h-0.25,N_{ri}})$$

$$- \left(\frac{2N_{ri}}{N_{ri} + 0.25}\right)\frac{F_0\Delta t}{\rho\Delta r}(C_{N_h-0.25,N_{ri}} - C_{ext})$$

$$- \frac{2}{M_h}(C_{N_h-1,N_{ri}+0.25} - C_{N_h,N_{ri}+0.25})$$

$$- \frac{2F_0\Delta t}{\rho\Delta h}(C_{N_h,N_{ri}+0.25} - C_{ext}) \tag{6.75}$$

6.8.2 Concentration-Dependent Diffusivity

Almost all the solutions for this problem have already been found for the case of a solid cylinder of finite length, with a finite rate of evaporation and concentration-dependent diffusivity. The effects of the internal surface must be given special attention.

Within the Cylinder, $N_{ri} + 1 \leqslant j \leqslant N_{re} - 1$ and $0 \leqslant n \leqslant N_h - 1$
The new concentration $CN_{n,j}$ is expressed in terms of the previous concentration and of the functions H and G:

$$CN_{n,j} = C_{n,j} + \frac{\Delta t}{j\Delta r}[H_{n,j+0.5} - H_{n,j-0.5}] + \frac{\Delta t}{\Delta h}[G_{n+0.5,j} - G_{n-0.5,j}] \tag{6.76}$$

with

$$H_{n,j+0.5} = (j + 0.5)D\frac{\partial C}{\partial r} \qquad \text{at } (n, j + 0.5) \tag{6.14}$$

$$G_{n+0.5,j} = D\frac{\partial C}{\partial h} \qquad \text{at } (n + 0.5, j) \tag{6.36}$$

Circular Surfaces, $j = N_{re}$ or $j = N_{ri}$ and n
For the external circular surface, defined by n and $j = N_{re}$, the new concentration is given by the relationship

$$CN_{n,N_{re}} = C_{n,N_{re}} - \left(\frac{2}{N_{re} - 0.25}\right)\frac{\Delta t}{\Delta r}H_{n,N_{re}-0.5}$$

$$- \left(\frac{2N_{re}}{N_{re} - 0.25}\right)\frac{\Delta t F_0}{\Delta r \rho}(C_{n,N_{re}} - C_{ext})$$

$$+ \frac{\Delta t}{\Delta h}[G_{n+0.5,N_{re}-0.25} - G_{n-0.5,N_{re}-0.25}] \qquad (6.77)$$

For the internal circular surface, defined by n and $j = N_{ri}$, the new concentration is expressed in a slightly different way:

$$CN_{n,N_{ri}} = C_{n,N_{ri}} - \left(\frac{2}{N_{ri} + 0.25}\right)\frac{\Delta t}{\Delta r}H_{n,N_{ri}+0.5}$$

$$- \left(\frac{2N_{ri}}{N_{ri} + 0.25}\right)\frac{\Delta t F_0}{\Delta r \rho}(C_{n,N_{ri}} - C_{ext})$$

$$+ \frac{\Delta t}{\Delta h}[G_{n+0.5,N_{ri}+0.25} - G_{n-0.5,N_{ri}+0.25}] \qquad (6.78)$$

Plane Parallel Surfaces, $n = N_h$ and j
The new concentration on the plane surface defined by $n = N_h$ and j, after elapse of time Δt is expressed by the equation already found for the solid cylinder of finite length.

$$CN_{N_h,j} = C_{N_h,j} - \frac{2\Delta t}{\Delta h}G_{N_h-0.5,j} - \frac{2\Delta t F_0}{\Delta h \rho}(C_{N_h,j} - C_{ext})$$

$$+ \frac{\Delta t}{j\Delta r}[H_{N_h-0.25,j+0.5} - H_{N_h-0.25,j-0.5}] \qquad (6.79)$$

Edges of the Cylinder
For the edge in contact with the external circular surface, defined by N_h and N_{re}, the new concentration is expressed by the relationship already found for the case of the solid cylinder of finite length:

$$CN_{N_h,N_{re}} = C_{N_h,N_{re}} - \left(\frac{2}{N_{re} - 0.25}\right)\frac{\Delta t}{\Delta h} H_{N_h-0.25,N_{re}-0.25}$$

$$- \left(\frac{2N_{re}}{N_{re} - 0.25}\right)\frac{\Delta t F_0}{\Delta r \rho}(C_{N_h-0.25,N_{re}} - C_{ext}) - \frac{2\Delta t}{\Delta r} G_{N_h-0.25,N_{re}-0.25}$$

$$- \frac{2\Delta t F_0}{\Delta h \rho}(C_{N_h,N_{re}-0.25} - C_{ext}) \qquad (6.80)$$

For the edge in contact with the internal circular surface, defined by N_h and N_{ri}, the new concentration is given by an expression which is slightly different from the preceding relation found for the external edge:

$$CN_{N_h,N_{ri}} = C_{N_h,N_{ri}} - \left(\frac{2}{N_{ri} + 0.25}\right)\frac{\Delta t}{\Delta h} H_{N_h-0.25,N_{ri}+0.25}$$

$$- \left(\frac{2N_{ri}}{N_{ri} + 0.25}\right)\frac{\Delta t F_0}{\Delta r \rho}(C_{N_h-0.25,N_{ri}} - C_{ext}) - \frac{2\Delta t}{\Delta h} G_{N_h-0.25,N_{ri}+0.25}$$

$$- \frac{2\Delta t F_0}{\Delta h \rho}(C_{N_h,N_{ri}+0.25} - C_{ext}) \qquad (6.81)$$

Amount of Substance Remaining in the Cylinder
The amount of substance remaining in the cylinder at any time can be obtained by integrating the concentrations at this time with respect to space.

6.9 Conclusions

Use of Numerical Models
Numerical models have a wider application than analytical solutions, as they are effective in all cases, and especially when no analytical solution can be found:

- When the diffusivity is concentration dependent
- When the radial and longitudinal diffusivities are different
- When the initial concentration is not uniform

Condition for Stability for Calculation
The dimensionless number M_r defined in Eq. (6.7) intervenes in the fundamental relationship between the new concentration $CN_{n,j}$ and the previous concentrations (Eq. (6.27)). Thus Eq. (6.27) can be rewritten as

$$CN_{n,j} = \frac{1}{M_r}[C_{n,j-1} + (M_r - 2)C_{n,j} + C_{n,j+1}]$$

$$+ \frac{1}{2jM_r}[C_{n,j+1} - C_{n,j-1}] + \frac{1}{M_h}[C_{n-1,j} - 2C_{n,j} + C_{n+1,j}] \qquad (6.27')$$

and Eq. (6.30) can be rewritten as

$$CN_{n,0} = \frac{1}{M_h}[C_{n+1,0} + (M_h - 2)C_{n,0} + C_{n-1,0}] + \frac{4}{M_r}[C_{n,0} - C_{n,1}] \quad (6.30')$$

The best value for M_h or M_r is

$$4 < M_h < 8 \qquad 4 < M_r < 8$$

Assumptions Made for Calculation
Only the concentrations in places defined by an integer are used. It is then necessary to calculate the value of concentration at position $(n + 0.5)$ or $(j + 0.5)$ by the simple relationships.

$$C_{n+0.5,j} = \tfrac{1}{2}[C_{n+1,j} + C_{n,j}]$$

$$C_{n,j+0.5} = \tfrac{1}{2}[C_{n,j+1} + C_{n,j}] \qquad (6.82)$$

The new concentration-dependent diffusivity can thus be obtained as a function of this mean concentration, as well as the functions G and H.

In the same way, the concentrations at $(N_h - 0.25)$ and $(N_r - 0.25)$ are calculated as follows:

$$C_{N_h-0.25,j} = \tfrac{3}{4} C_{N_h,j} + \tfrac{1}{4} C_{N_h-1,j} \qquad (6.55')$$

$$C_{n,N_r-0.25} = \tfrac{3}{4} C_{n,N_r} + \tfrac{1}{4} C_{n,N_r-1} \qquad (6.55'')$$

Symbols

C_0, C_j, C_N	Concentration of substance at the centre, at positon j and on the surface, respectively for a cylinder of infinite length, at time t
CN_j	New concentration of substance at positon j after time Δt, for a cylinder of infinite length
$C_{n,j}$	Concentration of substance at position (n,j) for a cylinder of finite length, at time t
$CN_{n,j}$	New concentration of substance at position (n,j) after time Δt, for a cylinder of finite length
C_{N_h, N_r}	Concentration at the corner of a cylinder of finite length
D	Diffusivity $(\text{cm}^2\,\text{s}^{-1})$
$D_{n,j}$	Diffusivity at position (n, j)
F_0	Evaporation rate of pure liquid $(\text{g cm}^{-2}\,\text{s}^{-1})$
$G_{n,j}$	Function ($D \dfrac{\partial C}{\partial h}$ at position (n, j))
$2H$	Height of finite cylinder
$H_{n,j}$	Function ($jD \dfrac{\partial C}{\partial r}$ at position (n, j)
j, n	Integers characterising position (radius, height)
M_r	Dimensionless number $\left(\dfrac{(\Delta r)^2}{D \Delta t} \right)$
M_h	Dimensionless number $\left(\dfrac{(\Delta h)^2}{D \Delta t} \right)$
N_h	Number of plane slices of thickness Δh
N_r	Number of circular slices of thickness Δr
N_{ri}, N_{re}	Minimum and maximum value of N_r for the hollow cylinder
Na	Dimensionless number $\left(= \dfrac{F_0 \Delta r}{\rho D} \right)$
Q_t	Amount of substance remaining in the cylinder at time t
R	Radius of the cylinder, with $R = N_r, \Delta r$
R_i and R_e	Minimum and maximum value of R, for the hollow cylinder
$\Delta h, \Delta r$	Increments of space
Δt	Increment of time
ρ	Density of the substance (g cm^{-3})

Chapter 7

Numerical Analysis for a Sphere

Many problems arise from spherical materials, essentially because various kinds of solids to be dried are spherical in shape. Moreover, the small grains of materials in powder form are very often considered as spherical, and when the shape of the grain is not known, the radius of a sphere having the same volume as that of the grain is sometimes used.

In the case of a sphere, the transfer of a substance is radial only. It is controlled by diffusion through the solid and evaporation from the surface. The rate of evaporation of the substance can be finite or infinite, and the diffusivity can be constant or concentration dependent.

Although generally solid spheres are encountered, hollow spheres are also considered. The usual case is where a liquid is surrounded by a spherical membrane. The concentration of diffusing substance at the internal surface can thus be considered as constant during the process, when there is enough liquid.

7.1   Solid Sphere with Infinite Rate of Evaporation

7.1.1   Constant Diffusivity

The case of a solid sphere with a liquid diffusing through the solid and evaporating from the surface is easy to study when the diffusivity is constant. It can thus be resolved by mathematical treatment. However, numerical analysis must be used when some difficulties arise from the surrounding atmosphere. Moreover, a mathematical treatment is not efficient when the initial concentration in the solid is not uniform.

A solid sphere of radius R is divided into N spherical membranes of constant thickness Δr (Fig. 7.1). The radius of the spherical membrane is defined by the integer j:

$$r = j\Delta r$$

$$R = N\Delta r \tag{7.1}$$

The following places must be considered: within the sphere, at its centre and at the surface.

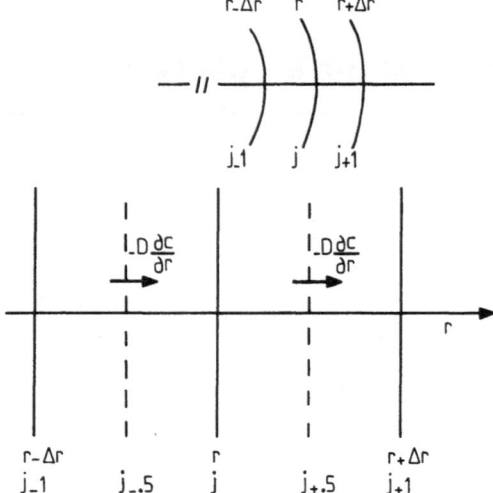

Fig. 7.1. Scheme for a solid sphere, with the integer j, and for a spherical membrane of radius $j\Delta r$.

Within the Sphere, $0 < j < N$

A spherical membrane of radius $j\Delta R$ and of thickness Δr is shown in Fig. 7.1. The matter balance during the increment of time Δt is calculated by taking into account the diffusion of the substance through the surfaces of radii $(j - 0.5)\Delta r$ and $(j + 0.5)\Delta r$:

$$\left[-AD\,\frac{\partial C}{\partial r} + AD\,\frac{\partial C}{\partial r}\right]\Delta t = 4\pi j^2 (\Delta r)^2 \Delta r [CN_j - C_j] \qquad (7.2)$$
$$\text{at } (j-0.5)\quad\ \text{at } (j+0.5)\qquad\quad \text{at } j$$

By expressing the area of the surfaces through which the substance diffuses, this equation becomes:

$$\left[-4\pi(j - 0.5)^2 (\Delta r)^2 D\,\frac{\partial C}{\partial r} + 4\pi(j + 0.5)^2 (\Delta r)^2 D\,\frac{\partial C}{\partial r}\right]\Delta t$$
$$\text{at } (j-0.5)\qquad\qquad\quad \text{at } (j+0.5)$$

$$= 4\pi j^2 (\Delta r)^3 [CN_j - C_j] \quad (7.2')$$

Upon replacing the gradient of concentration at position $(j + 0.5)$ by the chord slope between the positions $(j + 1)$ and j:

$$\frac{\partial C}{\partial r} = \frac{C_{j+1} - C_j}{\Delta r} \qquad \text{at } (j + 0.5) \qquad\qquad (7.3)$$

the new concentration CN_j at position j after time Δt is thus expressed in terms of the previous concentrations at the same and adjacent positions:

$$CN_j = C_j + \frac{1}{j^2 M}\left[(j + 0.5)^2 (C_{j+1} - C_j) - (j - 0.5)^2 (C_j - C_{j-1})\right] \quad (7.4)$$

with the dimensionless number M

$$M = \frac{(\Delta r)^2}{D\Delta t} \qquad\qquad (7.5)$$

Centre of the Sphere, $j = 0$, or $r = 0$

Eq. (7.4) cannot be used for the centre of the sphere. A small sphere of centre 0 and radius $\Delta r/2$ is considered (Fig. 7.2). The matter balance during the increment of time Δt is determined by considering the diffusion through the surface of this small sphere:

$$AD \frac{\partial C}{\partial r}\bigg|_{\text{at } j=0.5} \Delta t = \tfrac{4}{3}\pi(0.5)^3(\Delta r)^3[CN_0 - C_0] \tag{7.6}$$

By replacing the gradient of concentration at position $j = 0.5$ by the chord slope between the positions $j = 0$ and $j = 1$, the equation becomes

$$-4\pi(0.5)^2(\Delta r)^2 D \frac{C_0 - C_1}{\Delta r} \Delta t = \tfrac{4}{3}\pi(0.5)^3(\Delta r)^3[CN_0 - C_0] \tag{7.7}$$

The new concentration after time Δt is thus obtained as a function of the previous concentrations at positions 0 and 1:

$$CN_0 = C_0 - \frac{6}{M}[C_0 - C_1] \tag{7.8}$$

Surface of the Sphere, $r = R$, or $j = N$

When the rate of evaporation of the substance is infinite, the concentration of this substance at the surface falls to zero as soon as the drying process starts.

$$C_N = 0 \tag{7.9}$$

Amount of Matter Remaining in the Sphere

The amount of substance remaining in the solid sphere after time t can be obtained by integrating the concentrations at this time with respect to space:

$$Q_t = 4\pi \int_0^R C_r r^2 \, dr \tag{7.10}$$

This equation is rewritten as follows:

$$Q_t = 4\pi(\Delta r)^3 \sum_{j=1}^{N-1} j^2 C_j + \frac{\pi}{6}(\Delta r)^3 C_0 + 2\pi(N - 0.25)^2(\Delta r)^3 C_{N-0.25} \tag{7.11}$$

where the contribution of the small sphere of radius $\Delta r/2$ and of the spherical membrane of radius $(N - 0.25)\Delta r$ and thickness $\Delta r/2$ can be appreciated.

The concentrations at position $(N - 0.25)$ can be determined in many ways, one of them being

$$C_{N-0.25} = \tfrac{3}{4}C_N + \tfrac{1}{4}C_{N-1} \tag{7.12}$$

or more simply, as the concentration on the surface C_N is zero:

$$C_{N-0.25} = \tfrac{1}{4}C_{N-1} \tag{7.12'}$$

7.1.2 Concentration-Dependent Diffusivity

The solid sphere of radius R is divided into N spherical membranes of constant thickness Δr (Fig. 7.1). The matter balance is determined in various places: within the sphere, at the centre and on the surface.

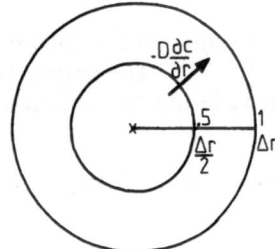

Fig. 7.2. Scheme for a small sphere of radius $\Delta r/2$.

Within the Sphere, $0 < j < N$

The matter balance is calculated during the time increment Δt within the spherical membrane of radius r and thickness Δr:

$$\left[-AD \frac{\partial C}{\partial r} + AD \frac{\partial C}{\partial r}\right]\Delta t = 4\pi j^2(\Delta r)^2 \Delta r[CN_j - C_j] \tag{7.2}$$

at $(j-0.5)$ at $(j+0.5)$ at j

By expressing the area of the two spherical surfaces of radii $(j - 0.5)$ and $(j + 0.5)$, this equation becomes

$$\left[-4\pi(j - 0.5)^2(\Delta r)^2 D \frac{\partial C}{\partial r} + 4\pi(j + 0.5)^2(\Delta r)^2 D \frac{\partial C}{\partial r}\right]\Delta t = 4\pi j^2(\Delta r)^3[CN_j - C_j]$$

at$(j-0.5)$ at$(j+0.5)$

$$(7.2')$$

On defining the function J,

$$J_{j-0.5} = (j - 0.5)^2 D \frac{\partial C}{\partial r} \qquad \text{at } (j - 0.5) \tag{7.13}$$

the new concentration after time Δt at position j is thus expressed in terms of the previous concentration at the same place and of the function J:

$$CN_j = C_j + \frac{\Delta t}{j^2 \Delta r}[J_{j+0.5} - J_{j-0.5}] \tag{7.14}$$

Centre of the Sphere, $j = 0$

The matter balance is determined in the small sphere of radius $\Delta r/2$ (Fig. 7.2) during the increment of time Δt, by considering the diffusion of the substance

$$AD \frac{\partial C}{\partial r} \Delta t = \tfrac{4}{3}\pi(0.5)^3(\Delta r)^3[CN_0 - C_0] \tag{7.6}$$

at $j=0.5$ at $j=0$

This equation is rewritten after expressing the area of this small sphere:

$$4\pi(0.5)^2(\Delta r)^2 D \frac{\partial C}{\partial r} \Delta t = \tfrac{4}{3}\pi(0.5)^3(\Delta r)^3[CN_0 - C_0] \tag{7.15}$$

By using the function $J_{0.5}$

$$J_{0.5} = (0.5)^2 D \frac{\partial C}{\partial r} \qquad \text{at } 0.5 \tag{7.13'}$$

the new concentration after time Δt at the centre of the sphere is obtained as a function of the previous concentration at this place:

$$CN_0 = C_0 + \frac{24\Delta t}{\Delta r} J_{0.5} \tag{7.16}$$

Surface of the Sphere, $j = N$
As the rate of evaporation is considered as infinite, the concentration of substance on the surface falls to zero as soon as the drying process starts:

$$C_N = 0 \tag{7.17}$$

Amount of Substance Remaining in the Sphere
The results obtained in the previous case with an infinite rate of evaporation and constant diffusivity can be used in the present case.

7.2   Solid Sphere with Finite Rate of Evaporation

7.2.1   Constant Diffusivity

The solid sphere divided into N spherical membranes of thickness Δr is considered (Fig. 7.1). The concentration of the substance is determined at various places within the solid, at the centre, and on the surface.

Within the Sphere, $0 < j < N$
From the matter balance calculated during the increment of time within the spherical membrane of radius $j\Delta r$ and thickness Δr, the new concentration after time Δt is thus obtained:

$$CN_j = C_j + \frac{1}{j^2 M} [(j + 0.5)^2(C_{j+1} - C_j) - (j - 0.5)^2(C_j - C_{j-1})] \tag{7.4}$$

with the dimensionless number M

$$M = \frac{(\Delta r)^2}{D\Delta t} \tag{7.5}$$

Centre of the Sphere, $j = 0$ (Fig. 7.2)
The matter balance is determined within the small sphere of radius $\Delta r/2$ during the time increment Δt. The new concentration at the centre of the sphere is thus expressed in terms of the previous concentrations obtained at the centre and at position Δr.

$$CN_0 = C_0 - \frac{6}{M} (C_0 - C_1) \tag{7.8}$$

Surface of the Sphere, with $j = N$
The spherical membrane of radius $(N - 0.25)\Delta r$ and thickness $\Delta r/2$, located next to the surface of the sphere, is considered (Fig. 7.3). The matter balance within this membrane is calculated during the increment of time Δt, by taking into

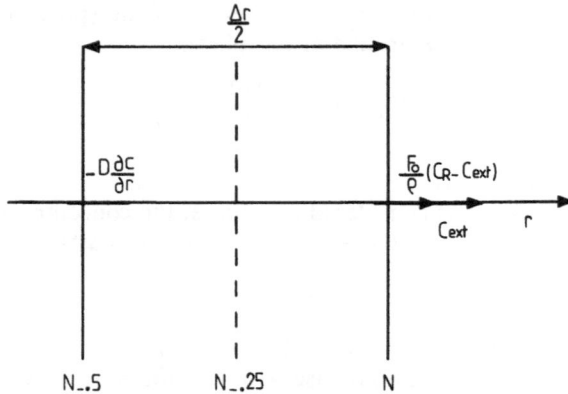

Fig. 7.3. Scheme for a spherical membrane thickness $\Delta r/2$ next to the surface, with finite rate of evaporation.

account the diffusion of the substance through the surface at position $(N - 0.5)$ and the evaporation at a finite rate from the surface.

$$\left[\underbrace{-AD\frac{\partial C}{\partial r}}_{\text{at } (N-0.5)} - \underbrace{A\frac{F_0}{\rho}(C_N - C_{\text{ext}})}_{\text{at } N}\right]\Delta t$$

$$= 4\pi(N - 0.25)^2(\Delta r)^2\left(\frac{\Delta r}{2}\right)[CN_{N-0.25} - C_{N-0.25}] \qquad (7.18)$$

This equation becomes, with the area of the surfaces at position $(N - 0.5)$ and N:

$$\left[\underbrace{-4\pi(N - 0.5)^2(\Delta r)^2 D\frac{\partial C}{\partial r}}_{\text{diffusion at }(N-0.5)} - \underbrace{4\pi N^2(\Delta r)^2\frac{F_0}{\rho}(C_N - C_{\text{ext}})}_{\text{evaporation at } N}\right]\Delta t$$

$$= 4\pi(N - 0.25)^2(\Delta r)^2\frac{\Delta r}{2}(CN_{N-0.25} - C_{N-0.25}) \quad (7.19)$$

Various assumptions can be made for the concentrations at positions $(N - 0.25)$.

First Assumption. The simple assumption is

$$CN_{N-0.25} - C_{N-0.25} = CN_N - C_N \qquad (7.20)$$

The new concentration after time Δt on the surface of the sphere becomes

$$CN_N = C_N + \left(\frac{N - 0.5}{N - 0.25}\right)^2\frac{2}{M}(C_{N-1} - C_N)$$

$$- \frac{2N^2}{(N - 0.25)^2}\frac{\Delta t F_0}{\Delta r\rho}(C_N - C_{\text{ext}}) \qquad (7.21)$$

The contribution of diffusion and evaporation to the change in concentration on the surface is clearly shown on the right-hand side of the preceding equation.

Second Assumption. A better assumption is made, with

$$CN_{N-0.25} - C_{N-0.25} = \tfrac{3}{4}(CN_N - C_N) + \tfrac{1}{4}(CN_{N-1} - C_{N-1}) \qquad (7.22)$$

The new concentration on the surface after time Δt is thus expressed by

$$CN_N = C_N - \tfrac{1}{3}(CN_{N-1} - C_{N-1}) + \left(\frac{N-0.5}{N-0.25}\right)^2 \frac{8}{3M}(C_{N-1} - C_N)$$

$$- \frac{8}{3}\left(\frac{N}{N-0.25}\right)^2 \frac{\Delta t F_0}{\Delta r \rho}(C_N - C_{\text{ext}}) \tag{7.23}$$

where CN_{N-1} is the new concentration at the adjacent place. This concentration must be calculated before the concentration CN_N, by using the relation already found when $0 < j < N$:

$$CN_{N-1} = C_{N-1} + \frac{1}{(N-1)^2 M}[(N-0.5)^2(C_N - C_{N-1})$$

$$- (N-1.5)^2(C_{N-1} - C_{N-2})] \tag{7.4'}$$

7.2.2 Concentration-Dependent Diffusivity

Fig. 7.1 shows the solid sphere divided into N spherical membranes of thickness Δr. The concentration of the substance is calculated at various places: within the sphere, at the centre and on the surface.

Within the Sphere, $0 < j < N$
The matter balance within the spherical membrane of radius $j\Delta r$ during the increment of time is written as follows (Fig. 7.1):

$$\left[-AD\frac{\partial C}{\partial r}_{\text{at }(j-0.5)} + AD\frac{\partial C}{\partial r}_{\text{at }(j+0.5)}\right]\Delta t = 4\pi j^2 (\Delta r)^2 \Delta r [CN_j - C_j]_{\text{at }j} \tag{7.2}$$

The new concentrations at position j after time Δt can thus be expressed in terms of the previous concentration at this same position and of the function J:

$$CN_j = C_j + \frac{\Delta t}{j^2 \Delta r}[J_{j+0.5} - J_{j-0.5}] \tag{7.14}$$

Centre of the Sphere, $j = 0$
The new concentration after time Δt at the centre of the sphere is calculated from the matter balance determined within the small sphere of radius $\Delta r/2$ during this time increment:

$$CN_0 = C_0 + \frac{24\Delta t}{\Delta r} J_{0.5} \tag{7.16}$$

with the function $J_{0.5}$.

$$J_{0.5} = (0.5)^2 D \frac{\partial C}{\partial r} \quad \text{at } j = 0.5 \tag{7.13'}$$

Surface of the Sphere, $j = N$
The matter balance is determined during the increment of time Δt within the spherical membrane of radius $(N - 0.25)\Delta r$ and thickness $\Delta r/2$, by considering

the diffusion through the surface of radius $(N - 0.25)\Delta r$ and evaporation from the external surface (Fig. 7.3).

$$\left[-AD \frac{\partial C}{\partial r} \bigg|_{\text{at } (N-0.5)} - A \frac{F_0}{\rho} (C_N - C_{\text{ext}}) \bigg|_{\text{at } N} \right] \Delta t$$

$$= 4\pi (N - 0.25)^2 (\Delta r)^2 \frac{\Delta r}{2} (CN_{N-0.25} - C_{N-0.25}) \quad (7.18)$$

By making the simple assumption

$$CN_{N-0.25} - C_{N-0.25} = CN_N - C_N \quad (7.20)$$

the new concentration on the surface after time Δt is thus expressed in terms of the previous concentration on the surface, as well as the function J and the rate of evaporation F_0:

$$CN_N = C_N - \frac{2}{(N - 0.25)^2} \frac{\Delta t}{\Delta r} J_{N-0.5} - \frac{2N^2}{(N - 0.25)^2} \frac{\Delta t F_0}{\Delta r \rho} (C_N - C_{\text{ext}})$$

$$(7.24)$$

With the better assumption

$$CN_{N-0.25} - C_{N-0.25} = \tfrac{3}{4}(CN_N - C_N) + \tfrac{1}{4}(CN_{N-1} - C_{N-1}) \quad (7.22)$$

the new concentration on the surface becomes

$$CN_N = C_N - \frac{1}{3}(CN_{N-1} - C_{N-1}) - \frac{8\Delta t}{3(N - 0.25)^2 \Delta r} J_{N-0.5}$$

$$- \frac{8N^2}{3(N - 0.25)^2} \frac{\Delta t F_0}{\Delta r \rho} (C_N - C_{\text{ext}}) \quad (7.25)$$

In Eq. (7.25), the new concentration on the surface CN_N is a function of the new concentration at the adjacent position $(N - 1)$. This concentration CN_{N-1} must be calculated beforehand, by using the relation already shown when $0 < j < N$:

$$CN_{N-1} = C_{N-1} + \frac{\Delta t}{j^2 \Delta r} [J_{N-0.5} - J_{N-1.5}] \quad (7.14)$$

Amount of Matter Remaining in the Sphere
The amount of matter remaining in the sphere after time t is obtained by integrating the concentrations at time t with respect to space:

$$Q_t = 4\pi \int_0^R C_{r,t} r^2 \, dr \quad (7.10)$$

This equation becomes

$$Q_t = 4\pi (\Delta r)^3 \sum_{j=1}^{N-1} j^2 C_j + \frac{\pi}{6} (\Delta r)^3 C_0 + 2\pi (N - 0.25)^2 (\Delta r)^3 C_{N-0.25}$$

$$(7.11)$$

where the concentration in the annulus of radius $(N - 0.25)\Delta r$ and thickness $\Delta r/2$ can be obtained by

$$C_{N-0.25} = \tfrac{3}{4} C_N + \tfrac{1}{4} C_{N-1} \quad (7.12)$$

7.3 Liquid-Filled Hollow Sphere with Infinite Rate of Evaporation

The hollow sphere is usually encountered as a spherical membrane which contains a liquid. As long as the liquid is in contact with the internal surface of the membrane, the concentration on this internal surface is constant. The liquid thus diffuses through the membrane and then evaporates from the surface.

Two cases are of interest for the transfer through the membrane, where the diffusivity is constant or concentration dependent.

In this section, the rate of evaporation is infinite, but the following case with a finite rate of evaporation is also of interest.

7.3.1 Constant Diffusivity

A hollow sphere located between the internal spherical surface of radius R_i and the external surface R_e is shown in Fig. 7.4. The sphere is divided into $(N_e - N_i)$ spherical membranes of thickness Δr. A position within the sphere is characterised by the radius of the sphere which passes through this position. Thus

$$R_i < r < R_e \qquad \text{(hollow sphere)}$$
$$N_i < j < N_e \qquad\qquad\qquad\qquad (7.26)$$

with

$$r = j\Delta r \qquad\qquad\qquad\qquad (7.1)$$
$$R_i = N_i\Delta r$$
$$R_e = N_e\Delta r \qquad\qquad\qquad\qquad (7.27)$$

Within the Sphere, $N_i + 1 \leqslant j \leqslant N_e - 1$
The matter balance during the increment of time Δt is determined within the spherical membrane of radius $j\Delta r$ and thickness Δr by considering the diffusion of substance through the surfaces at position $(j - 0.5)$ and $(j + 0.5)$. (Fig. 7.1).

From Eq. (7.2), the new concentration after time Δt at position j is expressed in terms of the previous concentration at this place and at the adjacent two places $(j - 1)$ and $(j + 1)$:

$$CN_j = C_j + \frac{1}{j^2 M}[(j + 0.5)^2(C_{j+1} - C_j) - (j - 0.5)^2(C_j - C_{j-1})] \quad (7.4)$$

with the dimensionless number M

$$M = \frac{(\Delta r)^2}{D\Delta t} \qquad\qquad\qquad\qquad (7.5)$$

Surfaces, $j = N_i$ and $j = N_e$
The internal surface is at constant concentration, being in contact with a liquid:

$$C_{N_i} = ct \qquad \text{(internal surface)} \qquad\qquad (7.28)$$

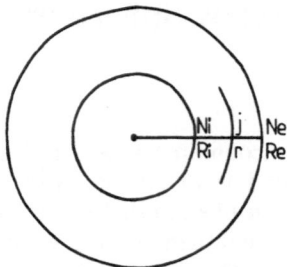

Fig. 7.4. Hollow sphere with internal and external surfaces at constant concentration.

The external surface is kept constant to zero, as the rate of evaporation is infinite:

$$C_{N_e} = 0 \quad \text{(external surface)} \tag{7.29}$$

Remark 7.1: Concentration Within the Hollow Sphere. Eq. (7.4) can also be written as follows:

$$CN_j = C_j + \frac{(j^2 + 0.25)}{j^2 M} [C_{j+1} - 2C_j + C_{j-1}] + \frac{1}{jM} [C_{j+1} - C_{j-1}] \tag{7.4'}$$

7.3.2 Concentration-Dependent Diffusivity

The hollow sphere located between the spherical surfaces of radii R_i and R_e is considered (Fig. 7.4). As shown in the previous section with constant diffusivity, the sphere is divided into spherical membranes of thickness Δr, and each of these membranes is characterised by the integer j such that the radius is $r = j\Delta r$.

Within the Sphere, $N_i + 1 \leq j \leq N_e - 1$
The matter balance is determined within the spherical membrane of radius $j\Delta r$ and of thickness Δr during the increment of time Δt, and Eq. (7.2) is obtained by considering the diffusion of substance through the surfaces of radii $(j - 0.5)\Delta r$ and $(j + 0.5)\Delta r$. The new concentration after time Δt at position j is thus expressed in terms of the previous concentration at the same place and of the function J:

$$CN_j = C_j + \frac{\Delta t}{j^2 \Delta r} [J_{j+0.5} - J_{j-0.5}] \tag{7.14}$$

with

$$J_{j+0.5} = (j + 0.5)^2 D \frac{\partial C}{\partial r} \quad \text{at } (j + 0.5) \tag{7.13}$$

Surfaces
The internal surface in contact with a liquid is maintained at a constant concentration:

$$C_{N_i} = ct \tag{7.28}$$

As the rate of evaporation is infinite, the concentration on the external surface falls to zero as soon as the process starts:

$$C_{N_e} = 0 \tag{7.29}$$

Amount of Matter Remaining in the Sphere

The amount of matter located in the hollow sphere at time t is obtained by integrating the concentration with respect to space:

$$Q_t = 4\pi \int_{R_i}^{R_e} r^2 C_r \, dr \tag{7.30}$$

This equation becomes with finite differences

$$Q_t = 4\pi(\Delta r)^3 \sum_{N_i+1}^{N_e-1} j^2 C_j + 2\pi(N_e - 0.25)^2(\Delta r)^3 C_{N_e-0.25}$$

$$+ 2\pi(N_i + 0.25)^2(\Delta r)^3 C_{N_i+0.25} \tag{7.31}$$

The concentrations at position $(N_e - 0.25)$ and $(N_i + 0.25)$ can be approximated by the relationships:

$$C_{N_e-0.25} = \tfrac{3}{4}C_{N_e} + \tfrac{1}{4}C_{N_e-1}$$

$$C_{N_i+0.25} = \tfrac{3}{4}C_{N_i} + \tfrac{1}{4}C_{N_i+1} \tag{7.32}$$

where C_{N_e} and C_{N_i} are given by Eqs (7.28) and (7.29).

7.4 Liquid-Filled Hollow Sphere with Finite Rate of Evaporation

This case is very common, where a liquid is encapsulated in a spherical membrane made of a polymer. The liquid diffuses through the membrane and evaporates from the surface. The condition at the surface is that the rate of evaporation is constantly equal to the rate of diffusing substance. Of course, the rate of evaporation is proportional to the difference of the concentrations of liquid on the surface and the concentration on the surface which is at equilibrium with the surrounding atmosphere, the coefficient of proportionality being the rate of evaporation of the pure liquid under the same conditions. A more complex case with a non-linear isotherm of absorption can also be considered with these numerical models, and a slight change in the final equation expressing the concentration at the surface is necessary.

The following two cases, with constant or concentration-dependent diffusivity, are considered.

7.4.1 Constant Diffusivity

The following three positions are studied: within the solid, and at the internal and external surfaces.

Within the Sphere, $N_i + 1 \leqslant j \leqslant N_e - 1$

From the matter balance already calculated within the spherical membrane of radius $j\Delta r$ and thickness Δr, the new concentration after time Δt at the position

j is obtained as a function of the previous concentration at the same place and adjacent places:

$$CN_j = C_j + \frac{1}{j^2 M}[(j + 0.5)^2(C_{j+1} - C_j) - (j - 0.5)^2(C_j - C_{j-1})] \quad (7.4)$$

with the dimensionless number M

$$M = \frac{(\Delta r)^2}{D\Delta t} \quad (7.5)$$

Internal Surface, $j = N_i$
The concentration of the substance is constant, as the liquid is in contact with the internal surface:

$$C_{N_i} = ct \quad (7.28)$$

External Surface, $j = N_e$
As already shown (Sect. 7.2.1), the spherical membrane of radius $(N_e - 0.25)\Delta r$ and of thickness $\Delta r/2$ is considered (Fig. 7.3). From the matter balance calculated during the increment of time Δt by considering the diffusion through the spherical surface of radius $(N - 0.5)\Delta r$ and the evaporation from the external surface, the new concentration at position $(N - 0.25)$ is obtained:

$$CN_{N_e} = $$
$$C_{N_e} + \left(\frac{N_e - 0.5}{N_e - 0.25}\right)^2 \frac{2}{M}(C_{N_e-1} - C_{N_e}) - \frac{2N_e^2}{(N_e - 0.25)^2}\frac{\Delta t F_0}{\Delta r \rho}(C_{N_e} - C_{ext}) \quad (7.33)$$

with the assumption

$$CN_{N_e-0.25} - C_{N_e-0.25} = CN_{N_e} - C_{N_e} \quad (7.20)$$

7.4.2 Concentration-Dependent Diffusivity

All calculations have been already made in Sect. 7.2.2, for positions within the solid and on the surfaces.

Within the Sphere, $N_i + 1 \leqslant j \leqslant N_e - 1$
From the matter balance calculated in the spherical membrane of radius $j\Delta r$ and thickness Δr during the increment of time Δt (Fig. 7.1), the new concentration at position j is obtained as a function of the previous concentration and the function J:

$$CN_j = C_j + \frac{\Delta t}{j^2 \Delta r}[J_{j+0.5} - J_{j-0.5}] \quad (7.14)$$

with

$$J_{j+0.5} = (j + 0.5)^2 D \frac{\partial C}{\partial r} \quad \text{at } (j + 0.5) \quad (7.13)$$

Internal Surface, $j = N_i$

As long as the liquid is in contact with the internal surface, the concentration on this surface is constant:

$$CN_i = ct \tag{7.28}$$

External Surface, $j = N_e$

The matter balance during the increment of time Δt is determined in the spherical membrane of radius $(N_e - 0.25)\Delta r$ and thickness $\Delta r/2$. The new concentration after time Δt is thus obtained in terms of the previous concentration at the same place, as well as of the function J and the concentration in equilibrium with the surrounding atmosphere:

$$CN_{N_e} = C_{N_e} - \frac{2}{(N_e - 0.25)^2}\frac{\Delta t}{\Delta r} J_{N_e - 0.25} - \frac{2N_e^2}{(N_e - 0.25)^2}\frac{\Delta t F_0}{\Delta r \rho}(C_{N_e} - C_{ext}) \tag{7.34}$$

with the assumption

$$CN_{N_e - 0.25} - C_{N_e - 0.25} = CN_{N_e} - C_{N_e} \tag{7.20}$$

By making the better assumption

$$CN_{N_e - 0.25} - C_{N_e - 0.25} = \tfrac{3}{4}(CN_{N_e} - C_{N_e}) + \tfrac{1}{4}(CN_{N_e - 1} - C_{N_e - 1}) \tag{7.22}$$

the new concentration on the surface CN_{N_e} becomes

$$CN_{N_e} = C_{N_e} - \frac{1}{3}(CN_{N_e - 1} - C_{N_e - 1}) - \frac{8\Delta t}{3(N_e - 0.25)^2 \Delta r} J_{N_e - 0.25}$$

$$- \frac{8N_e^2}{3(N_e - 0.25)^2}\frac{\Delta t F_0}{\Delta r \rho}(C_{N_e} - C_{ext}) \tag{7.35}$$

Amount of Substance Remaining in the Sphere

The amount of substance remaining in the sphere is determined by integrating the concentration with respect to space, at any time.

Remark 7.2: Evaporation or Condensation at the Internal Surface. When the substance evaporates from or condenses onto the internal surface, the concentration at this surface can be obtained from the matter balance obtained within the spherical membrane of radius $(N_i + 0.25)\Delta r$ and of thickness Δr (Fig. 7.5).

Constant Diffusivity. By making the assumption

$$CN_{N_i + 0.25} - C_{N_i + 0.25} = CN_{N_i} - C_{N_i} \tag{7.36}$$

the new concentration after time Δt on the internal surface is written as follows:

$$CN_{N_i} = C_{N_i} + \left(\frac{N_i + 0.5}{N_i + 0.25}\right)^2 \frac{2}{M}(C_{N_i + 1} - C_{N_i})$$

$$- \frac{2N_i^2}{(N_i + 0.25)^2}\frac{\Delta t F_0}{\Delta r \rho}(C_{N_i} - C_{ext}) \tag{7.37}$$

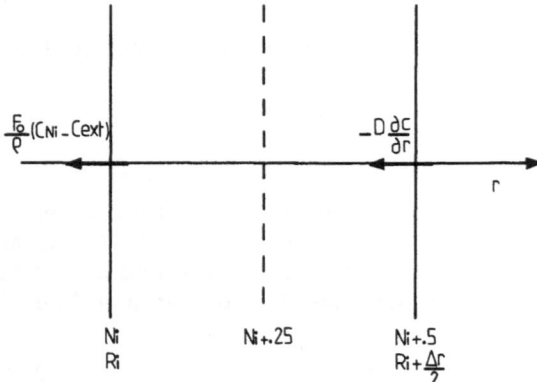

Fig. 7.5. Hollow sphere with evaporation or condensation at the internal surface.

Concentration-Dependent Diffusivity. With the same assumption (Eq. (7.36)), the new concentration after time Δt at the internal surface becomes

$$CN_{N_i} = C_{N_i} - \frac{2}{(N_i + 0.25)^2} \frac{\Delta t}{\Delta r} J_{N_i+0.25} - \frac{2N_i^2}{(N_i + 0.25)^2} \frac{\Delta t F_0}{\Delta r \rho} (C_{N_i} - C_{\text{ext}})$$

(7.38)

Of course, evaporation or condensation takes place according to the respective values of the concentration on the internal surface and the surface which is at equilibrium with the surrounding atmosphere:

$$C_{N_i} > C_{\text{ext}} \quad \text{(evaporation)}$$

$$C_{N_i} < C_{\text{ext}} \quad \text{(condensation)}$$

(7.39)

7.5 Conclusions

The following conclusions must be considered.

Use of Numerical Models
Analytical solutions are obtained only in simple cases, when the diffusivity is constant and the initial profile of concentration at the beginning of the process is uniform.

In all cases, numerical methods are efficient, whatever the diffusivity and the initial profile of concentration.

Conditions of Stability for Calculation
As shown in Eq. (7.4'), the dimensionless number M must be greater than 2.

Following Eq. (7.8) expressing the new concentration after time Δt at the centre of the sphere,

$$CN_0 = C_0 - \frac{6}{M} (C_0 - C_1)$$

(7.8)

where the dimensionless number M must be greater than 6.

Assumptions Made for Calculation
As the positions defined by an integer have a known concentration, the concentrations in other places must be calculated. For instance, the concentration in the position $(N - 0.25)\,\Delta r$ can be calculated as follows:

$$C_{N-0.25} = \tfrac{3}{4}C_N + \tfrac{1}{4}C_{N-1} \tag{7.12}$$

Symbols

A	Area of the spherical surface through which the substance diffuses
C_j	Concentration of substance at position j
CN_j	New concentration of substance at position j after time Δt
D	Diffusivity $(cm^2\,s^{-1})$
D_j	Diffusivity at position j
F_0	Rate of evaporation of the pure liquid $(g\,cm^2\,s^{-1})$
j	Integer characterising the position $r = j\Delta r$
J_j	Function $\left(j^2 D \dfrac{\partial C}{\partial r}\right)$
M	Dimensionless number $((\Delta r)^2/D\Delta t)$
N	Number of spherical membranes of constant thickness Δr
N_i, N_e	Minimum and maximum values of N for the hollow sphere
Q_t	Amount of substance remaining in the sphere
r, R	Position, radius
R_i, R_e	Internal and external radii for the hollow sphere
$\Delta r, \Delta t$	Increments of space and of time, respectively
ρ	Density of the substance $(g\,cm^{-3})$

Chapter 8

Diffusion–Evaporation in Two and Three Dimensions: Isotropic and Anisotropic Media

8.1 Introduction

The process of drying of media having two or three dimensions is of great interest, and numerical models can be successfully applied to these various applications.

The case where drying is controlled by diffusion through the solid and by evaporation from the surface is considered, for both isotropic and anisotropic media.

8.1.1 Isotropic Medium

In an isotropic medium, diffusion properties in the neighbourhood of any point are the same relative to all directions. The rate of transfer of diffusing substance through a section of area A is proportional to the gradient of concentration measured normal to the section (Fig. 8.1).

$$\mathrm{Ra}_x = -AD\frac{\partial C}{\partial x} \tag{8.1}$$

For two-dimensional transfer, the fundamental differential equation is (Fig. 8.2), when the diffusivity is constant:

$$\frac{\partial C}{\partial t} = D\left[\frac{\partial^2 C}{\partial x^2} + \frac{\partial^2 C}{\partial y^2}\right] \tag{8.2}$$

When the diffusivity is concentration dependent:

$$\frac{\partial C}{\partial t} = \frac{\partial}{\partial x}\left(D\frac{\partial C}{\partial x}\right) + \frac{\partial}{\partial y}\left(D\frac{\partial C}{\partial y}\right) \tag{8.3}$$

For the three-dimensional transfer of substance, the differential equation is (Fig. 8.3), when the diffusivity is constant:

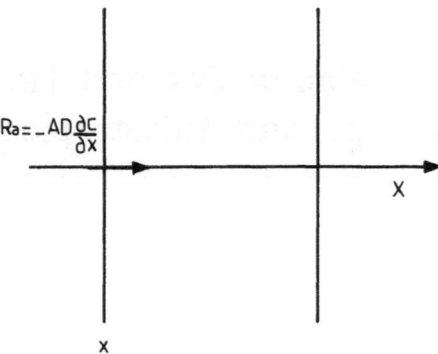

Fig. 8.1. Transfer in one dimension.

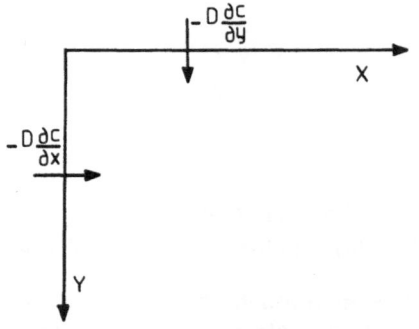

Fig. 8.2. Transfer in two dimensions.

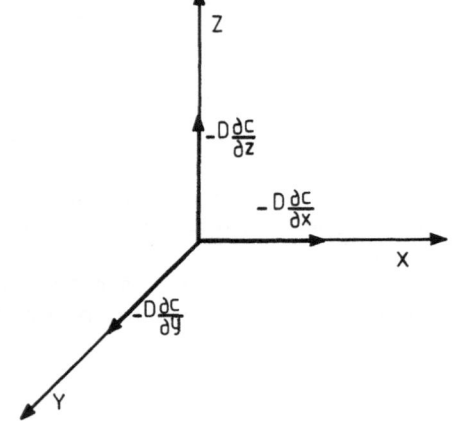

Fig. 8.3. Transfer in three dimensions.

$$\frac{\partial C}{\partial t} = D\left[\frac{\partial^2 C}{\partial x^2} + \frac{\partial^2 C}{\partial y^2} + \frac{\partial^2 C}{\partial z^2}\right] \tag{8.4}$$

When the diffusivity is concentration dependent:

$$\frac{\partial C}{\partial t} = \frac{\partial}{\partial x}\left(D\frac{\partial C}{\partial x}\right) + \frac{\partial}{\partial y}\left(D\frac{\partial C}{\partial y}\right) + \frac{\partial}{\partial z}\left(D\frac{\partial C}{\partial z}\right) \tag{8.5}$$

8.1.2 Anisotropic Medium

Many types of media exist for which the diffusion properties depend on the direction in which they are measured. The molecules in various solids such as textile fibres and polymer films have a preferential direction of orientation. Another important example is found with wood. These media are anisotropic and they exhibit different diffusion properties in different directions.

For three-dimensional transfer of substance, the rates of transfer are

$$-\text{Ra}_x = D_{11}\frac{\partial C}{\partial x} + D_{12}\frac{\partial C}{\partial y} + D_{13}\frac{\partial C}{\partial z}$$

$$-\text{Ra}_y = D_{21}\frac{\partial C}{\partial x} + D_{22}\frac{\partial C}{\partial y} + D_{23}\frac{\partial C}{\partial z}$$

$$-\text{Ra}_z = D_{31}\frac{\partial C}{\partial x} + D_{32}\frac{\partial C}{\partial y} + D_{33}\frac{\partial C}{\partial z} \tag{8.6}$$

where $D_{12}\partial C/\partial y$ is the contribution to the rate of transfer in the x-direction due to the component of the gradient of concentration in the y-direction.

By writing the matter balance

$$\frac{\partial C}{\partial t} = \frac{\partial \text{Ra}_x}{\partial x} + \frac{\partial \text{Ra}_y}{\partial y} + \frac{\partial \text{Ra}_z}{\partial z} \tag{8.7}$$

and substituting from Eq. (8.6) for the rates Ra in Eq. (8.7), the following is obtained:

$$\frac{\partial C}{\partial t} = D_{11}\frac{\partial^2 C}{\partial x^2} + D_{22}\frac{\partial^2 C}{\partial y^2} + D_{33}\frac{\partial^2 C}{\partial z^2} + (D_{12} + D_{21})\frac{\partial^2 C}{\partial x\partial y}$$

$$+ (D_{23} + D_{32})\frac{\partial^2 C}{\partial y\partial z} + (D_{31} + D_{13})\frac{\partial^2 C}{\partial x\partial z} \tag{8.8}$$

if the diffusivities are constant.

A transformation to rectangular coordinates x', y' and z' can be found which reduces this equation to

$$\frac{\partial C}{\partial t} = D_1\frac{\partial^2 C}{\partial x'^2} + D_2\frac{\partial^2 C}{\partial y'^2} + D_3\frac{\partial^2 C}{\partial z'^2} \tag{8.9}$$

These new axes x', y' and z' are called the principal axes of diffusion, and D_1, D_2 and D_3 the principal diffusivities. When the diffusivities are concentration dependent we have

$$\frac{\partial C}{\partial t} = \frac{\partial}{\partial x'}\left(D_1\frac{\partial C}{\partial x'}\right) + \frac{\partial}{\partial y'}\left(D_2\frac{\partial C}{\partial y'}\right) + \frac{\partial}{\partial z'}\left(D_3\frac{\partial C}{\partial z'}\right) \tag{8.10}$$

For pieces of wood (e.g. beams, boards) as well as for living trees, three principal directions exist: longitudinal, radial and tangential. The longitudinal direction is vertical for a tree, the radial direction passes through the centre of the circular cross-section, and the tangential direction is perpendicular to the radial direction in the circular cross-section (Fig. 8.4).

Eq. (8.10) can thus be written in three dimensions:

$$\frac{\partial C}{\partial t} = \frac{\partial}{\partial L}\left(D_L\frac{\partial C}{\partial L}\right) + \frac{\partial}{\partial R}\left(D_R\frac{\partial C}{\partial R}\right) + \frac{\partial}{\partial T}\left(D_T\frac{\partial C}{\partial T}\right) \tag{8.11}$$

and in two dimensions, L and R for instance:

$$\frac{\partial C}{\partial t} = \frac{\partial}{\partial L}\left(D_L\frac{\partial C}{\partial L}\right) + \frac{\partial}{\partial R}\left(D_R\frac{\partial C}{\partial R}\right) \tag{8.12}$$

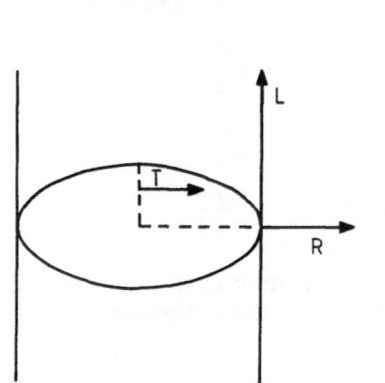

Fig. 8.4. Cross-section of a tree, showing the three principal directions of diffusion: longitudinal, radial and tangential.

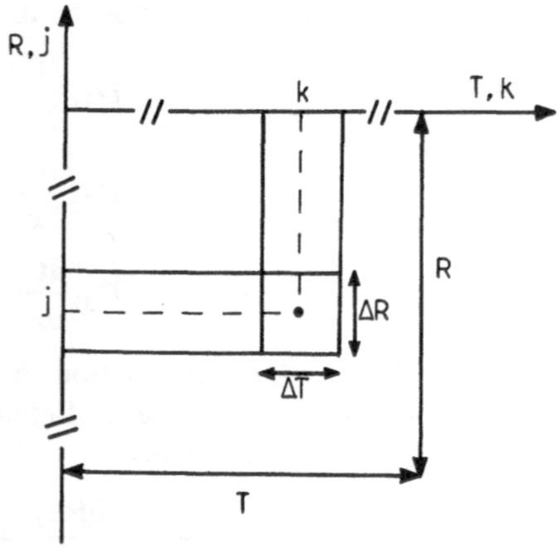

Fig. 8.5. Small parallelepiped of dimensions ΔR, ΔT and thickness U.

8.2   Two Dimensions with Infinite Rate of Evaporation

The problem considered is that of two-dimensional transfer through parallelepipedic boards in the following two cases:

1. Where the board is very long in one direction with regard to the other two; the aim is to determine the profiles of concentration of the substance within the cross-section perpendicular to this direction. Of course, these calculations can be made at the places where the transfer along this direction is negligible.

2. Where the substance is transferred only in two directions, transfer along the third direction being stopped by some means.

The two directions considered are two principal directions of transfer with two principal diffusivities (Sect. 8.1).

As the rate of evaporation is considered infinite, the concentration of evaporating substance on the surfaces falls to zero as soon as the process starts.

Two cases are examined, with constant diffusivity and concentration-dependent diffusivity.

8.2.1   Constant Diffusivity

Each of the sides (R, T) of the rectangular cross-section is divided into N_T and N_R slices of constant thickness ΔT and ΔR. Each position is characterised by two integers j and k with (Fig. 8.5).

$$T = N_T \Delta T$$

$$R = N_R \Delta R$$

$$T' = k \Delta T$$

$$R' = j \Delta R \tag{8.13}$$

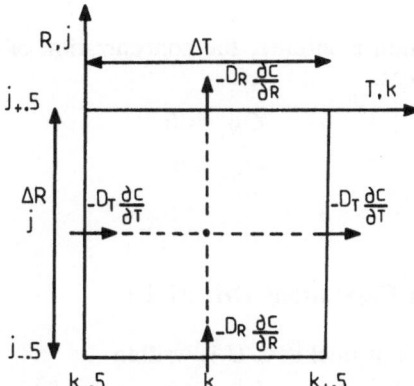

Fig. 8.6. Transfer by diffusion within the parallelepiped of dimensions ΔR, ΔT and thickness U.

The concentration of diffusing substance is calculated within the cross-section and on the surface.

Within the Cross-Section
A small rectangle of dimensions ΔT and ΔR is shown in Fig. 8.6, and the matter balance is calculated during the increment of time Δt by taking into account the diffusion of substance along the two directions $R(j)$ and $T(k)$, where the two integers j and k characterise the position along the axes R and T. The matter balance per unit thickness U is written as follows:

$$\Delta RU\left[-D_T\frac{\partial C}{\partial T} + D_T\frac{\partial C}{\partial T}\right]\Delta t + \Delta TU\left[-D_R\frac{\partial C}{\partial R} + D_R\frac{\partial C}{\partial R}\right]\Delta t$$

$\quad\;$ at $(k-0.5)$ $\;$ at $(k+0.5)$ $\qquad\qquad\qquad$ at $(j-0.5)$ $\;$ at $(j+0.5)$

$\qquad$ (along T-axis) $\qquad\qquad\qquad\qquad\quad$ (along R-axis)

$$= \Delta R\Delta TU[CN_{j,k} - C_{j,k}] \quad (8.14)$$

The new concentration after time Δt is thus expressed as a function of the previous concentrations obtained at the same place and at the adjacent places:

$$CN_{j,k} = C_{j,k} + \frac{1}{M_T}[C_{j,k-1} - 2C_{j,k} + C_{j,k+1}]$$

$$+ \frac{1}{M_R}[C_{j-1,k} - 2C_{j,k} + C_{j+1,k}] \quad (8.15)$$

with the dimensionless numbers M_T and M_R:

$$M_T = \frac{(\Delta T)^2}{D_T\Delta t} \quad (8.16)$$

$$M_R = \frac{(\Delta R)^2}{D_R\Delta t} \quad (8.16')$$

assuming that the concentration gradient at position $(k-0.5)$ is equal to the chord slope of concentration between the positions k and $(k-1)$,

$$\frac{\partial C}{\partial T} = \frac{C_k - C_{k-1}}{\Delta T} \quad \text{at} \quad (k-0.5) \quad (8.17)$$

Surfaces

As the rate of evaporation is infinite, the concentration of evaporating substance is zero on the two surfaces:

$$C_{N_T} = 0 \qquad\qquad (8.18)$$

$$C_{N_R} = 0 \qquad\qquad (8.18')$$

8.2.2   Concentration-Dependent Diffusivity

The matter balance per unit thickness U is written

$$\Delta R U \left[\underbrace{-D_T \frac{\partial C}{\partial T}}_{\text{at } (k-0.5)} + \underbrace{D_T \frac{\partial C}{\partial T}}_{\text{at } (k+0.5)} \right] \Delta t + \Delta T U \left[\underbrace{-D_R \frac{\partial C}{\partial R}}_{\text{at } (j-0.5)} + \underbrace{D_R \frac{\partial C}{\partial R}}_{\text{at } (j+0.5)} \right] \Delta t$$

$$= \Delta R \Delta T U [CN_{j,k} - C_{j,k}] \quad (8.14)$$

On defining the function G,

$$G_{T_{j,k}} = D_T \frac{\partial C}{\partial T} \qquad \text{at} \qquad j, k \qquad\qquad (8.19)$$

$$G_{R_{j,k}} = D_R \frac{\partial C}{\partial R} \qquad \text{at} \qquad j, k \qquad\qquad (8.20)$$

the new concentration after time Δt is thus expressed in terms of the previous concentration at the same place and of the functions G_T and G_R:

$$CN_{j,k} = C_{j,k} - \frac{\Delta t}{\Delta T} [G_{T_{j,k-0.5}} - G_{T_{j,k+0.5}}]$$

$$- \frac{\Delta t}{\Delta R} [G_{R_{j-0.5,k}} - G_{R_{j+0.5,k}}] \qquad\qquad (8.21)$$

The diffusivity at position $(k - 0.5)$ can be obtained by many ways, for instance by

$$D_{k-0.5} = \tfrac{1}{2} (D_k + D_{k-1}) \qquad\qquad (8.22)$$

or better,

$$D_{j,k-0.5} = D_0 \exp \left[\text{cte} - \frac{C_{j,k} + C_{j,k-1}}{2} \right] \qquad\qquad (8.23)$$

where D_0 is constant.

Remark 8.1: Symmetry. The problem is symmetrical, whatever the diffusivity. The centre of the rectangle being characterised by the integers $N_R/2$ and $N_T/2$, it is obvious that

$$C_{(N_r/2)-1,k} = C_{(N_r/2)+1,k} \qquad\qquad (8.24)$$

$$C_{j,(N_t/2)-1} = C_{j,(N_t/2)+1} \qquad\qquad (8.24')$$

8.3   Two Dimensions with Finite Rate of Evaporation

The rate of evaporation on each surface is proportional to the difference of the actual concentration on the surface and the concentration on the surface necessary to maintain equilibrium with the surrounding atmosphere, the coefficient of proportionality being the rate of evaporation of the pure liquid under the same conditions:

$$\frac{F_0}{\rho}\,(C_{\text{surface}} - C_{\text{ext}})$$

The following places are considered: within the solid, on the surfaces and at the corner defined by the intersection of these two surfaces.

8.3.1   Constant Diffusivity

The matter balance is determined at various places, with constant diffusivity.

Within the Solid, $N_r/2 \leqslant j \leqslant N_R - 1$ and $N_t/2 \leqslant k \leqslant N_T - 1$
The new concentration after time Δt at position j, k is given by

$$CN_{j,k} = C_{j,k} + \frac{1}{M_T}[C_{j,k-1} - 2C_{j,k} + C_{j,k+1}]$$

$$+ \frac{1}{M_R}[C_{j-1,k} - 2C_{j,k} + C_{j+1,k}] \qquad (8.15)$$

On the Surfaces
The matter balance is calculated during the time increment Δt within the slice of thickness $\Delta R/2$ next to the surface N_R (Fig. 8.7), by considering the diffusion

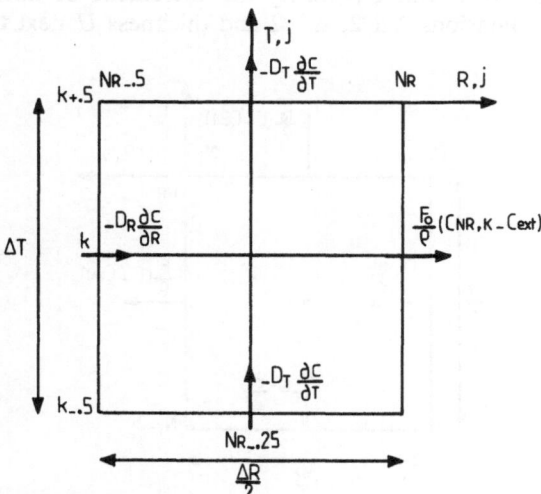

Fig. 8.7. Two-dimensional transfer by diffusion within the solid and evaporation from a surface $(j = N_R)$.

along the T-axis and the R-axis and evaporation from the surface:

$$\frac{\Delta R}{2} U\left[-D_T \frac{\partial C}{\partial T} + D_T \frac{\partial C}{\partial T}\right] + \Delta T U\left[-D_R \frac{\partial C}{\partial R} - \frac{F_0}{\rho}(C_{N_R,k} - C_{ext})\right]$$

at $(k-0.5)$ at $(k+0.5)$ at $(N_R-0.5)$ at N_R

$$= \frac{\Delta R \Delta T U}{2\Delta t}\left[CN_{N_R-0.25,k} - C_{N_R-0.25,k}\right]$$

$$(8.25)$$

By assuming that

$$CN_{N_R-0.5,k} - C_{N_R-0.5,k} = CN_{N_R,k} - C_{N_R,k} \qquad (8.26)$$

the new concentration on the surface N_R after time Δt is given by

$$CN_{N_R,k} = C_{N_R,k} + \frac{2}{M_R}(C_{N_R-1,k} - C_{N_R,k})$$

$$- \frac{2F_0\Delta t}{\rho\Delta R}[C_{N_R,k} - C_{ext}]$$

$$+ \frac{1}{M_T}[C_{N_R-0.25,k-1} - 2C_{N_R-0.25,k} + C_{N_R-0.25,k+1}] \qquad (8.27)$$

The new concentration on the other surface N_T is obtained in the same way by considering the slice of thickness $\Delta T/2$ next to this surface:

$$CN_{j,N_T} = C_{j,N_T} + \frac{2}{M_T}(C_{j,N_T-1} - C_{j,N_T}) - \frac{2F_0\Delta t}{\rho\Delta T}(C_{j,N_T} - C_{ext})$$

$$+ \frac{1}{M_R}[C_{j-1,N_T-0.25} - 2C_{j,N_T-0.25} + C_{j+1,N_T-0.25}] \qquad (8.28)$$

At the Corner, N_R and N_T (Fig. 8.8)
The matter balance is calculated during the increment of time Δt within the parallelepiped of dimensions $\Delta R/2$, $\Delta T/2$ and thickness U next to the corner, by

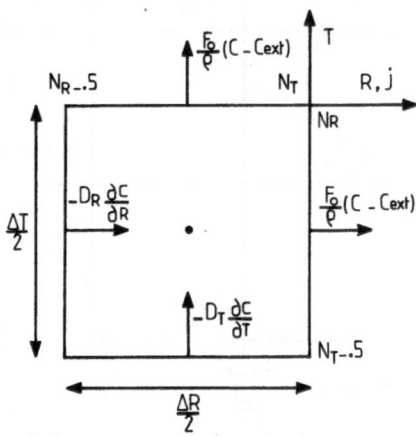

Fig. 8.8. Two-dimensional transfer by diffusion within the solid and evaporation from two faces near an edge ($j = N_R$, $k = N_T$).

considering the diffusion through the solid and evaporation from the surfaces:

$$\frac{\Delta T}{2}U\left[-D_R\frac{\partial C}{\partial R} - \frac{F_0}{\rho}(C_{N_R,N_T-0.25} - C_{ext})\right]\Delta t$$

$$+\frac{\Delta R}{2}U\left[-D_T\frac{\partial C}{\partial T} - \frac{F_0}{\rho}(C_{N_R,-0.25,N_T} - C_{ext})\right]\Delta t$$

$$= \frac{\Delta R}{2}\frac{\Delta T}{2}U[CN_{N_R-0.25,N_T-0.25} - C_{N_R-0.25,N_T-0.25}] \quad (8.29)$$

By assuming that

$$CN_{N_R-0.25,N_T-0.25} - C_{N_R-0.25,N_T-0.25} = CN_{N_R,N_T} - C_{N_R,N_T} \quad (8.30)$$

the new concentration at the corner after time Δt is given by

$$CN_{N_R,N_T} = C_{N_R,N_T} - \frac{2}{M_R}(C_{N_R,N_T-0.25} - C_{N_R-1,N_T-0.25})$$

$$-\frac{2\Delta t F_0}{\Delta R\rho}(C_{N_R,N_T-0.25} - C_{ext}) - \frac{2\Delta t F_0}{\Delta T\rho}(C_{N_R-0.25,N_T} - C_{ext})$$

$$-\frac{2}{M_T}(C_{N_R-0.25,N_T} - C_{N_R-0.25,N_T-1}) \quad (8.31)$$

With the other assumptions

$$C_{N_R-0.25,N_T} = C_{N_R,N_T-0.25} = C_{N_R,N_T} \quad (8.32)$$

the new concentration at the corner becomes

$$CN_{N_R,N_T} = C_{N_R,N_T} - \frac{2}{M_R}(C_{N_R,N_T} - C_{N_R-1,N_T})$$

$$-\frac{2\Delta t F_0}{\rho}\left(\frac{1}{\Delta R} + \frac{1}{\Delta T}\right)(C_{N_R,N_T} - C_{ext})$$

$$-\frac{2}{M_T}(C_{N_R,N_T} - C_{N_R,N_T-1}) \quad (8.33)$$

Remark 8.2: Other Assumptions. A better assumption than that made in Eq. (8.30) is

$$CN_{N_R-0.25,N_T-0.25} - C_{N_R-0.25,N_T-0.25}$$

$$= \tfrac{3}{4}(CN_{N_R,N_T} - C_{N_R,N_T}) + \tfrac{1}{4}(CN_{N_R-1,N_T-1} - C_{N_R-1,N_T-1}) \quad (8.34)$$

8.3.2 Concentration-Dependent Diffusivity

The concentration is determined within the solid, on the surface and at the corner.

Within the Solid, $N_R/2 \leqslant j \leqslant N_R-1$ and $N_T/2 \leqslant k \leqslant N_T-1$
The matter balance is evaluated during the increment of time Δt within the parallelepiped of thickness U and other dimensions ΔR and ΔT (Fig. 8.6):

$$U\Delta R\left[-D_T\frac{\partial C}{\partial T} + D_T\frac{\partial C}{\partial T}\right]\Delta t + U\Delta T\left[-D_R\frac{\partial C}{\partial R} + D_R\frac{\partial C}{\partial R}\right]\Delta t$$

$$\;_{k-0.5}\qquad\;_{k+0.5}\qquad\qquad\quad_{j-0.5}\qquad\;_{j+0.5}$$

$$= U\Delta R\Delta T[CN_{j,k} - C_{j,k}] \quad (8.14)$$
$$_{j,k}$$

The new concentration after time Δt at position $(j,\ k)$ is thus expressed in terms of the previous concentration at the same place and of the functions G_R and G_T:

$$CN_{j,k} = C_{j,k} + \frac{\Delta t}{\Delta T}[G_{T_{j,k+0.5}} - G_{T_{j,k-0.5}}] + \frac{\Delta t}{\Delta R}[G_{R_{j+0.5,k}} - G_{R_{j-0.5,k}}] \,(8.21)$$

the functions G_R and G_T being defined by

$$G_{R_{j+0.5,k}} = D_R\frac{\partial C}{\partial R} \quad \text{at} \quad (j + 0.5,\ k) \qquad\qquad (8.20)$$

$$G_{T_{j,k+0.5}} = D_T\frac{\partial C}{\partial T} \quad \text{at} \quad (j, k + 0.5) \qquad\qquad (8.19)$$

Surfaces $j = N_R$, k
The parallelepiped of thickness U and other dimensions $\Delta R/2$ and ΔT next to the surface N_R is shown in Fig 8.7. The matter balance within this parallelepiped during the increment of time Δt, by taking into account the diffusion and evaporation is

$$\frac{\Delta R}{2}U\left[-D_T\frac{\partial C}{\partial T} + D_T\frac{\partial C}{\partial T}\right] + \Delta TU\left[-D_R\frac{\partial C}{\partial R} - \frac{F_0}{\rho}(C_{N_R,k} - C_{ext})\right]$$

$$\phantom{\frac{\Delta R}{2}U}\text{at } (k-0.5) \text{ at } (k+0.5)\qquad\qquad \text{at } (N_R-0.5)\qquad\;\; \text{at } N_R$$

$$= \frac{\Delta R\Delta TU}{2\Delta t}\left[CN_{N_R-0.25,k} - C_{N_R-0.25,k}\right]$$

$$(8.25)$$

with the assumption

$$CN_{N_R-0.25,k} - C_{N_R-0.25,k} = CN_{N_R,k} - C_{N_R,k} \qquad\qquad (8.26)$$

The new concentration on the surface N_R after time Δt can thus be calculated by the relationship

$$CN_{N_R,k} = C_{N_R,k} + \frac{\Delta t}{\Delta T}[-G_{T_{N_R-0.25,k-0.5}} + G_{T_{N_R-0.25,k+0.5}}]$$

$$- \frac{2\Delta t}{\Delta R}\left[G_{R_{N_R-0.5,k}} + \frac{F_0}{\rho}(C_{N_R,k} - C_{ext})\right] \qquad\qquad (8.35)$$

The new concentration on the other surface N_T after time Δt can be obtained by a similar equation:

$$CN_{j,N_T} = C_{j,N_T} + \frac{\Delta t}{\Delta R}[-G_{R_{j-0.5,N_T-0.25}} + G_{R_{j+0.5,N_T-0.25}}]$$

$$-\frac{2\Delta t}{\Delta T}\left[G_{T_{j,N_T-0.5}} + \frac{F_0}{\rho}(C_{j,N_T} - C_{\text{ext}})\right] \tag{8.36}$$

Corner, $j = N_R$ and $k = N_T$

The parallelepiped of thickness U and other dimensions $\Delta R/2$ and $\Delta T/2$ next to the corner is considered (Fig. 8.8). The matter balance during the increment of time Δt in this parallelepiped, by considering the diffusion and evaporation from two surfaces, is written as follows:

$$\frac{\Delta T}{2}U\left[-D_R\frac{\partial C}{\partial R} - \frac{F_0}{\rho}(C_{N_R,N_T-0.25} - C_{\text{ext}})\right]\Delta t$$

$$+\frac{\Delta R}{2}U\left[-D_T\frac{\partial C}{\partial T} - \frac{F_0}{\rho}(C_{N_R-0.25,N_T} - C_{\text{ext}})\right]\Delta t$$

$$= \frac{\Delta R}{2}\frac{\Delta T}{2}U[CN_{N_R-0.25,N_T-0.25} - C_{N_R-0.25,N_T-0.25}] \tag{8.29}$$

By assuming that

$$CN_{N_R-0.25,N_T-0.25} - C_{N_R-0.25,N_T-0.25} = CN_{N_R,N_T} - C_{N_R,N_T} \tag{8.30}$$

the new concentration at the corner after time Δt is given by

$$CN_{N_R,N_T} = C_{N_R,N_T} - \frac{2\Delta t}{\Delta T}\left[G_{T_{N_R-0.25,N_T-0.5}} + \frac{F_0}{\rho}(C_{N_R-0.25,N_T} - C_{\text{ext}})\right]$$

$$-\frac{2\Delta t}{\Delta R}\left[G_{R_{N_R-0.5,N_T-0.25}} + \frac{F_0}{\rho}(C_{N_R,N_T-0.25} - C_{\text{ext}})\right] \tag{8.37}$$

8.4 Three Dimensions with Infinite Rate of Evaporation

8.4.1 Constant Diffusivity

The parallelepiped cut along the principal axes L, R and T is divided into small parallelepipeds of dimensions ΔL, ΔR and ΔT. Each position is characterised by the integers i, j, k, by following the scheme

L, D_L	(longitudinal axis)	$i,$	N_L
R, D_R	(radial axis)	$j,$	N_R
T, D_T	(tangential axis)	$k,$	N_T (8.38)

The principal diffusivities are D_L, D_R and D_T.

Each dimension of the parallelepiped is cut into N_L, N_R and N_T parallel slices of thicknesses ΔL, ΔR and ΔT.

The concentrations are calculated in two positions, within the solid and on the surface.

Within the Solid

$$N_L - 1 \leqslant i \leqslant \frac{N_L}{2}$$

$$N_R - 1 \leqslant j \leqslant \frac{N_R}{2}$$

$$N_T - 1 \leqslant k \leqslant \frac{N_T}{2} \tag{8.39}$$

A small parallelepiped of dimensions ΔL, ΔR and ΔT is considered (Fig. 8.9). The matter balance during the increment of time Δt is calculated within this parallelepiped, by considering the transfer of substance by diffusion along the principal axes:

$$\Delta R \Delta T \left[-D_L \frac{\partial C}{\partial L} + D_L \frac{\partial C}{\partial L} \right] \Delta t + \Delta L \Delta T \left[-D_R \frac{\partial C}{\partial R} + D_R \frac{\partial C}{\partial R} \right] \Delta t$$

$$\text{at } i-0.5 \qquad \text{at } i+0.5 \qquad\qquad \text{at } j-0.5 \qquad \text{at } j+0.5$$

$$+ \Delta L \Delta R \left[-D_T \frac{\partial C}{\partial T} + D_T \frac{\partial C}{\partial T} \right] \Delta t = \Delta L \Delta R \Delta T [CN_{i,j,k} - C_{i,j,k}]$$

$$\text{at } k-0.5 \qquad \text{at } k+0.5$$

$$\tag{8.40}$$

The new concentration after time Δt at the position (i, j, k) can thus be easily expressed in terms of the previous concentration at the same place and at the adjacent places:

$$CN_{i,j,k} = C_{i,j,k} + \frac{1}{M_L}[C_{i-1,j,k} - 2C_{i,j,k} + C_{i+1,j,k}]$$

$$+ \frac{1}{M_R}[C_{i,j-1,k} - 2C_{i,j,k} + C_{i,j+1,k}]$$

$$+ \frac{1}{M_T}[C_{i,j,k-1} - 2C_{i,j,k} + C_{i,j,k+1}] \tag{8.41}$$

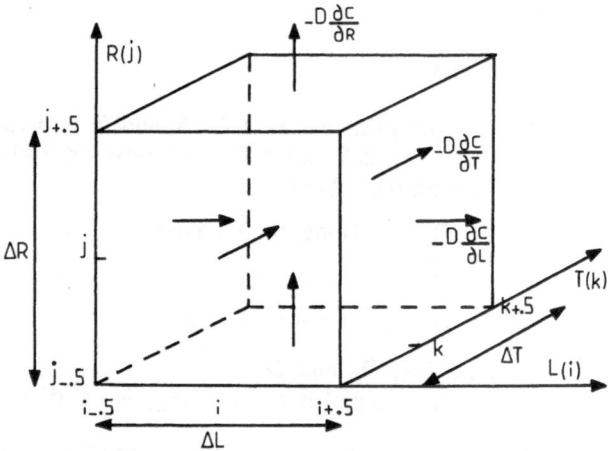

Fig. 8.9. Three-dimensional transfer by diffusion within the solid. Parallelepiped of dimensions ΔR, ΔL, ΔT.

with the dimensionless numbers

$$M_L = \frac{(\Delta L)^2}{D_L \Delta t}$$

$$M_R = \frac{(\Delta R)^2}{D_R \Delta t}$$

$$M_T = \frac{(\Delta T)^2}{D_T \Delta t} \qquad (8.16)$$

Surfaces, $i = N_L$ or $j = N_R$ or $k = N_T$
As the rate of evaporation is infinite, the concentration on each surface is constantly equal to zero:

$$C_{N_L,j,k} = C_{i,N_R,k} = C_{i,j,N_T} = 0 \qquad (8.42)$$

8.4.2 Concentration-Dependent Diffusivity

The problem is about the same as for the previous case, with a difference resulting from the concentration dependency of the principal diffusivities. The concentrations are easily obtained within the solid and on the surfaces.

Within the Solid

$$N_L - 1 \leq i \leq \frac{N_L}{2}$$

$$N_R - 1 \leq j \leq \frac{N_R}{2}$$

$$N_T - 1 \leq k \leq \frac{N_T}{2} \qquad (8.39)$$

The small parallelepiped shown in Fig. 8.9 is considered. The matter balance during the time increment Δt, by considering the diffusion with concentration-dependent diffusivities, can be written as follows:

$$\Delta R \Delta T[-G_{L_{i-0.5,j,k}} + G_{L_{i+0.5,j,k}}]\Delta t$$

$$+\Delta L \Delta T[-G_{R_{i,j-0.5,k}} + G_{R_{i,j+0.5,k}}]\Delta t$$

$$+\Delta L \Delta R[-G_{T_{i,j,k-0.5}} + G_{T_{i,j,k+0.5}}]\Delta t$$

$$= \Delta L \Delta R \Delta T[CN_{i,j,k} - C_{i,j,k}] \qquad (8.43)$$

The new concentration after time Δt at position (i, j, k) can thus be expressed in terms of the previous concentration at the same place and of the function G at various places:

$$CN_{i,j,k} = C_{i,j,k} + \frac{\Delta t}{\Delta L}[G_{L_{i+0.5,j,k}} - G_{L_{i-0.5,j,k}}]$$

$$+\frac{\Delta t}{\Delta R}[G_{R_{i,j+0.5,k}} - G_{R_{i,j-0.5,k}}]$$

$$+\frac{\Delta t}{\Delta T}[G_{T_{i,j,k+0.5}} - G_{T_{i,j,k-0.5}}] \qquad (8.44)$$

Surfaces, $i = N_L$, $j = N_R$ or $k = N_T$
The concentration on each surface is constantly equal to zero, as the rate of evaporation is infinite:

$$C_{N_L,j,k} = C_{i,N_R,k} = C_{i,j,N_T} = 0 \qquad (8.42)$$

Amount of Matter Remaining in the Solid
The amount of matter remaining in the solid at any time can be calculated by integrating the concentration at this time with respect to space:

$$Q_t = \sum_{i=0}^{N_L} \sum_{j=0}^{N_R} \sum_{k=0}^{N_T} C_{i,j,k} \Delta L \Delta R \Delta T \qquad (8.45)$$

8.5   Three Dimensions with Finite Rate of Evaporation

This case is very common for anisotropic solids such as wood, and the diffusivity can be constant or concentration dependent.

8.5.1   Constant Diffusivity

The following positions in the parallelepiped are of interest: within the solid, the surfaces, the lines of intersection of two surfaces, and the corners. They are considered successively.

Within the Solid

$$N_L - 1 \leqslant i \leqslant \frac{N_L}{2}$$

$$N_R - 1 \leqslant j \leqslant \frac{N_R}{2}$$

$$N_T - 1 \leqslant k \leqslant \frac{N_T}{2} \qquad (8.39)$$

The problem has already been solved for an infinite rate of evaporation (Sect. 8.4.1).

From the matter balance written in the parallelepiped of dimensions ΔL, ΔR and ΔT (Fig. 8.9), the new concentration after time Δt at the position (i, j, k) is obtained as a function of the previous concentration at the same place and adjacent places:

$$CN_{i,j,k} = C_{i,j,k} + \frac{1}{M_L}[C_{i-1,j,k} - 2C_{i,j,k} + C_{i+1,j,k}]$$

$$+ \frac{1}{M_R}[C_{i,j-1,k} - 2C_{i,j,k} + C_{i,j+1,k}]$$

$$+ \frac{1}{M_T}[C_{i,j,k-1} - 2C_{i,j,k} + C_{i,j,k+1}] \qquad (8.41)$$

with dimensionless numbers

$$M_L = \frac{(\Delta L)^2}{D_L \Delta t}$$

$$M_R = \frac{(\Delta R)^2}{D_R \Delta t}$$

$$M_T = \frac{(\Delta T)^2}{D_T \Delta t} \tag{8.16}$$

Surfaces, $i = N_L$, j, k (Fig. 8.10)
A parallelepiped of dimensions $\Delta L/2$, ΔR and ΔT located next to the surface $i = N_L$ is considered. The matter balance during the time increment Δt is calculated, by considering the transfer of substance by diffusion (longitudinal, radial and tangential) through the solid and by evaporation from the surface:

$$\Delta R \Delta T \left[-D_L \frac{\partial C}{\partial L} - \frac{F_0}{\rho}(C_{N_L,j,k} - C_{ext}) \right] \Delta t + \frac{\Delta L}{2} \Delta R \left[-D_T \frac{\partial C}{\partial T} + D_T \frac{\partial C}{\partial T} \right] \Delta t$$

$$\underset{N_L-0.5 \text{ (longitudinal)}}{\qquad\qquad} \overset{N_L}{\qquad} \qquad\qquad\qquad \underset{k-0.5 \text{ (tangential)}}{\qquad} \overset{k+0.5}{\qquad}$$

$$+ \frac{\Delta L}{2} \Delta T \left[-D_R \frac{\partial C}{\partial R} + D_R \frac{\partial C}{\partial R} \right] \Delta t = \frac{\Delta L}{2} \Delta R \Delta T [CN_{N_L-0.25,j,k} - C_{N_L-0.25,j,k}]$$

$$\underset{j-0.5 \text{ (radial)}}{\qquad} \overset{j+0.5}{\qquad}$$

$$\tag{8.46}$$

With the assumption

$$CN_{N_L-0.25,j,k} - C_{N_L-0.25,j,k} = CN_{N_L,j,k} - C_{N_L,j,k} \tag{8.47}$$

the new concentration after time Δt on the surface can be expressed in terms of the previous concentration at various places:

$$CN_{N_L,j,k} = C_{N_L,j,k} + \frac{2}{M_L}[C_{N_L-1,j,k} - C_{N_L,j,k}] - \frac{2\Delta t F_0}{\Delta L \rho}(C_{N_L,j,k} - C_{ext})$$

$$+ \frac{1}{M_T}[C_{N_L,j,k-1} - 2C_{N_L,j,k} + C_{N_L,j,k+1}]$$

$$+ \frac{1}{M_R}[C_{N_L,j-1,k} - 2C_{N_L,j,k} + C_{N_L,j+1,k}] \tag{8.48}$$

The new concentration can be determined on the other surfaces as follows.

Surface, i, $j = N_R$, k

$$CN_{i,N_R,k} = C_{i,N_R,k} + \frac{2}{M_R}[C_{i,N_R-1,k} - C_{i,N_R,k}] - \frac{2\Delta t F_0}{\Delta R \rho}(C_{i,N_R,k} - C_{ext})$$

$$+ \frac{1}{M_L}[C_{i+1,N_R,k} - 2C_{i,N_R,k} + C_{i-1,N_R,k}]$$

$$+ \frac{1}{M_T}[C_{i,N_R,k-1} - 2C_{i,N_R,k} + C_{i,N_R,k+1}] \tag{8.49}$$

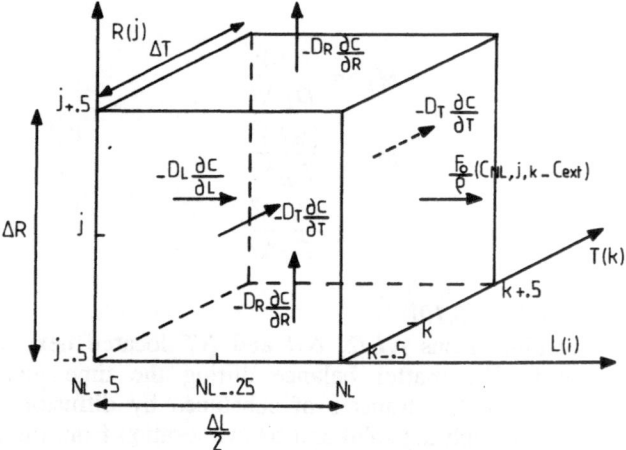

Fig. 8.10. Three-dimensional transfer by diffusion within the solid and evaporation from the surface $(i = N_L)$.

Surface, i, j, $k = N_T$

$$CN_{i,j,N_T} = C_{i,j,N_T} + \frac{2}{M_T}[C_{i,j,N_T-1} - C_{i,j,N_T}] - \frac{2\Delta t F_0}{\Delta T \rho}(C_{i,jN_T} - C_{ext})$$

$$+ \frac{1}{M_L}[C_{i-1,j,N_T} - 2C_{i,j,N_T} + C_{i+1,j,N_T}]$$

$$+ \frac{1}{M_R}[C_{i,j-1,N_T} - 2C_{i,j,N_T} + C_{i,j+1,N_T}] \tag{8.50}$$

Edge, $i = N_L$, $j = N_R$ and k

A parallelepiped of dimensions $\Delta L/2$, $\Delta R/2$ and ΔT next to the edge formed by the surfaces N_L and N_R is considered (Fig. 8.11). The matter balance during the small time Δt is calculated by considering the diffusion through the solid and the evaporation from the two surfaces:

$$\frac{\Delta R}{2}\Delta T\left[-D_L\frac{\partial C}{\partial L} - \frac{F_0}{\rho}(C_{N_L,N_R-0.25,k} - C_{ext})\right]\Delta t$$

$$\quad N_L-0.5 \; _{\text{(longitudinal)}} \; ^{N_L}$$

$$+ \frac{\Delta L}{2}\Delta T\left[-D_R\frac{\partial C}{\partial R} - \frac{F_0}{\rho}(C_{N_L-0.25,N_R,k} - C_{ext})\right]\Delta t$$

$$\quad N_R-0.5 \; _{\text{(radial)}} \; ^{N_R}$$

$$+ \frac{\Delta L}{2}\frac{\Delta R}{2}\left[-D_T\frac{\partial C}{\partial T} + D_T\frac{\partial C}{\partial T}\right]\Delta T$$

$$\quad k-0.5 \; _{\text{tangential}} \; ^{k+0.5}$$

$$= \frac{\Delta L}{2}\frac{\Delta R}{2}\Delta T[CN_{N_L-0.25,N_R-0.25,k} - C_{N_L-0.25,N_R-0.25,k}]$$

$$\tag{8.51}$$

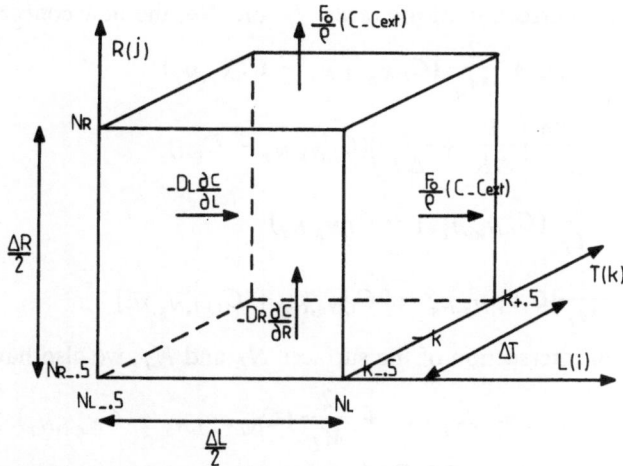

Fig. 8.11. Three-dimensional transfer by diffusion within the solid and evaporation from two faces near an edge ($i = N_L$ and $j = N_R$).

With the assumption

$$CN_{N_L-0.25,N_R-0.25,k} - C_{N_L-0.25,N_R-0.25,k} = CN_{N_L,N_R,k} - C_{N_L,N_R,k} \quad (8.52)$$

the new concentration on the line of intersection of the two faces N_L and N_R after time Δt is expressed by

$$CN_{N_L,N_R,k} = C_{N_L,N_R,k} + \frac{2}{M_L}(C_{N_L-1,N_R-0.25,k} - C_{N_L,N_R-0.25,k})$$

$$- \frac{2\Delta t F_0}{\Delta L \rho}(C_{N_L,N_R-0.25,k} - C_{ext}) + \frac{2}{M_R}(C_{N_L-0.25,N_R-1,k} - C_{N_L-0.25,N_R,k})$$

$$- \frac{2\Delta t F_0}{\Delta R \rho}(C_{N_L-0.25,N_R,k} - C_{ext})$$

$$+ \frac{1}{M_T}[C_{N_L-0.25,N_R-0.25,k-1} - 2C_{N_L-0.25,N_R-0.25,k} + C_{N_L-0.25,N_R-0.25,k+1}]$$

$$(8.53)$$

With further assumptions such as

$$C_{N_L-0.25,N_R-0.25,k} = C_{N_L-0.25,N_R,k} = C_{N_L,N_R,k} \quad (8.54)$$

the expression for the new concentration on the line of intersection of the two faces is reduced to

$$CN_{N_L,N_R,k} = C_{N_L,N_R,k} + \frac{2}{M_L}(C_{N_L-1,N_R,k} - C_{N_L,N_R,k})$$

$$- \frac{2\Delta t F_0}{\rho}\left(\frac{1}{\Delta L} + \frac{1}{\Delta R}\right)(C_{N_L,N_R,k} - C_{ext})$$

$$+ \frac{2}{M_R}(C_{N_L,N_R-1,k} - C_{N_L,N_R,k})$$

$$+ \frac{1}{M_T}[C_{N_L,N_R,k-1} - 2C_{N_L,N_R,k} + C_{N_L,N_R,k+1}] \quad (8.55)$$

For the line of intersection of the faces N_R and N_T, the new concentration is

$$CN_{i,N_R,N_T} = C_{i,N_R,N_T} + \frac{2}{M_R}(C_{i,N_R-1,N_T} - C_{i,N_R,N_T})$$

$$-\frac{2\Delta t F_0}{\rho}\left(\frac{1}{\Delta R} + \frac{1}{\Delta T}\right)(C_{i,N_R,N_T} - C_{\text{ext}})$$

$$+\frac{2}{M_T}(C_{i,N_R,N_T-1} - C_{i,N_R,N_T})$$

$$+\frac{1}{M_L}[C_{i-1,N_R,N_T} - 2C_{i,N_R,N_T} + C_{i+1,N_R,N_T}] \qquad (8.55')$$

and for the line of intersection of the surfaces N_L and N_T, we also have

$$CN_{N_L,j,N_T} = C_{N_L,j,N_T} + \frac{2}{M_L}(C_{N_L-1,j,N_T} - C_{N_L,j,N_T})$$

$$-\frac{2\Delta t F_0}{\rho}\left(\frac{1}{\Delta L} + \frac{1}{\Delta T}\right)(C_{N_L,j,N_T} - C_{\text{ext}})$$

$$+\frac{2}{M_T}(C_{N_L,j,N_T-1} - C_{N_L,j,N_T})$$

$$+\frac{1}{M_R}[C_{N_L,j-1,N_T} - 2C_{N_L,j,N_T} + C_{N_L,j+1,N_T}] \qquad (8.55'')$$

Corners

A parallelepiped of dimensions $\Delta L/2$, $\Delta R/2$ and $\Delta T/2$ next to the corner formed by the surfaces N_L, N_R and N_T is considered (Fig. 8.12). The coordinates of the centre of this parallelepiped are $N_L - 0.25$, $N_R - 0.25$ and $N_T - 0.25$.

The matter balance during the increment of time Δt is calculated within this parallelepiped by considering the diffusion through the solid and the evaporation from the three faces:

$$\frac{\Delta L}{2}\frac{\Delta T}{2}\left[-D_R\frac{\partial C}{\partial R}\bigg|_{\substack{N_R-0.25 \\ (\text{radial})}}^{N_R} - \frac{F_0}{\rho}(C_{N_L-0.25,N_R,N_T-0.25} - C_{\text{ext}})\right]\Delta t$$

$$+\frac{\Delta R}{2}\frac{\Delta T}{2}\left[-D_L\frac{\partial C}{\partial L}\bigg|_{\substack{N_L-0.5 \\ (\text{longitudinal})}}^{N_L} - \frac{F_0}{\rho}C_{N_L,N_R-0.25,N_T-0.25} - C_{\text{ext}})\right]\Delta t$$

$$+\frac{\Delta L}{2}\frac{\Delta R}{2}\left[-D_T\frac{\partial C}{\partial T}\bigg|_{\substack{Nt-0.25 \\ (\text{tangential})}}^{N_T} - \frac{F_0}{\rho}(C_{N_L-0.25,N_R-0.25,N_T} - C_{\text{ext}})\right]$$

$$=\frac{\Delta L}{2}\frac{\Delta R}{2}\frac{\Delta T}{2}[CN_{N_L-0.25,N_R-0.25,N_T-0.25} - C_{N_L-0.25,N_R-0.25,N_T-0.25}]$$

$$(8.56)$$

With the assumption

$$CN_{N_L-0.25,N_R-0.25,N_T-0.25} - C_{N_L-0.25,N_R-0.25,N_T-0.25}$$

$$= CN_{N_L,N_R,N_T} - C_{N_L,N_R,N_T} \qquad (8.57)$$

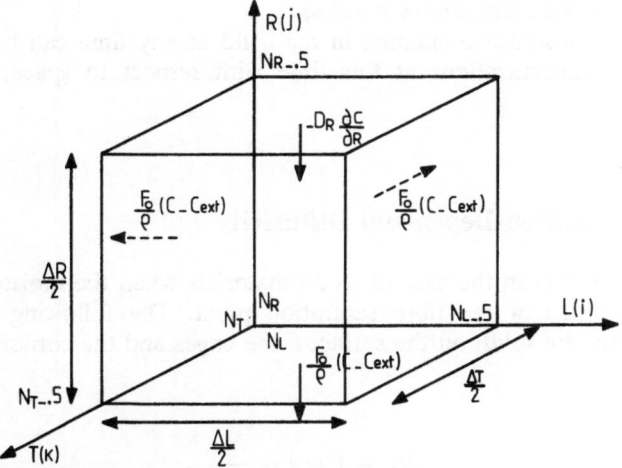

Fig. 8.12. Three-dimensional transfer by diffusion within the solid and evaporation from three surfaces near a corner ($i = N_L$, $j = N_R$, $k = N_T$).

the new concentration at the corner is given by

$$CN_{N_L,N_R,N_T} = C_{N_L,N_R,N_T} + \frac{2}{M_R}\left(C_{N_L-0.25,N_R-1,N_T-0.25} - C_{N_L-0.25,N_R,N_T-0.25}\right)$$

$$+ \frac{2}{M_L}\left(C_{N_L-1,N_R-0.25,N_T-0.25} - C_{N_L,N_R-0.25,N_T-0.25}\right)$$

$$+ \frac{2}{M_T}\left(C_{N_L-0.25,N_R-0.25,N_T-1} - C_{N_L-0.25,N_R-0.25,N_T}\right)$$

$$- \frac{2\Delta t F_0}{\Delta R\rho}\left(C_{N_L-0.25,N_R,N_T-0.25} - C_{\text{ext}}\right)$$

$$- \frac{2\Delta t F_0}{\Delta L\rho}\left(C_{N_L,N_R-0.25,N_T-0.25} - C_{\text{ext}}\right)$$

$$- \frac{2\Delta t F_0}{\Delta T\rho}\left(C_{N_L-0.25,N_R-0.25,N_T} - C_{\text{ext}}\right) \tag{8.58}$$

With the other assumptions

$$C_{N_L-0.25,N_R,N_T-0.25} = C_{N_L-0.25,N_R-0.25,N_T-0.25} = C_{N_L-0.25,N_R-0.25,N_T} = C_{N_L,N_R,N_T}$$
$$\tag{8.59}$$

the expression of the new concentration after time Δt at the corner is reduced to

$$CN_{N_L,N_R,N_T} = C_{N_L,N_R,N_T} + \frac{2}{M_R}\left(C_{N_L,N_R-1,N_T} - C_{N_L,N_R,N_T}\right)$$

$$+ \frac{2}{M_L}\left(C_{N_L-1,N_R,N_T} - C_{N_L,N_R,N_T}\right)$$

$$+ \frac{2}{M_T}\left(C_{N_L,N_R,N_T-1} - C_{N_L,N_R,N_T}\right)$$

$$- \frac{2\Delta t F_0}{\rho}\left(\frac{1}{\Delta L} + \frac{1}{\Delta R} + \frac{1}{\Delta T}\right)[C_{N_L,N_R,N_T} - C_{\text{ext}}] \tag{8.60}$$

Amount of Matter Remaining in the Solid
The amount of substance remaining in the solid at any time can be obtained by integrating the concentrations at this time with respect to space, as shown in Eq. (8.45).

8.5.2 Concentration-Dependent Diffusivity

This problem appears in the case of wood materials when the moisture content is below that obtained at the fibre saturation point. The following positions are examined: within the solid, on the surfaces, the edges and the corners.

Within the Solid

$$N_L - 1 \le i \le \frac{N_L}{2}$$

$$N_R - 1 \le j \le \frac{N_R}{2}$$

$$N_T - 1 \le k \le \frac{N_T}{2} \tag{8.39}$$

As shown in Sect. 8.4.2, the new concentration after time Δt at the position (i, j, k) is expressed as a function of the previous concentrations obtained at the same position and adjacent places:

$$CN_{i,j,k} = C_{i,j,k} + \frac{\Delta t}{\Delta L}[G_{L_{i+0.5,j,k}} - G_{L_{i-0.5,j,k}}] + \frac{\Delta t}{\Delta R}[G_{R_{i,j+0.5,k}} - G_{R_{i,j-0.5,k}}]$$

$$+ \frac{\Delta t}{\Delta T}[G_{T_{i,j,k+0.5}} - G_{T_{i,j,k-0.5}}] \tag{8.44}$$

with the functions G_L, G_R and G_T shown in Eq. (8.20).

Surface, $i = N_L$, j, k (Fig. 8.10)
The matter balance is calculated during the time increment Δt within the parallelepiped of dimensions $\Delta L/2$, ΔR and ΔT, located next to the surface N_L:

$$-\Delta R \Delta T \left[G_{L_{N_L-0.5,j,k}} + \frac{F_0}{\rho}(C_{N_L,j,k} - C_{\text{ext}}) \right] \Delta t$$

$$+ \frac{\Delta L}{2} \Delta R [-G_{T_{N_L-0.25,j,k-0.5}} + G_{T_{N_L-0.25,j,k+0.5}}] \Delta t$$

$$+ \frac{\Delta L}{2} \Delta T [-G_{R_{N_L-0.25,j-0.5,k}} + G_{R_{N_L-0.25,j+0.5,k}}]$$

$$= \frac{\Delta L}{2} \Delta R \Delta T [CN_{N_L-0.25,j,k} - C_{N_L-0.25,j,k}] \tag{8.61}$$

With the assumption

$$CN_{N_L-0.25,j,k} - C_{N_L-0.25,j,k} = CN_{N_L,j,k} - C_{N_L,j,k} \tag{8.47}$$

the new concentration on the surface N_L after time Δt is then obtained as a function of the previous concentration and function G:

$$CN_{N_L,j,k} = C_{N_L,j,k} - \frac{2\Delta t}{\Delta L}\left[GL_{N_L-0.5,j,k} + \frac{F_0}{\rho}(C_{N_L,j,k} - C_{\text{ext}})\right]$$

$$+ \frac{\Delta t}{\Delta T}[GT_{N_L-0.25,j,k+0.5} - GT_{N_L-0.25,j,k-0.5}]$$

$$+ \frac{\Delta t}{\Delta R}[GR_{N_L-0.25,j+0.5,k} - GR_{N_L-0.25,j-0.5,k}] \qquad (8.62)$$

Surface, i, $j = N_R$, k
The new concentration on this surface after time Δt is also obtained by

$$CN_{i,N_R,k} = C_{i,N_R,k} - \frac{2\Delta t}{\Delta R}\left[GR_{i,N_R-0.5,k} + \frac{F_0}{\rho}(C_{i,N_R,k} - C_{\text{ext}})\right]$$

$$+ \frac{\Delta t}{\Delta L}[GL_{i+0.5,N_R-0.25,k} - GL_{i-0.5,N_R-0.25,k}]$$

$$+ \frac{\Delta t}{\Delta T}[GT_{i,N_R-0.25,k+0.5} - GT_{i,N_R-0.25,k-0.5}] \qquad (8.63)$$

Surface, i, j, $k = N_T$

$$CN_{i,j,N_T} = C_{i,j,N_T} - \frac{2\Delta t}{\Delta T}\left[GT_{i,j,N_T-0.5} + \frac{F_0}{\rho}(C_{i,j,N_T} - C_{\text{ext}})\right]$$

$$+ \frac{\Delta t}{\Delta L}[GL_{i+0.5,j,N_T-0.25} - GL_{i-0.5,j,N_T-0.25}]$$

$$+ \frac{\Delta t}{\Delta R}[GR_{i,j+0.5,N_T-0.25} - GR_{i,j-0.5,N_T-0.25}] \qquad (8.64)$$

Edge, $i = N_L$, $j = N_R$ and k
A parallelepiped of dimensions $\Delta L/2$, $\Delta R/2$ and ΔT, located next to edge formed by the two surfaces, is considered (Fig. 8.11). The matter balance during the time increment Δt is rewritten by considering the transfer of substance by diffusion with the function G, and the transfer by evaporation from the two surfaces:

$$-\frac{\Delta R}{2}\Delta T\left[GL_{N_L-0.5,N_R-0.25,k} + \frac{F_0}{\rho}(C_{N_L,N_R-0.25,k} - C_{\text{ext}})\right]\Delta t$$

$$-\frac{\Delta L}{2}\Delta T\left[GR_{N_L-0.25,N_R-0.5,k} + \frac{F_0}{\rho}(C_{N_L-0.25,N_R,k} - C_{\text{ext}})\right]\Delta t$$

$$+\frac{\Delta L}{2}\frac{\Delta R}{2}[-GT_{N_L-0.25,N_R-0.25,k-0.5} + GT_{N_L-0.25,N_R-0.25,k+0.5}]\Delta t$$

$$= \frac{\Delta L}{2}\frac{\Delta R}{2}\Delta T[CN_{N_L-0.25,N_R-0.25,k} - C_{N_L-0.25,N_R-0.25,k}] \qquad (8.65)$$

With the assumption

$$CN_{N_L-0.25,N_R-0.25,k} - C_{N_L-0.25,N_R-0.25,k} = CN_{N_L,N_R,k} - C_{N_L,N_R,k} \qquad (8.52)$$

the new concentration after time Δt at the edge is

$$
CN_{N_L,N_R,k} = C_{N_L,N_R,k} - \frac{2\Delta t}{\Delta L}\left[G_{L_{N_L-0.5,N_R-0.25,k}} + \frac{F_0}{\rho}\left(C_{N_L,N_R-0.25,k} - C_{ext} \right) \right]
$$
$$
- \frac{2\Delta t}{\Delta R}\left[G_{R_{N_L-0.25,N_R-0.5,k}} + \frac{F_0}{\rho}\left(C_{N_L-0.25,N_R,k} - C_{ext} \right) \right]
$$
$$
+ \frac{\Delta t}{\Delta T}[G_{T_{N_L-0.25,N_R-0.25,k+0.5}} - G_{T_{N_L-0.25,N_R-0.25,k-0.5}}] \tag{8.66}
$$

Edge, i, j = N_R and k = N_T
The parallelepiped of dimensions ΔL, $\Delta R/2$ and $\Delta T/2$ located next to the edge formed by the two surfaces is considered. The new concentration after time Δt at this line of intersection is thus obtained:

$$
CN_{i,N_R,N_T} = C_{i,N_R,N_T} - \frac{2\Delta t}{\Delta R}\left[G_{R_{i,N_R-0.5,N_T-0.25}} + \frac{F_0}{\rho}\left(C_{i,N_R,N_T-0.25} - C_{ext} \right) \right]
$$
$$
- \frac{2\Delta t}{\Delta T}\left[G_{T_{i,N_R-0.25,N_T-0.5}} + \frac{F_0}{\rho}\left(C_{i,N_R-0.25,N_T} - C_{ext} \right) \right]
$$
$$
+ \frac{\Delta t}{\Delta L}[G_{L_{i+0.5,N_R-0.25,N_T-0.25}} - G_{L_{i-0.5,N_R-0.25,N_T-0.25}}] \tag{8.67}
$$

For the edge $i = N_L$, j, $k = N_T$, the new concentration after time Δt is obtained as follows:

$$
CN_{N_L,j,N_T} = C_{N_L,j,N_T} - \frac{2\Delta t}{\Delta L}\left[G_{L_{N_L-0.5,j,N_T-0.25}} + \frac{F_0}{\rho}\left(C_{N_L,j,N_T-0.25} - C_{ext} \right) \right]
$$
$$
- \frac{2\Delta t}{\Delta T}\left[G_{T_{N_L-0.25,j,N_T-0.5}} + \frac{F_0}{\rho}\left(C_{N_L-0.25,j,N_T} - C_{ext} \right) \right]
$$
$$
+ \frac{\Delta t}{\Delta R}[G_{R_{N_L-0.25,j+0.5,N_T-0.25}} - G_{R_{N_L-0.25,j-0.5,N_T-0.25}}] \tag{8.68}
$$

Corner, i = N_L, j = N_R and k = N_T
A parallelepiped of dimensions $\Delta L/2$, $\Delta R/2$ and $\Delta T/2$, located next to the corner, is considered (Fig. 8.12). The matter balance during the increment of time Δt is calculated within this parallelepiped by considering the diffusion through the solid and the evaporation from the three surfaces:

$$
- \frac{\Delta L}{2}\frac{\Delta T}{2}\left[G_{R_{N_L-0.25,N_R-0.5,N_T-0.25}} + \frac{F_0}{\rho}\left(C_{N_L-0.25,N_R,N_T-0.25} - C_{ext} \right) \right]\Delta t
$$
$$
- \frac{\Delta R}{2}\frac{\Delta T}{2}\left[G_{L_{N_L-0.5,N_R-0.25,N_T-0.25}} + \frac{F_0}{\rho}\left(C_{N_L,N_R-0.25,N_T-0.25} - C_{ext} \right) \right]\Delta t
$$
$$
- \frac{\Delta L}{2}\frac{\Delta R}{2}\left[G_{T_{N_L-0.25,N_R-0.25,N_T-0.5}} + \frac{F_0}{\rho}\left(C_{N_L-0.25,N_R-0.25,N_T} - C_{ext} \right) \right]
$$
$$
= \frac{\Delta L}{2}\frac{\Delta R}{2}\frac{\Delta T}{2}[CN_{N_L-0.25,N_R-0.25,N_T-0.25} - C_{N_L-0.25,N_R-0.25,N_T-0.25}]
$$
$$
\tag{8.69}
$$

With the assumption

$$CN_{N_L-0.25,N_R-0.25,N_T-0.25} - C_{N_L-0.25,N_R-0.25,N_T-0.25}$$

$$= CN_{N_L,N_R,N_T} - C_{N_L,N_R,N_T} \qquad (8.57)$$

the new concentration after time Δt at the corner becomes

$$CN_{N_L,N_R,N_T} = C_{N_L,N_R,N_T}$$

$$- \frac{2\Delta t}{\Delta R}\left[G_{R_{N_L-0.25,N_R-0.5,N_T-0.25}} + \frac{F_0}{\rho}\,(C_{N_L-0.25,N_R,N_T-0.25} - C_{\text{ext}})\right]$$

$$- \frac{2\Delta t}{\Delta L}\left[G_{L_{N_L-0.5,N_R-0.25,N_T-0.25}} + \frac{F_0}{\rho}\,(C_{N_L,N_R-0.25,N_T-0.25} - C_{\text{ext}})\right]$$

$$- \frac{2\Delta t}{\Delta T}\left[G_{T_{N_L-0.25,N_R-0.25,N_T-0.5}} + \frac{F_0}{\rho}\,(C_{N_L-0.25,N_R-0.25,N_T} - C_{\text{ext}})\right]$$

$$\qquad (8.70)$$

8.6 Conclusions

The following conclusions are worth noting.

Use of Numerical Models
Numerical models are of help for calculating the concentration within parallelepipeds as well as the kinetics of drying, especially when three principal diffusivities exist.

Conditions of Stability for Calculation
The dimensionless numbers M_L, M_R and M_T found when the diffusivity is
constant, must be higher than two.

Assumptions Made for Calculation
The concentrations are calculated only at places defined by integers. At other
places, the concentrations are obtained by a simple way, depending whether
transfer is two- or three-dimensional.

- Two-dimensional transfer:

$$C_{j+0.5,k} = \tfrac{1}{2}[C_{j+1,k} + C_{j,k}]$$

$$C_{N_R-0.25,k} = \tfrac{3}{4}C_{N_R,k} + \tfrac{1}{4}C_{N_R-1,k}$$

$$C_{N_R-0.25,N_T-0.25} = \tfrac{3}{4}C_{N_R,N_T} + \tfrac{1}{4}C_{N_R-1,N_T-1}$$

- Three-dimensional transfer:

$$C_{i+0.5,j,k} = \tfrac{1}{2}[C_{i,j,k} + C_{i+1,j,k}]$$

$$C_{N_L-0.25,j,k} = \tfrac{3}{4}C_{N_L,j,k} + \tfrac{1}{4}C_{N_L-1,j,k}$$

$$C_{N_L-0.25,N_R-0.25,k} = \tfrac{3}{4}C_{N_L,N_R,k} + \tfrac{1}{4}C_{N_L-1,N_R-1,k}$$

$$C_{N_L-0.25,N_R-0.25,N_T-0.25} = \tfrac{3}{4}C_{N_L,N_R,N_T} + \tfrac{1}{4}C_{N_L-1,N_R-1,N_T-1}$$

$$C_{N_L-0.25,N_R-0.25,N_T-0.5} = \tfrac{3}{8}[C_{N_L,N_R,N_T} + C_{N_L,N_R,N_T-1}]$$
$$+ \tfrac{1}{8}[C_{N_L-1,N_R-1,N_T} + C_{N_L-1,N_R-1,N_T-1}]$$

In the case of concentration-dependent diffusivities, the diffusivities at these various positions are thus easily obtained by using the right relationship between the diffusivity and concentration.

Symbols

A	Area through which the substance diffuses
$C_{j,k}$	Concentration at position (j, k) with two-dimensional transfer
$C_{i,j,k}$	Concentration at position (i, j, k) with three-dimensional transfer
D	Diffusivity ($cm^2\,s^{-1}$)
D_L, D_R, D_T	Principal diffusivities (longitudinal, radial, tangential)
F_0	Rate of evaporation ($g\,cm^{-2}\,s^{-1}$)
$G_L(G_R, G_T)$	Function $\left(G_L = D_L\dfrac{\partial C}{\partial L}\right)$
i, j, k	Integers representing longitudinal, radial and tangential positions, respectively
M_L, M_R, M_T	Dimensionless numbers
N_L, N_R, N_T	Number of slices
U	Unit thickness
$\Delta L, \Delta R, \Delta T$	Finite dimensions of a parallelepiped
Δt	Increment of time
ρ	Density of the evaporating substance ($g\,cm^{-3}$)

Chapter 9

Drying of Paints

9.1 Introduction

One of the purposes of this chapter is to attain a further insight into the nature of the process involved in the drying of paints, since drying is an important step in paint preparation. An additional objective is to build up a model capable of simulating the whole drying process by considering the various steps involved, and thus enabling the prediction of the time necessary to dry the coating with a given thickness and solvent concentration. Another purpose is to gain a fuller understanding of the process when the paint is composed not only of one layer but of several superimposed layers.

The case considered is that of a paint which dries by solvent evaporation without chemical reaction, and the solvent plays the role of diluent. This general problem is rather complicated because of the following facts:

1. The process of drying combines two diverse factors, i.e., diffusion of the solvent through the polymer and solid ingredients, and evaporation of the solvent from the surface. Drying is thus a diffusion-controlled process [1].

2. The rate of evaporation depends on
   - Volatility of the solvent
   - Temperature which affects both the volatility of the solvent and its diffusivity
   - Agitation of the surrounding atmosphere, which plays an important role [2]

3. The transfer within the solid is controlled by diffusion under transient conditions. This fact has already been demonstrated for a liquid evaporating from a polymer, either a rubber or plasticised PVC [3, 4]. The diffusivity was found to depend on the temperature and the concentration of the liquid located in the polymer.

4. Following drying and loss of solvent, a shrinkage of the coating can be significant especially when the coating is flexible. This shrinkage is also observed during the desorption of a liquid from a swollen rubber [1]. During some drying stages, the coating solidifies, becoming hard and stiff, and sometimes being constrained by adhesion to the substrate.

9.2 One Layer

9.2.1 Theory

Assumptions

The following assumptions were made in order to enable a clear description of all the known facts regarding the process [5, 6]:

1. The transfer of solvent is controlled by transient diffusion within the paint, and evaporation from the surface.
2. The diffusivity is concentration dependent.
3. The rate of evaporation of the solvent is taken to be either constant or proportional to the concentration of solvent on the surface.
4. One-dimensional diffusion through the paint thickness is considered.
5. (a) The initial solvent concentration is uniform throughout the paint where there is only one layer of paint [5].
    (b) The solvent concentration is not uniform throughout the thickness of the first layer when a second layer is deposited and dried [6].
6. A frame of reference is fixed with respect to the paint thickness, despite shrinkage of the coating during drying.
7. The pressure of solvent vapour in the surrounding atmosphere is zero, as well as the concentration of liquid on the surface which is at equilibrium with this atmosphere.

Mathematical Treatment

The equation for one-dimensional diffusion with concentration-dependent diffusivity is

$$\frac{\partial C}{\partial t} = \frac{\partial}{\partial x}\left(D\,\frac{\partial C}{\partial x}\right) \tag{9.1}$$

where C is the concentration of solvent in the coating and D the concentration-dependent diffusivity.

The initial and boundary conditions are as follows [5, 6]:

$$t = 0 \qquad 0 < x < L \qquad C_{in} \neq 0 \tag{9.2}$$

$$t > 0 \qquad -D\,\frac{\partial C}{\partial x} = \frac{F_0}{\rho}\,(C_s - C_{ext}) \tag{9.3}$$

where F_0 is the rate of evaporation of the pure solvent, ρ is the density of this liquid, C_s is the concentration of liquid on the surface and C_{ext} the concentration of liquid on the surface which is at equilibrium with the surrounding atmosphere.

Eq. (9.3) has two precise meanings:

1. The rate of evaporation is constantly equal to the rate of diffusing substance which is brought to the surface by diffusion.
2. The rate of evaporation is proportional to the difference between the concentration of solvent on the surface C_s, and the concentration of solvent

on the surface which is at equilibrium with the surrounding atmosphere C_{ext}. The coefficient of proportionality F_0 is the rate of evaporation of the pure liquid $(g\, cm^{-2}\, s^{-1})$.

In the present case,

$$C_{ext} = 0 \qquad (9.4)$$

during the whole process, as the surrounding atmosphere is well stirred and the vapour pressure of the solvent is zero.

Numerical Analysis

No analytical solution can be found for these equations, and a numerical method with finite differences is used to solve the problem [5, 6]. The paint thickness is divided into N constant increments of space Δx. The space–time region is covered by a grid of rectangles of sides Δx and Δt, Δt being the time increment (Fig. 9.1).

By using the integers n and j, the coordinates are expressed by

$$x = n\Delta x \qquad (9.5)$$

$$L = N\Delta x \qquad (9.5')$$

$$t = j\Delta t \qquad (9.6)$$

and the concentration of solvent is denoted as

$$C_{n,j} \qquad \text{or simply} \qquad C_n$$

$$\text{and } CN_n \qquad \text{instead of} \qquad C_{n,j+1} \qquad (9.7)$$

From the matter balance evaluated during the time increment Δt within the slice of thickness Δx located at position n, the new concentration after time Δt is expressed in terms of the previous concentrations at this position and adjacent positions:

$$CN_n = \frac{1}{M_n}[C_{n-1} + (M_n - 2)C_n + C_{n+1}] \qquad (9.8)$$

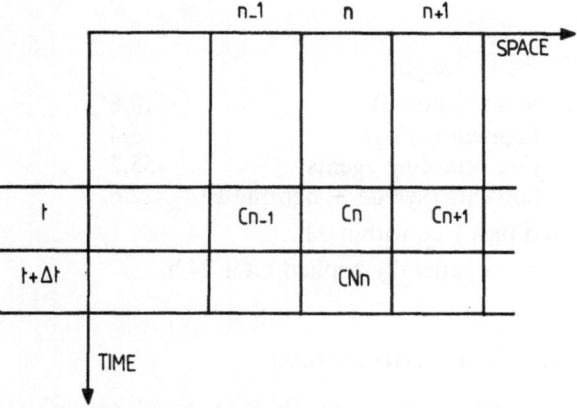

Fig. 9.1. Space–time diagram for numerical analysis: variation of the concentration of solvent with the thickness of the coating and time.

with the dimensionless number M_n:

$$M_n = \frac{(\Delta x)^2}{D_n \Delta t} \tag{9.9}$$

For the coating plane adjacent to the plane substrate, the thickness $\Delta x/2$ is taken, and the new concentration is thus obtained by (Fig. 9.2)

$$CN_n = \frac{1}{M_n} [2C_{N-1} + (M_N - 2)C_N] \tag{9.10}$$

For the surface of the paint, the slice of thickness $\Delta x/2$ next to the surface is considered, and the matter balance is calculated by considering the diffusion of solvent within the paint and the evaporation of solvent from the surface (Fig. 9.3). The new concentration is thus expressed in terms of the previous concentrations and of the evaporation rate, by

$$CN_0 = C_0 - \frac{1}{3}(CN_1 - C_1) + \frac{8}{3M_1}(C_1 - C_0) - \frac{8}{3}\frac{F_0\Delta t}{\rho\Delta x}C_0 \tag{9.11}$$

Of course, the new concentration CN_1 at position 1 next to the surface must be calculated beforehand, as the new concentration on the surface CN_0 is a function of CN_1. This new concentration CN_1 can be easily obtained using Eq. (9.8).

The amount of solvent remaining in the coating at time t can be obtained by integrating the concentration of solvent with respect to space, or rather by summing up the solvent concentrations throughout the thickness of the coating.

$$Q_t = \left[\frac{1}{2}C_0 + \frac{1}{2}C_n + \sum_1^{N-1} C_N\right]\Delta x \tag{9.12}$$

9.2.2 Experiment

Materials

The coating used was a fire retardant coating PI 248 (JIVAL, France) with the following properties:

- Density: 1.25
- Dry extract: 72.4% by weight
- Components: Resin (pliolite)   10.8
              Pigment (TiO_2)   6.4
              Fire retarding agents   55.2
              Solvents (xylene + benzine F)   27.6
- The coating dried tack free within 8 h.
- The second layer was generally applied after 24 h.

Coating Preparation and Measurement

Coatings of various thicknesses were applied to glass plates, and exposed to motionless air at various constant temperatures ranging from 20 to 46 °C. The

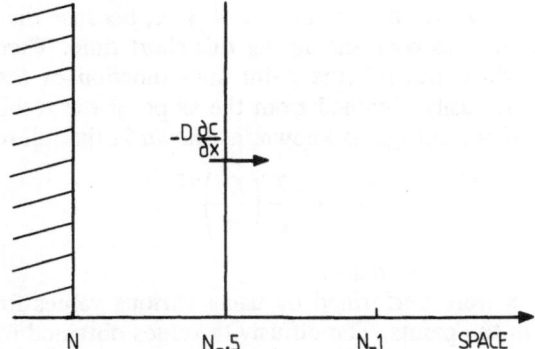

Fig. 9.2. Slice next to the plane substrate.

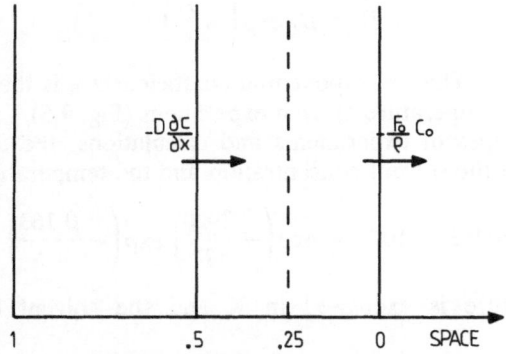

Fig. 9.3. Slice next to the surface.

drying process was followed by recording the weight loss of the coating on a weighing scale. The influence of the solvent concentration on the rate of drying was determined by using paints having various solvent concentrations ranging from 22 to 35 wt %.

9.2.3 Results

Before evaluating the validity of the numerical model by comparing the experimental and theoretical kinetics of drying, it is necessary to determine the values of diffusivity for various solvent concentrations in the paint and various temperatures in order to establish relationships between diffusivity and these parameters.

Determination of the Parameters

Diffusivity
A variety of methods may be used to determine the diffusivity of a liquid through a given material, as shown in previous studies of toluene in elastomeric materials [1], or for the plasticiser migrating through polyvinylchloride (PVC) [7, 8].

When the diffusivity is concentration dependent, the best method involves the use of the short test technique [7]. The principle of this method consists of measuring the diffusivity over a short period of time, because the concentration of liquid can be considered as constant during this short time. Then by plotting the amount of solvent which has left the paint as a function of the square root of time, the diffusivity is easily obtained from the slope of the resulting straight line if the total amount of solvent Q_∞ is known, as shown in the following equation:

$$\frac{Q_t}{Q_\infty} = \frac{2}{L}\left(\frac{Dt}{\pi}\right)^{0.5} \tag{9.13}$$

where L is the thickness of the paint.

These experiments were performed by using various values for the concentration of the solvent in the paints. The diffusivity values obtained for these different paints are expressed as a function of the concentration of solvent (Fig. 9.4), conforming to the relationship:

$$D = D_0 \exp\left(-\frac{A}{C}\right) \tag{9.14}$$

where A is a constant. The pre-exponential coefficient D_0 is then plotted against the reciprocal of the temperature of each experiment (Fig. 9.5).

From these two types of experiments and calculations, the diffusivity can be expressed in terms of the solvent concentration and the temperature

$$D = 9.3 \times 10^{-3} \times \exp\left(-\frac{2920}{T}\right)\exp\left(-\frac{0.163}{C}\right) \tag{9.15}$$

where the temperature is expressed in K and the solvent concentration in $cm^3\ cm^{-3}$.

Rate of Evaporation
The rate of evaporation was found to vary with the temperature by following an expression similar to the Clausius–Clapeyron law, as the rate of evaporation is proportional to the vapour pressure of the solvent (Fig. 9.6).

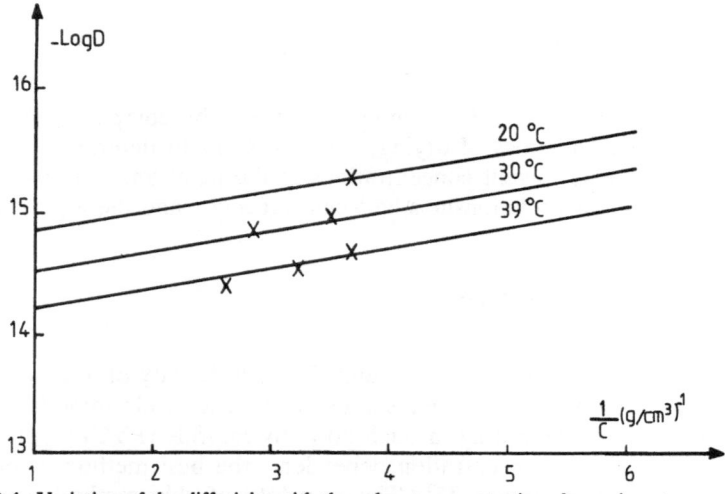

Fig. 9.4. Variation of the diffusivity with the solvent concentration, for various temperatures.

The following equation obtained from experiments expresses the value of the rate of evaporation as a function of the temperature (K), F being expressed in $g \, cm^{-2} \, s^{-1}$.

$$F = 1.41 \exp\left(- \frac{7610}{RT}\right) \tag{9.16}$$

Validity of the Model

The numerical model was tested by comparing the kinetics of drying obtained either by experiments or by calculation. These comparisons were made by using various temperatures ranging from 30 to 46 °C, and various concentrations of solvent.

The results are illustrated in

- Fig. 9.7, at 30 °C for two solvent concentrations
- Fig. 9.8, at 39 °C for three diverse solvent concentrations
- Fig. 9.9, at 46 °C for two solvent concentrations

All the values for the diffusivity and rate of evaporation were determined using the experimental data. In all cases examined, obtained by varying the temperature and the solvent concentration, the theoretical values calculated with the numerical model are very well superimposed on the experimental values, proving the validity of the model.

Effect of Parameters

Three parameters are of interest in the drying process of paints:

- the temperature at which drying is conducted
- the initial solvent concentration within the paint
- the thickness of the paint layer

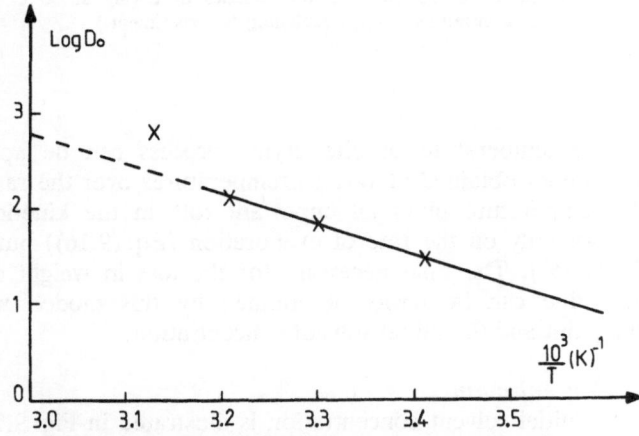

Fig. 9.5. Variation of the coefficient D_0 with temperature.

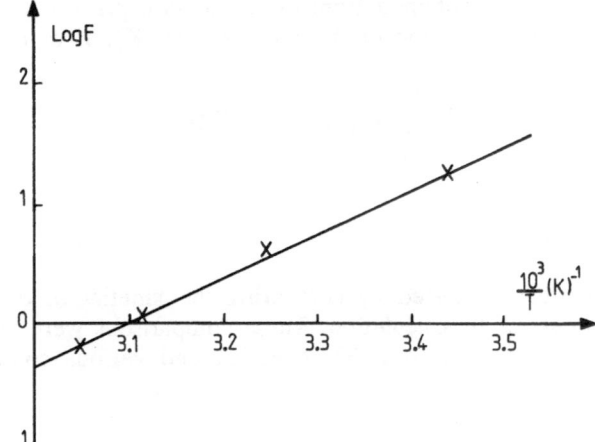

Fig. 9.6. Variation of the rate of evaporation of the solvent with temperature.

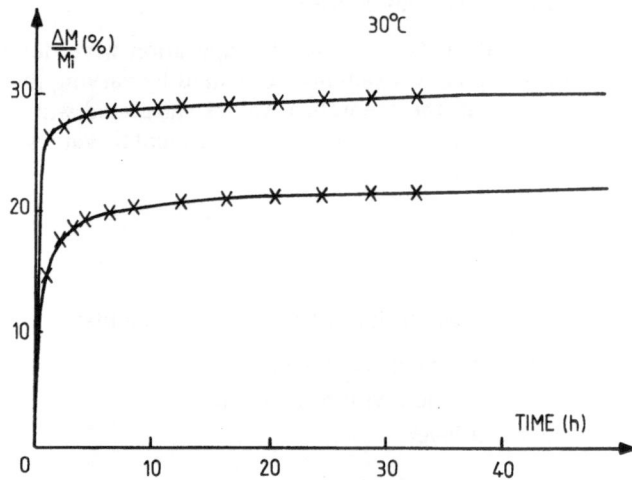

Fig. 9.7. Validity of the model, estimated from the kinetics of drying at 30 °C for two solvent concentrations. (——, calculation; ×, experiment.)

Temperature

The effect of the temperature on the drying process can be appreciated by comparing the kinetics obtained at various temperatures over the range 30–46 °C (Fig. 9.7–9.9). Temperature plays an important role in the kinetics of drying, because it acts not only on the rate of evaporation (Eq. (9.16)) but also on the diffusivity (Eq. (9.15)). The time necessary for the loss in weight of solvent to reach a given value can be easily determined by this model, whatever the thickness of the paint and the initial solvent concentration.

Initial Solvent Concentration

The effect of the initial solvent concentration is illustrated in Figs 9.7–9.9, where various concentrations of solvent are used for various temperatures.

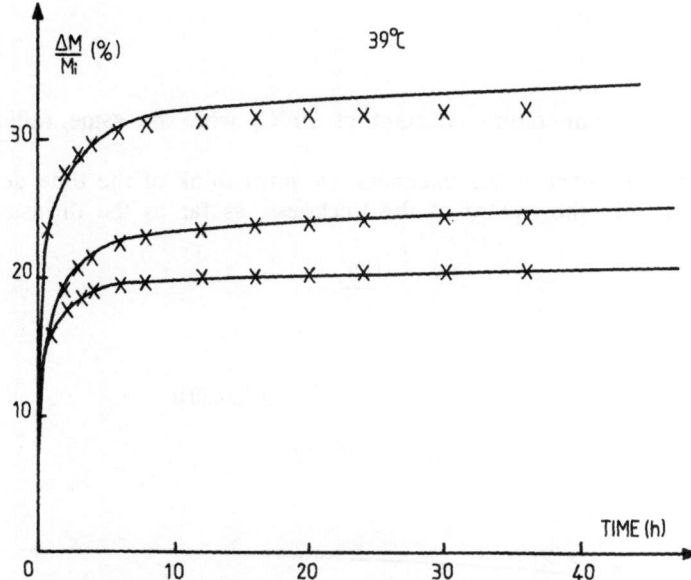

Fig. 9.8. Validity of the model, estimated from the kinetics of drying at 39 °C, for three solvent concentrations. (——, calculation; ×, experiment.)

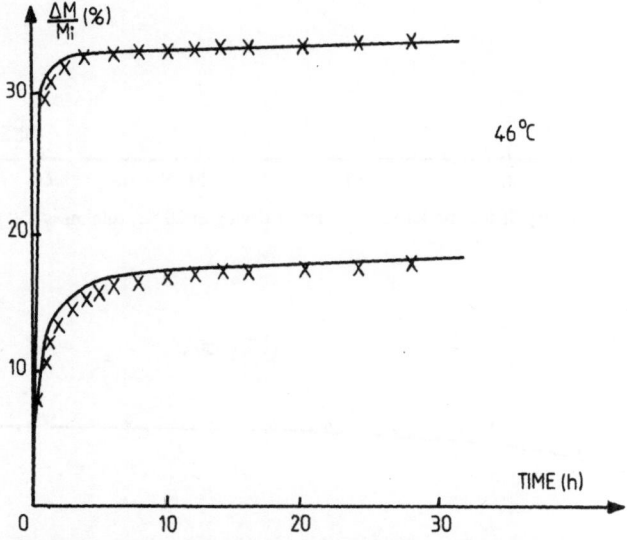

Fig. 9.9. Validity of the model, estimated from the kinetics of drying at 46 °C, for two solvent concentrations. (——, calculation; ×, experiment.)

Thickness of the Paint Layer

The thickness of the paint layer is a parameter of interest, especially for fire retardant coatings. The effect of the thickness of the layer is shown in Figs 9.10–9.12, by considering the following values:

- 450 μm (Fig. 9.10)
- 530 μm (Fig. 9.11)
- 670 μm (Fig. 9.12)

and keeping the temperature constant at 20 °C, with the same initial solvent concentration.

To examine the effect of the thickness, we must think of the time dependency of the process with the square of the thickness, as far as the diffusion only is concerned:

$$\frac{Dt}{L^2} \tag{9.17}$$

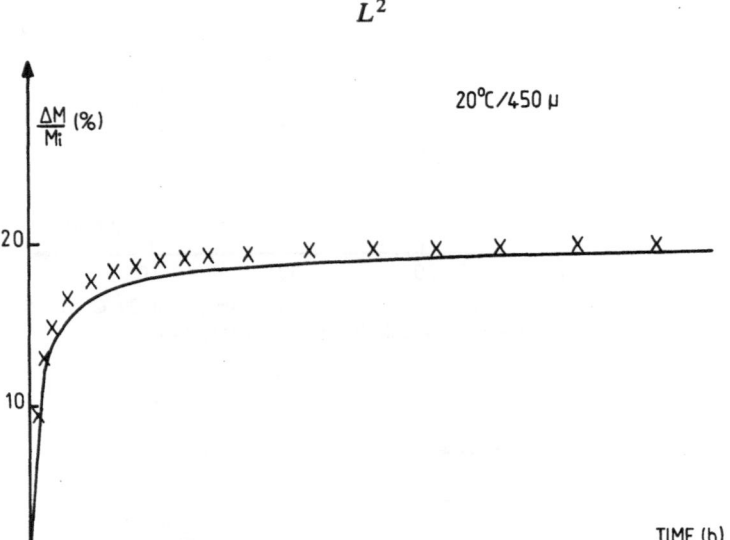

Fig. 9.10. Validity of the model for kinetics of drying at 20 °C, thickness 450 μm.

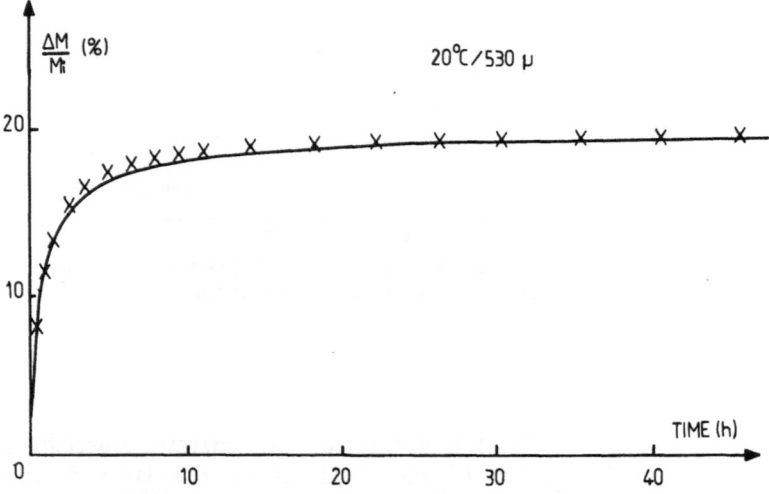

Fig. 9.11. Validity of the model for kinetics of drying at 20 °C, thickness 530 μm.

9.3 Two Layers

9.3.1 Theory

The process is about the same as for one layer of paint, in the sense that diffusion of liquid through the solid and evaporation from the surface take place. However, a difference exists between these two cases, due to the transport of solvent into the first layer when the second layer is deposited.

Assumptions

All the assumptions made for the drying process of one layer of a coating are also considered with two layers: on the whole, the process is controlled by diffusion of solvent through the paint and evaporation from the surface.

Another assumption is made during the short time when the second layer is deposited. A part of the solvent of the second layer penetrates quickly the first layer which is partly dried, and an approximately constant gradient of concentration through the two layers is thus obtained by this hydrodynamic process. This process is described in Fig. 9.13, where the previous gradient of concentration in the first layer is shown, as well as the new gradient of solvent expanded just after spreading the second layer.

9.3.2 Experiment

The first layer was deposited on a glass substrate and partly dried under given conditions of temperature and time. The second layer was then spread on the first layer, and the two layers left to dry under selected conditions of time and temperature.

The conditions selected for preparing various two-layer coatings are described in Table 9.1.

The coatings were then weighed at intervals.

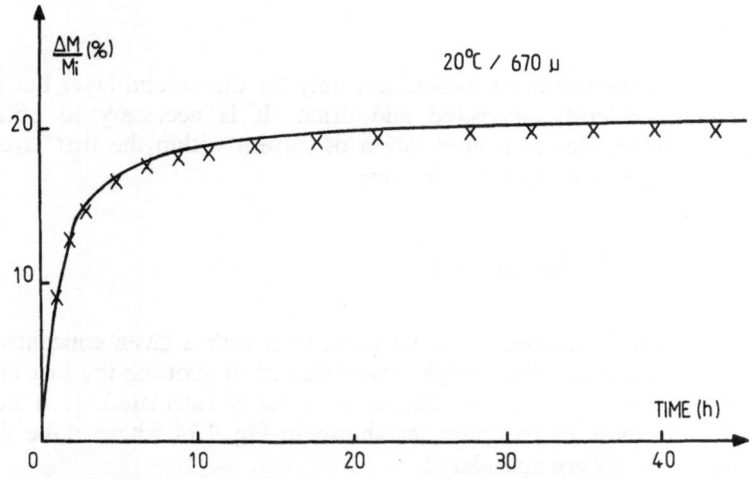

Fig. 9.12. Validity of the model for kinetics of drying at 20 °C, thickness 670 μm.

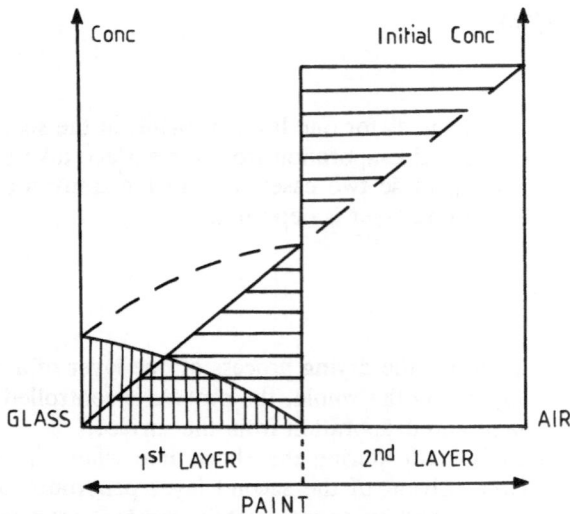

Fig. 9.13. Scheme for solvent transfer between the two layers when the second layer is spread over the first, showing profiles of concentration of solvent: |||, remaining in the partly dried first layer; ≡, resulting from transfer from the second layer to the first; ———, resulting from both of these contributions.

Table 9.1. Preparation of two-layer coatings

| Time of drying (h) | First layer | | Second layer | |
	Thickness (μm)	Δm (1) (g g^{-1})	Thickness (μm)	Δm (2) (g g^{-1})
1	506	0.136	610	0.336
14	500	0.043	495	0.272
24	510	0.034	600	0.331

Δm (1) is the amount of solvent remaining in the first layer when the second layer is spread over.
Δm (2) represents the total amount of solvent in the second layer when applied over the first one.

9.3.3 Results

The process of drying has been studied not only for the second layer but also for the first layer previously deposited and dried. It is necessary to obtain full knowledge of the profiles of concentration of solvent within the first layer when the second layer is spread over the first one.

Determination of the Parameters

Diffusivity
The diffusivity was determined by using short tests with a given concentration of solvent. From the slope of the straight line obtained by plotting the loss in weight with the square root of time, the diffusivity is easily calculated. It is the same whatever the thickness of the paint, as shown in Fig. 9.14 where three different values of the thickness are considered.

Of course, the diffusivity depends on the value of the concentration of solvent.

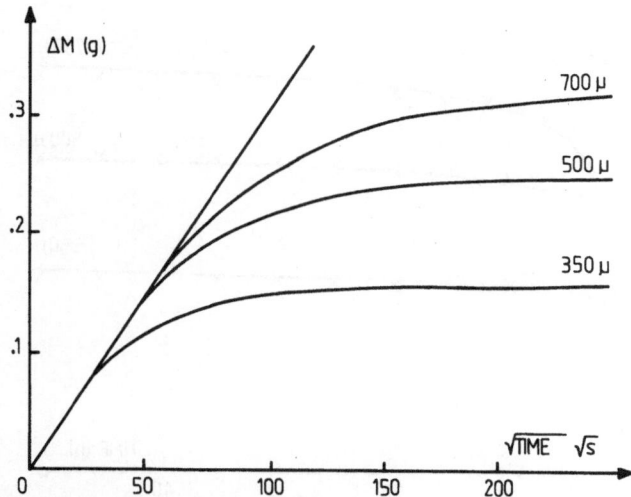

Fig. 9.14. Determination of the diffusivity by using short tests. Loss in weight of solvent as a function of the square root of time.

Rate of Evaporation

The rate of evaporation of the pure solvent is determined by using the same method as for the first layer.

Validity of the Model for the First Layer

The curves expressing the kinetics of drying obtained from experiments and from calculation are very well superimposed, proving the validity of the model for the first layer. These kinetics were determined for various values of the thickness ranging from 350 to 700 μm (Fig. 9.15).

The profiles of concentration of solvent developed through the thickness of the first layer can be calculated by using the model. Some profiles are shown in Fig. 9.16 for various times of drying. As seen from these curves, the concentration of solvent next to the surface falls rapidly to a low value around zero. After 20 h of drying, a small but significant amount of solvent remains in the coating.

It is difficult to make experiments in order to determine the profiles of concentration of a liquid throughout a solid, as shown previously in the case of plasticiser and liquid transfers within a plasticised PVC sheet in contact with this liquid [9, 10]. However, a proof of the validity of these profiles of concentration exists, as the kinetics of drying are obtained by integrating the concentration of the solvent with respect to space throughout the thickness of the paint.

Validity of the Model for the Second Layer

The validity of the model was tested for the second layer by comparing the kinetics of drying obtained from experiments and calculation. As shown in Fig. 9.17, compatibility was found between experimental and theoretical results for the kinetics of drying for the three samples described in Table 9.1. The three

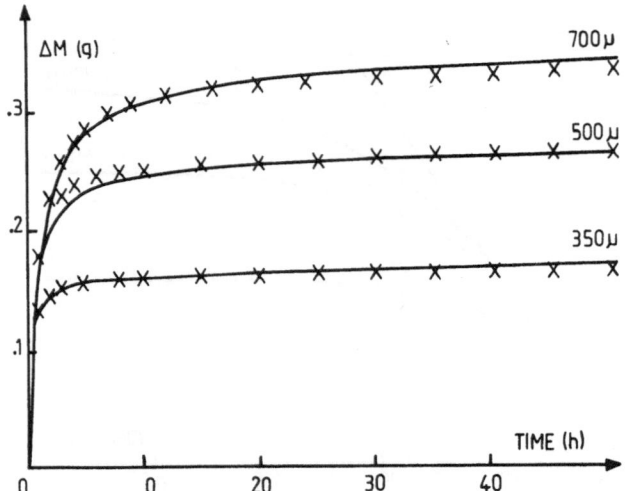

Fig. 9.15. Validity of the model for the first layer. Kinetics of drying obtained by experiments (×) and calculation (——).

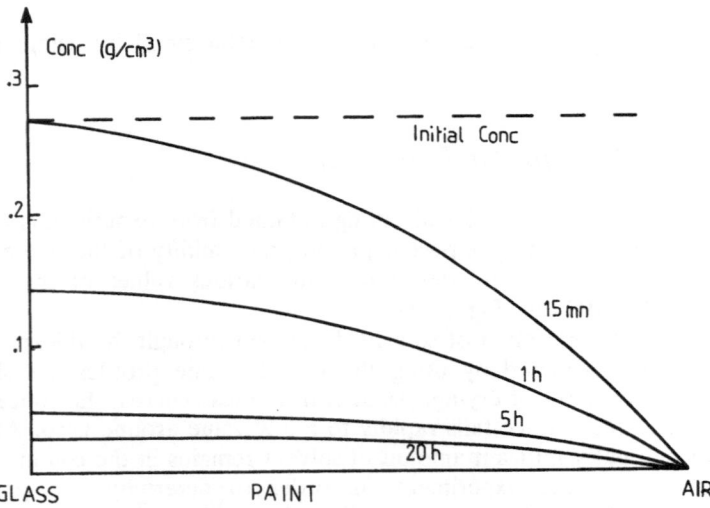

Fig. 9.16. Profiles of concentration during the drying process of one-layer coating, for various times of drying.

curves in Fig. 9.17, plotted for increasing times of drying for the first layer, are quite different. Of course, the longer the time of drying for the first layer, the shorter the time of drying for the second layer.

The kinetics were obtained by assuming that a fast transfer of solvent from the second layer to the first one takes place as soon as the second layer is deposited. The solvent is transferred so that a linear gradient of concentration is attained through the thickness of the two layers.

The profiles of concentration of solvent through the two layers of paint calculated using the numerical model are shown in Figs 9.18–9.20. The following results are worth noting:

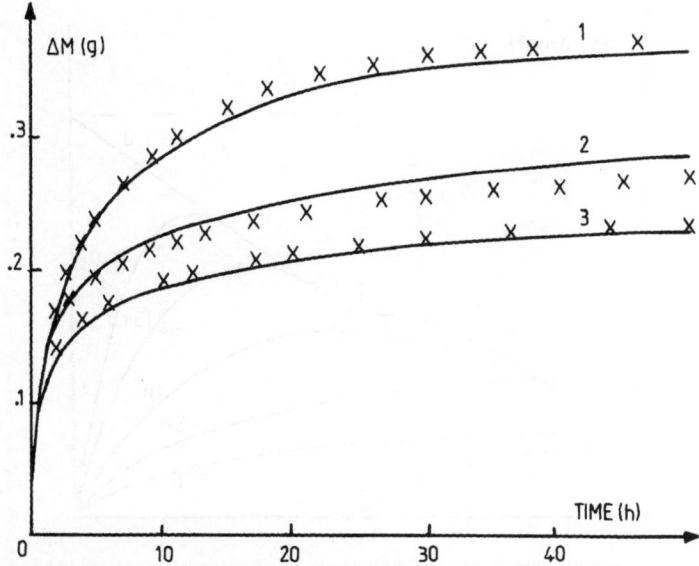

Fig. 9.17. Kinetics of drying of two-layer coatings, with various times of drying for the first layer. 1: 1 h; 2: 14 h; 3: 24 h. (——, calculation; ×, experiment.)

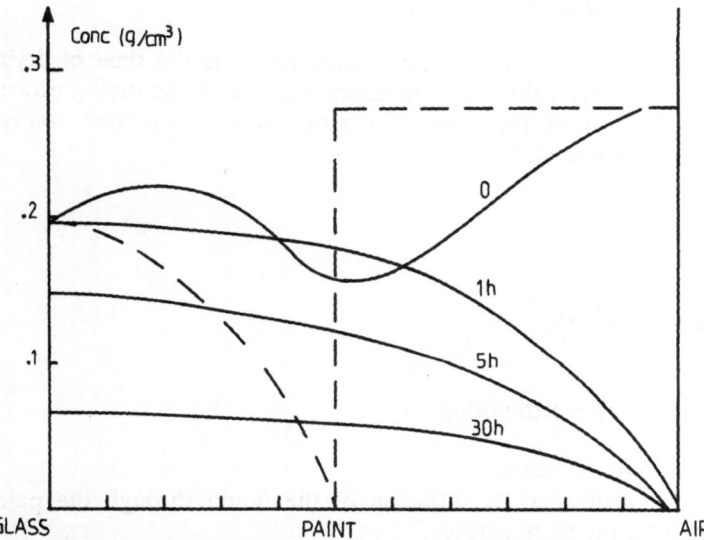

Fig. 9.18. Profiles of concentration during the drying process of a two-layer coating, when the first layer is dried for 1 h.

1. The concentration of solvent next to the surface falls to a very low value as soon as the process starts for the two-layer coating.

2. After 30 h of drying of the two-layer coating, the solvent seems to be entrapped within the paint. This fact is especially due to the dependency of diffusivity with the solvent concentration. A low concentration of solvent on the surface of the paint is responsible for a low rate of transfer of the solvent by diffusion.

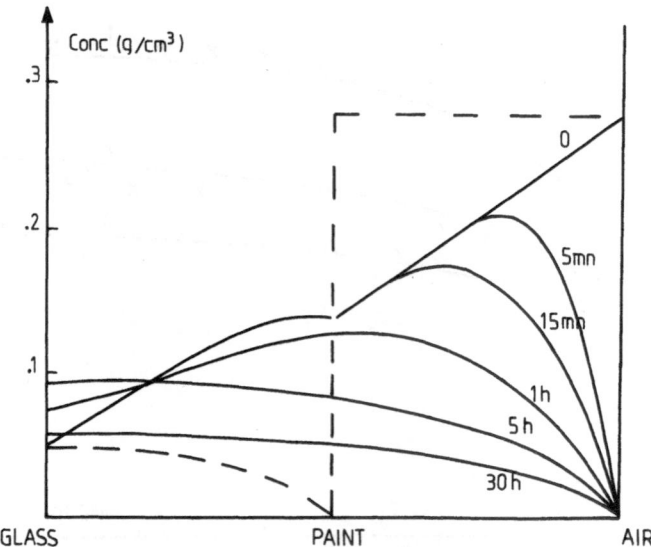

Fig. 9.19. Profiles of concentration during the drying process of a two-layer coating, when the first layer is dried for 14 h.

Effect of the Parameters

The parameter which was especially considered [6] is the time of drying for the first coating. Of course, this plays an important role in the drying process of both layers. Optimisation of the time of drying for the two-layer coating can be obtained using the model.

9.4   Conclusions

Two conclusions are worth noting.

Process of Drying of Paints
The process is controlled by diffusion of the liquid through the paint and by evaporation from the paint surface.

The main parameter is not the evaporation rate but the diffusivity. In the original papers [5, 6], the rate of evaporation was sometimes taken as infinite, and a slight change in the kinetics of drying was observed, proving the unimportant effect of the rate of evaporation on the whole process.

The diffusivity greatly depends on the concentration of the diffusing substance, as well as the temperature, and the higher the concentration of solvent (or the higher the temperature), the faster the rate of diffusion. This concentration dependency of diffusivity is responsible for an important fact which becomes very significant at the end of the process. As diffusivity decreases with solvent concentration, the rate of diffusion of the solvent is very low next to the surface

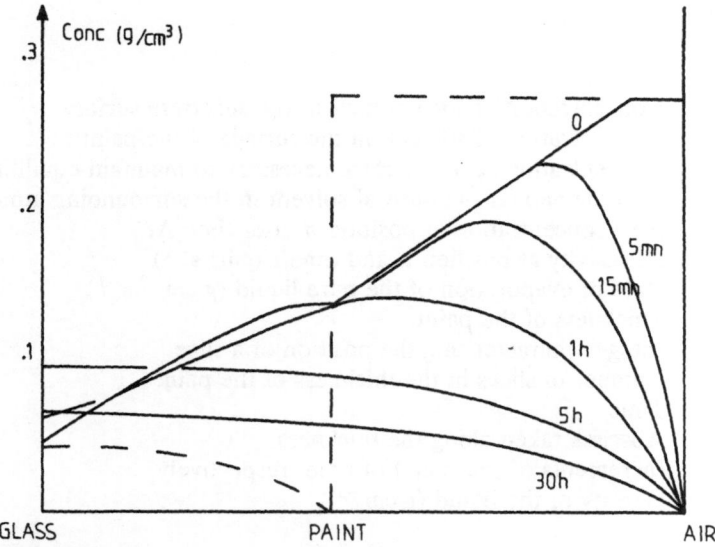

Fig. 9.20. Profiles of concentration during the drying process of a two-layer coating, when the first layer is dried for 24 h.

of the paint where the solvent concentration tends to zero. Resulting from this fact, part of the solvent remains entrapped within the paint.

The model can be used in order to determine operational conditions able to reduce this inconvenience.

Assumptions Made in This Study

The assumption bears essentially on the diffusivity, due to the fact that this parameter is concentration dependent.

The dimensionless number M_n

$$M_n = \frac{(\Delta x)^2}{D_n \Delta t} \qquad (9.9)$$

was used in the two papers [5, 6], for calculating the new concentration of solvent after time Δt at position n:

$$CN_n = \frac{1}{M_n} [C_{n-1} + (M_n - 2)C_n + C_{n+1}] \qquad (9.8)$$

Of course, being concentration dependent, diffusivity varies with time and position, and D_n must be calculated for each new position and time.

A better equation can be used, as shown in Chap. 5:

$$CN_n = C_n + \frac{\Delta t}{\Delta x} (G_{n-0.5} - G_{n+0.5}) \qquad (9.18)$$

with the function G

$$G_{n-0.5} = D_{n-0.5} \frac{\partial C}{\partial x} \qquad \text{at } (n - 0.5) \qquad (9.19)$$

Symbols

C_N	Concentration of solvent next to the substrate surface
C_s, C_0	Concentration of solvent on the surface of the paint
C_{ext}	Concentration on the surface necessary to maintain equilibrium with the vapour pressure of solvent in the surrounding atmosphere
CN_n	New concentration at position n after time Δt
D_n	Diffusivity at position n and time t (cm^2 s^{-1})
F_0	Rate of evaporation of the pure liquid (g cm^{-2} s^{-1})
L	Thickness of the paint
n	Integer characterising the position of a slice
N	Number of slices in the thickness of the paint
t	Time
x	Abscissa taken along the thickness
Δx, Δt	Increments of space and of time, respectively
ρ	Density of the liquid (g cm^{-3})

References

1. Khatir Y, Bouzon J, Vergnaud JM. Liquid sorption by rubber sheets and evaporation; models and experiments. Polym Test 1986; 6: 253–267
2. Abdul M, Bathnagar VM, Vergnaud JM. Effect of air velocity on pyrolysis of fire retardant coatings exposed to air heated at controlled temperatures. Fire Safety 1984/85; 8: 135–140
3. Aboutaybi A, Bouzon J, Vergnaud JM. Modelling the process of drying of plasticized PVC previously immersed in n-hexane. Europ Polym J 1989; 25: 1013–1018
4. Aboutaybi A, Bouzon J, Vergnaud JM. Drying at room temperature of plasticized PVC previously immersed in liquids. Modelling and experiments. Europ Polym J 1990; 26: 285–291
5. Blandin HP, David JC, Illien JP, Malizevicz M, Vergnaud JM. Modelling of drying of coatings: effect of the thickness, temperature and concentration of solvent. Prog Organic Coat 1987; 15: 163–172
6. Blandin HP, David JC, Illien JP, Malizevicz M, Vergnaud JM. Modelling the drying process of coatings with various layers. J Coat Technol 1987; 59: 27–32
7. Taverdet JL, Vergnaud JM. Study of transfer process of liquid into and plasticizer out of plasticized PVC by using short tests. J Appl Polym Sci 1984; 29: 3391–3400
8. Taverdet JL, Vergnaud JM. Modelization of matter transfers between plasticized PVC and liquids in case of a maximum for liquid-time curves. J Appl Polym Sci 1986; 31: 111–122
9. Messadi D, Hivert M, Vergnaud JM. A new approach to the study of plasticizer migration from PVC into methanol. J Appl Polym Sci 1981; 26: 667–677
10. Messadi D, Vergnaud JM. Plasticizer transfer from plasticized PVC into ethanol–water mixtures. J Appl Polym Sci 1982; 27: 3945–3955

Chapter 10

Drying of Wet Earth for Adobe Construction

10.1 Introduction

Primitive adobe dwellings constructed from thick walls of crude earth (clay) still predominate in poor or developing countries where bricks and concrete are not produced. This type of earth wall is very inexpensive, but it does not perform very well because of its poor mechanical properties.

Adobe construction on a large scale depends on the supply of a large quantity of earth blocks of good quality at low cost. This is achieved by pressing moist earth firmly into a mould; simultaneous vibration allows the pressure to be reduced as well as the time required for moulding [1, 2]. The last step of the process is the drying of the blocks in still air.

As the mechanical properties of these earth blocks were found to depend greatly on their state of drying, studies of the drying process are of interest for its development. Moreover, if the rate of drying is too high, the solid may develop cracks which may be responsible for poor durability of the final product. In previous works [3] describing the desorption of a solvent from rubber, the drying process was found to be controlled by diffusion of the liquid through the solid and evaporation from the surface. The diffusion was found to obey Fick's laws under transient conditions with a concentration-dependent diffusivity when the solid was a polymer [4, 5].

In the case of moist earth, the transport of moisture through the solid could result from a pressure gradient, as capillary forces play an important role. The driving force for the water transport in a porous solid could thus be a pressure gradient. In spite of this fact, it was assumed that the water could be transported through the earth in response to a gradient of concentration of water, and thus diffusion would play an important role.

The purpose in this chapter is to develop a method for determining the values of the main parameters, i.e., the diffusivity and the evaporation rate, and to build up a model capable of describing the drying process. A technique studied previously [4], based on short tests using thin sheets of the material, was applied to the problem of earth drying [6]. A model based on numerical analysis with finite differences was utilised for this problem. The numerical model exhibits many advantages over the analytical solution, because it can be used in the cases where the analytical solutions are not efficient:

- Where the diffusivity is concentration dependent
- Where the initial concentration of moisture is not uniform
- Where the shape of the material is complex
- Where there are special conditions for the surrounding atmosphere

10.2 Modelling the Drying of Wet Earth

10.2.1 Theory

Assumptions

The following assumptions are made:

1. A one-dimensional transport through a thin plane sheet of material is considered.
2. The drying process is controlled by diffusion of water through the solid and evaporation from the surface.
3. Transient diffusion takes place, the diffusivity being constant or concentration-dependent.
4. The rate of evaporation of water from the surface is proportional to the water concentration at the surface and to the rate of evaporation of pure water under the same conditions of temperature.
5. The upper surface of the solid is exposed to still air at room temperature with a relative humidity of 40%.

Mathematical Treatment

The transport through the solid is described by Fick's second equation in one dimension:

$$\frac{\partial C}{\partial t} = \frac{\partial}{\partial x}\left(D\frac{\partial C}{\partial x}\right) \tag{10.1}$$

The rate of evaporation at the surface is constantly equal to the rate at which water is supplied to the evaporating surface by internal diffusion.

$$-D\frac{\partial C}{\partial x} = \frac{F_0}{\rho}(C_0 - C_{ext}) \tag{10.2}$$

where C_0 is the actual concentration of water on the surface and C_{ext} is the concentration on the surface required to maintain equilibrium with the surrounding atmosphere.

In the case of constant diffusivity, and when the initial concentration of water is uniform, the amount of water Q_t leaving the sheet of thickness $2L$ up to time t is expressed as a fraction of Q_∞, the corresponding quantity after infinite time [7, 8]:

$$\frac{Q_\infty - Q_t}{Q_\infty} = \sum_{n=1}^{\infty} \frac{S^2}{\beta_n^2(\beta_n^2 + S^2 + S)} \exp\left(-\frac{\beta_n^2 D}{L^2}t\right) \qquad (10.3)$$

where the β_ns are the positive roots of

$$\beta \tan \beta = S \qquad (10.4)$$

with the dimensionless number S

$$S = \frac{LF_0}{\rho D} \qquad (10.5)$$

The β_ns are given in Table 2.2 for various values of S.

Numerical Analysis

An explicit method with finite differences is able to solve the problem of drying. The sheet occupying the space $0 < x < 2L$ is divided into equal intervals Δx and the time into increments Δt, so that the space–time region is covered by a grid with rectangles of sides Δx and Δt (Fig. 10.1).

Within the Solid
The matter balance is calculated during the time increment Δt within the slice n by considering the amount of water which enters and leaves, as shown in Chap. 2 [8]. From this matter balance, the new concentration after time Δt at position n can be expressed in terms of the previous concentrations obtained at the same place and at the adjacent two places:

$$CN_n = \frac{1}{M}[C_{n-1} + (M - 2)C_n + C_{n+1}] \qquad (10.6)$$

with the dimensionless number M:

$$M = \frac{(\Delta x)^2}{D\Delta t} \qquad (10.7)$$

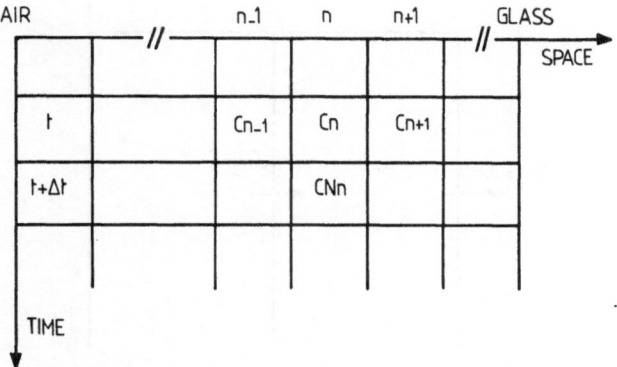

Fig. 10.1 Space–time diagram for numerical analysis.

On the Surface

A slice of thickness $\Delta x/2$ located next to the external surface is considered (Fig. 10.2). The matter balance during the time Δt within this slice, by considering the diffusion of water through the plane at 0.5 and evaporation from the surface, is as follows:

$$\left[-D \frac{\partial C}{\partial x} - \frac{F_0}{\rho} C_0 \right] \Delta t = \frac{\Delta x}{2} [CN_{0.25} - C_{0.25}] \tag{10.8}$$

By assuming that

$$CN_{0.25} - C_{0.25} = \tfrac{3}{4}(CN_0 - C_0) + \tfrac{1}{4}(CN_1 - C_1) \tag{10.9}$$

and that the gradient of concentration at the plane at 0.5 is equal to the mean gradient of concentration between the planes at 0 and 1, we obtain

$$CN_0 = C_0 - \frac{1}{3}(CN_1 - C_1) + \frac{8}{3M}(C_1 - C_0) - \frac{8F_0\Delta t}{3\rho\Delta x} C_0 \tag{10.10}$$

Amount of Water Remaining in the Sheet

The amount of water remaining in the sheet at time t is obtained by integrating the water concentration with respect to space:

$$Q_t = \left[\frac{1}{2} C_0 + \frac{1}{2} C_N + \sum_{n=1}^{n-1} C_n \right] \Delta x \tag{10.11}$$

10.2.2 Experiment

Samples of Earth

A mixture of earth (clay, 416 g with 0.34% water) and water (26 g) was prepared and pressed into slabs in a mould, according to the process previously described [2]. During the compression, the material was vibrated in order to reduce the pressure required for achieving perfect compactness.

Sheets of $10 \times 10 \times 1.5$ cm were prepared and put on glass plates.

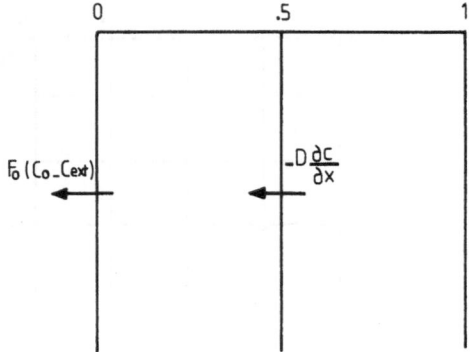

Fig. 10.2 Water transport in the slice next to the surface, with diffusion through the plane at 0.5 and evaporation from the plane at 0.

Test Conditions for Drying
The sheets were exposed to still air at room temperature, and the weight of the sheets constantly recorded during the drying period.

10.2.3 Results

Measurement of the Diffusivity and Rate of Evaporation

The amount of water evaporated from a sheet was plotted as a function of the square root of time (Fig. 10.3). From the slope of the straight line obtained, the diffusivity was calculated by the following equation:

$$\frac{Q_t}{Q_\infty} = \frac{2}{L}\left(\frac{Dt}{\pi}\right)^{0.5}$$
(10.12)

where $2L$ is the thickness of the sheet.
   The rate of evaporation was calculated using two techniques:

1. From the rate of loss in weight of the earth sample, at the beginning of the process, when the process can be assumed to be controlled by evaporation (Fig. 10.4) [3], the value obtained being $F_0 C_0 / \rho$.
2. From the rate of loss in weight of pure water under the same conditions of temperature and surrounding atmosphere.

   The values of the diffusivity and evaporation rate obtained under the operational conditions of the test are:

- Diffusivity $D = 1.56 \times 10^{-5}$ (cm^2 s^{-1})
- Evaporation rate $F_0 = 5.8 \times 10^{-6}$ (g cm^{-2} s^{-1})

Validity of the Model

The validity of the numerical model was tested by comparing the kinetics of drying either from experiments or by calculation. A good compatibility between

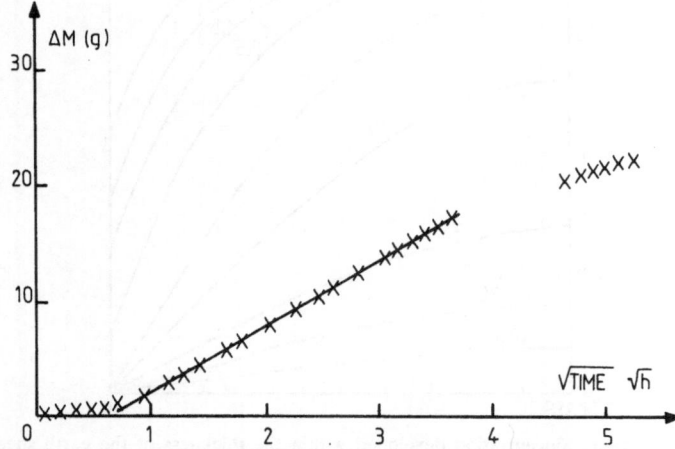

Fig. 10.3 Loss in weight as a function of the square root of time, for determining the diffusivity.

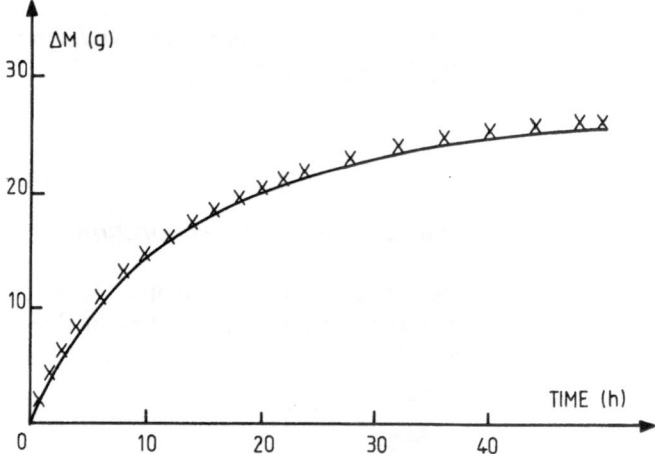

Fig. 10.4 Validity of the model for kinetics for drying for the earth sheet. (———, calculation; ×, experiment.)

these calculated and experimental results can be appreciated in Fig. 10.4, proving the validity of the model for a sheet exposed to still air on one surface, the other being in contact with a glass plate.

The model affords a fuller insight into the process, by giving the profiles of water concentration developed within the earth at various times (Fig. 10.5). The following conclusions can be drawn from these curves:

1. A proof for the validity of the profiles of concentration exists, as the kinetics of drying are obtained by integrating the concentration with respect to space.

2. The concentration of moisture decreases more quickly on the external surface because of the evaporation.

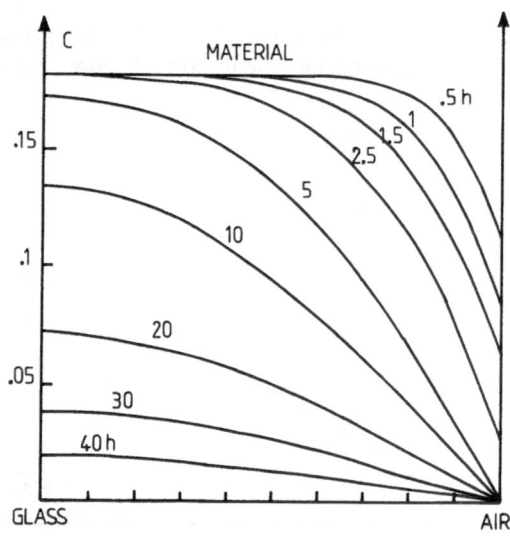

Fig. 10.5 Profiles of water concentration developed within the thickness of the earth sheet at various times.

3. After 40 h of drying, a small amount of moisture remains in the sheet, with the highest concentration next to the glass plate.

10.3 Conclusions

The numerical model describes the process of drying of earth sheets. This model has also been successfully applied to the case of thick wet earth blocks, where the evaporation took place from two parallel surfaces or from five surfaces.

Of course, the conditions of the surrounding atmosphere can also be accounted for in the model, with the term C_{ext} which represents the concentration of moisture on the surface at equillibrium with the surrounding atmosphere.

Symbols

C	Concentration of moisture in the earth sheet
C_0	Concentration on the external surface
CN_n	New concentration at n after time Δt
D	Diffusivity ($cm^2 s^{-1}$)
F_0	Rate of evaporation of pure water ($g\,cm^{-2}\,s^{-1}$)
$2L$	Thickness of the sheet
M	Dimensionless number
Q_t, Q_∞	Amount of water which has evaporated after time t and after infinite time, respectively
$\Delta t, \Delta x$	Increments of time and space, respectively
ρ	Density of water ($g\,cm^{-3}$)

References

1. Accetta A. Compactage dynamique. French patent no. 82.09078, 1982
2. Accetta A. Centrale de production de blocs à bâtir. French patent no. 84.08892, 1984
3. Khatir Y, Bouzon J, Vergnaud JM. Liquid sorption by rubber sheets and evaporation. Models and experiments. Polym Test (1986); 6: 253–265
4. Taverdet JL, Vergnaud JM. Study of transfer process of liquid into and plasticizer out of a plasticized PVC by using short tests. J Appl Polym Sci 1984; 29: 3391–3400
5. Taverdet JL, Vergnaud JM. Modelization of matter transfer between platicized PVC and liquids in case of a maximum for liquid-time curves. J Appl Polym Sci 1986; 31: 111–122
6. Accetta A, Armand JY, Vergnaud JM. Model and short tests for studying the drying of wet earth (clay) sheets for adobe construction. Durability Build Mater 1987; 5: 27–34
7. Crank J. The mathematics of diffusion, 2nd edn. Clarendon Press, Oxford, 1975, pp. 44–68
8. Vergnaud JM. Liquid transport processes in polymeric materials. Modelling and industrial applications. Prentice Hall, 1990

Chapter 11

Drying of Rubbers

11.1 Introduction

Testing of rubber materials involves exposure to various types of environmental factors: temperature, which plays a part in all environment tests, moisture, steam, liquids and gases.

The effects of humidity are of interest and they are studied in the same way as heat ageing.

Very little testing is carried out on the effect of gases, apart from exposure to air and ozone.

The most important problem for rubbers is their resistance to liquids. Tests in which rubbers are in contact with liquids are often called "swelling tests" because of the resulting change in volume of the test piece. The tests are also referred to as "oil ageing" because standard grades of mineral oil are the liquids most often specified.

Generally, rubber samples are in contact with a variety of mineral oils in motor cars, especially petrol. Some additional problems may appear with the use of unleaded petrol, having a greater amount of aromatic hydrocarbons or other components.

When a hydrocarbon, and especially an aromatic hydrocarbon, is in contact with a rubber, the action of the liquid may result in absorption of liquid by the rubber, and extraction of soluble constituents from the rubber. As absorption is generally greater than extraction, an increase in volume results. Volume change is a very good measure of the general resistance of a rubber to a given liquid, a high degree of swelling clearly indicating that the rubber is not suitable in that environment. In addition, the degree of swelling can be related to the state of cure of the rubber [1].

After the absorption stage, there follows the complementary stage of evaporation from the rubber. Although this enables recovery of the rubber, extraction of additive and plasticiser may take place to the detriment of the main characteristics of the rubber.

The main interest in the drying process is to be able to predict by calculation the conditions under which it must be conducted. In other words, as testing the process on the sample may be destructive, a better way of working is by modelling the process and determining the parameters needed for calculation by conducting a short test on a sample representative of the real industrial sample.

Following this principle, a model based on a numerical method with finite differences is built up and tested in various cases. The short test technique is described when the rubber sample is in the form of a thin plane sheet, in order to obtain the parameters necessary for calculation. The numerical model is then used for various shapes such as: plane sheet, cylinder of finite length, tubes and annuli.

The process of drying of rubbers is controlled by diffusion of the liquid within the solid and evaporation from the surface. Depending on the nature of the liquid, the rubber may contain a large amount of liquid, up to three times the volume of the dried rubber, so that considerable shrinkage is observed during the drying process, and the diffusivity is very often concentration dependent. Up to now, the numerical models in use have not taken into account this change in dimensions of the sample, and a reference framework is taken with for instance the dimensions of the original rubber sample when free of liquid.

The use of thin plane sheet has been recommended for measuring the parameters of interest, for the following three reasons [2]:

- Determining the diffusivity from the kinetics of desorption is very simple for a thin plane sheet, when the effect of the edges is negligible.
- Preparing a plane sheet of rubber is easy.
- The time of drying, as far as the process is controlled by diffusion, is proportional to the square of the thickness (Dt/L^2).

Three parameters are necessary to describe the process of drying of rubbers: the rate of evaporation of the liquid, the diffusivity of the liquid within the rubber, and the maximum amount of liquid initially contained by the rubber.

11.2 Thin Sheets

The rubber sheet is so thin that one-dimensional transfer through the thickness only is considered, the effect of the edges being neglected. The process of drying is assumed to be controlled by transient diffusion within the rubber and by evaporation from the surface.

The liquid chosen for the study was toluene [1, 2].

11.2.1 Theory

Assumptions

The following assumptions are made to clarify the problem:

1. A thin plane sheet of rubber is studied with a one-dimensional transfer of liquid.
2. Transient diffusion takes place within the rubber.

3. The rate of evaporation from the surface is proportional to the liquid concentration on the surface. The liquid concentration on the surface required to maintain equilibrium with the surrounding atmosphere is zero.
4. In spite of the change in dimensions during the process, a frame of reference is considered for calculation with the sample free of liquid.

Mathematical Treatment

The equation for one-dimensional diffusion

$$\frac{\partial C}{\partial t} = \frac{\partial}{\partial x}\left(D\,\frac{\partial C}{\partial x}\right) \tag{11.1}$$

with initial and boundary conditions

$$t = 0, \quad -L < x < L, \quad C = C_{in} \tag{11.2}$$

$$t > 0, \quad -D\,\frac{\partial C}{\partial x} = \frac{F_0}{\rho}\,(C_s - C_{ext}) \tag{11.3}$$

has a solution when the diffusivity is constant and the initial concentration in the rubber is uniform [3]:

$$\frac{C_{x,t}}{C_{in}} = \sum_{n=1}^{\infty} \frac{2S\cos\dfrac{\beta_n x}{L}}{(\beta_n^2 + L^2 + L)\cos\beta_n}\exp\left(-\frac{\beta_n^2}{L^2}Dt\right) \tag{11.4}$$

where the β_ns are the positive roots of

$$\beta\tan\beta = S \tag{11.5}$$

with the dimensionless number S:

$$S = \frac{LF_0}{D\rho} \tag{11.6}$$

The total amount of substance evaporating from the sheet up to time t is expressed as a fraction of Q_∞, the corresponding quantity after infinite time, by

$$\frac{Q_\infty - Q_t}{Q_\infty} = \sum_{n=1}^{\infty} \frac{2S^2\exp\left(-\dfrac{\beta_n^2}{L^2}Dt\right)}{\beta_n^2(\beta_n^2 + S^2 + S)} \tag{11.7}$$

Numerical Analysis

The sheet of thickness $2L$ is divided into N equal intervals of thickness Δx, and constant time increments Δt are considered, so that the space–time region is covered by a grid of rectangles of sides Δx and Δt (Fig. 11.1). Each position is defined by an integer n, with

$$x = n\Delta x$$

$$2L = N\Delta x \tag{11.8}$$

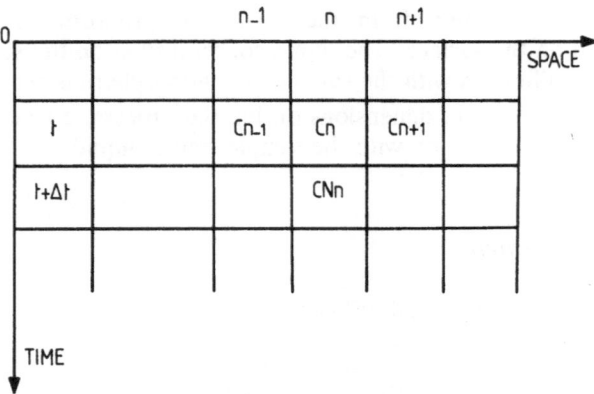

Fig. 11.1. Space–time diagram for the process of drying of a thin plane rubber sheet.

The concentration of liquid is calculated within the rubber and at the surface of the sheet.

Within the Rubber
The matter balance is determined during the time increment Δt within the slice of thickness Δx at position n, by considering the diffusion through the planes $(n - 0.5)$ and $(n + 0.5)$ (Chap. 5).
The new concentration after time Δt at position n is thus expressed as a function of the previous concentration at the same place and adjacent places:

$$CN_n = \frac{1}{M}[C_{n-1} + (M - 2)C_n + C_{n+1}] \tag{11.9}$$

where the dimensionless number M is

$$M = \frac{(\Delta x)^2}{D\Delta t} \tag{11.10}$$

Surfaces (Fig. 11.2)
The matter balance during the time Δt within the slice of thickness $\Delta x/2$, located next to a surface, can be written by considering the diffusion through the plane 0.5 and evaporation from the surface at 0:

$$\left[-D\frac{\partial C}{\partial x}\Big|_{\text{at } 0.5} - \frac{F_0}{\rho}C_0\Big|_{\text{at } 0}\right]\Delta t = \frac{\Delta x}{2}[CN_{0.25} - C_{0.25}] \tag{11.11}$$

where $CN_{0.25}$ and $C_{0.25}$ represent the new concentration after time Δt at position 0.25, and the previous concentration at the same place, respectively.
With the assumption

$$CN_{0.25} - C_{0.25} = \tfrac{3}{4}(CN_0 - C_0) + \tfrac{1}{4}(CN_1 - C_1) \tag{11.12}$$

the new concentration on the surface CN_0 can thus be expressed in terms of the previous concentrations at position 0 and 1, as well as the new concentration at position 1:

$$CN_0 = C_0 - \frac{1}{3}(CN_1 - C_1) + \frac{8}{3M}(C_1 - C_0) - \frac{8}{3}\frac{F_0\Delta t}{\rho\Delta x}C_0 \tag{11.13}$$

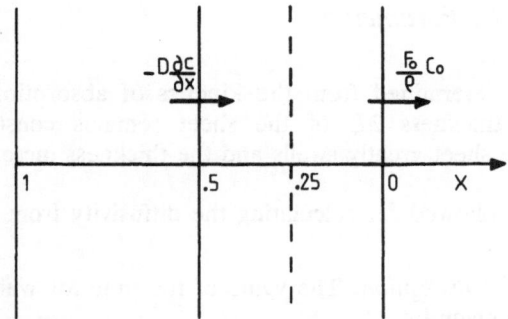

Fig. 11.2. Matter balance within the slice of thickness $\Delta x/2$ located next to the surface.

Of course the new concentration CN_1 must be calculated beforehand, it is easily obtained by using Eq. (11.9).

Amount of Liquid Remaining
The amount of liquid which remains in the sheet at any time can be obtained by integrating the concentration of liquid with respect to space:

$$Q_t = \left[C_0 + \sum_{n=1}^{N-1} C_n \right] \Delta x \qquad (11.14)$$

11.2.2 Experiment

Preparation of Samples

Rubber sheets of thickness of 0.15 cm were cured by pressing rubber in a steel mould operated by a power press under a pressure of 50 bars. The operational conditions such as time (10 min) and temperature (170 °C) were determined from earlier studies [4, 5].

Test Conditions

The rubber sheets were immersed in toluene, and then weighed at intervals.
   The samples were then exposed either to motionless air or to well-stirred air. The sheets were vertical and the temperature constant at 20 °C. The weight of the samples was constantly recorded.

11.2.3 Results

After measuring the parameters from experiments, the validity of the numerical model was tested by comparing the theoretical and experimental kinetics of drying.

Determination of the Parameters

Diffusivity

The diffusivity was determined from the kinetics of absorption of toluene, by assuming that the thickness $2L$ of the sheet remains constant as diffusion proceeds. In fact the sheet greatly swells and the thickness increases as the liquid enters the sheet.

Three ways were followed for calculating the diffusivity from the experimental kinetics of absorption.

1. *Use of half-life of absorption.* The value of the time for which $Q_t/Q_\infty = 0.5$, is approximately given by

$$D = 0.049\left(\frac{4L^2}{t}\right)$$ (11.15)

   with an error about 0.001% (Chap. 2).

2. *Use of initial rate of absorption.* In the first stages of absorption in the sheet, the amount of liquid which enters the sheet is proportional to the square root of time (Chap. 2):

$$\frac{Q_t}{Q_\infty} = \frac{2}{L}\left(\frac{Dt}{\pi}\right)^{0.5} \quad \text{when} \quad \frac{Q_t}{Q_\infty} < 0.5$$ (11.16)

   and the diffusivity is calculated from the slope of the straight line obtained by plotting the amount of liquid in the sheet as a function of the square root of time.

3. *Use of final rate of absorption.* In the later stages of absorption, only the first term in the series expressing the kinetics of absorption, can be retained (Chap. 2):

$$\frac{Q_\infty - Q_t}{Q_\infty} = \frac{8}{\pi^2}\exp\left(-\frac{\pi^2 Dt}{4L^2}\right) \quad \text{with} \quad \frac{Q_t}{Q_\infty} > 0.6$$ (11.17)

   The diffusivity can thus be calculated from the slope of the straight line obtained by plotting the logarithm of the left-hand side as a function of time.

4. *Use of final rate of desorption.* The first term of the series in Eq. (11.7) dominates when $Q_t/Q_\infty > 0.6$. Expressed in a logarithmic form, Eq. (11.7) is reduced to

$$\ln\left(\frac{Q_\infty - Q_t}{Q_\infty}\right) = -\frac{\beta_1^2}{L^2}Dt + \ln\left(\frac{2S^2}{\beta_1^2(\beta_1^2 + S^2 + S)}\right)$$ (11.18)

   It is easy to determine the diffusivity from the slope of the straight line obtained by plotting the quantity on the left-hand side as a function of time. The values of β_1 are given in Table 2.2.

Rate of Evaporation

The rate of evaporation was determined by using the rate of desorption at the beginning of the drying process, when this process can be assumed to be controlled by evaporation:

$$\frac{F_0 C_0}{\rho} = \frac{dQ}{dt} \quad \text{for } t \to 0$$ (11.19)

The rate of evaporation was also obtained by evaporating the pure liquid under the same conditions for the temperature and surrounding atmosphere.

The values of the diffusivity for absorption and desorption, and of the rate of evaporation are shown in Table 11.1.

Some conclusions can be drawn from these values:

1. The values for the diffusivity are about the same for the process of absorption and desorption.
2. The values of diffusivity obtained for absorption do not depend on the method of calculation, or on the amount of liquid absorbed.
3. These above results for the diffusivity show that in this case the diffusivity is constant, and does not depend on the concentration of liquid.
4. The values of the rate of evaporation measured by the two techniques described are about the same. However, the method using the pure liquid is basically better, as it is easy to follow a constant rate of evaporation.
5. The effect of the agitation of air on the rate of evaporation is of interest.

Validity of the Model

The validity of the model was tested by comparing the kinetics of drying obtained from experiments and calculation, in the case of motionless air and well-stirred air.

Motionless Air
The kinetics of drying determined by experiments and calculation are shown in Fig. 11.3, using the values of the parameters shown in Table 11.1. Experimental results and theoretical results calculated using the numerical model or the analytical expression are totally compatible throughout the whole drying process.

Well-Stirred Air
In the case of stirred air, the experimental curve corresponds well to results calculated with the help of the model and the analytical expression (Fig. 11.4).

11.2.4 Conclusions

The following conclusions are worth nothing.

Process
The process of drying is controlled not only by evaporation but also by diffusion of the liquid through the solid, and diffusion plays an important role in the process.

Calculation
The analytical solution obtained by taking into account the evaporation and diffusion can describe the kinetics of drying when the diffusivity is constant and the initial concentration of liquid in the sample is uniform.

The numerical model may be used for every case, whatever the diffusivity and the initial concentration of liquid.

Table 11.1. Diffusivity and rate of evaporation

Process	Equation	$D \times 10^7$ (cm^2 s^{-1})	$\dfrac{Q_\infty}{Q_i} 100$	$F \times 10^5$ (g cm^{-2} s^{-1})
Absorption	11.15	5	338	
Absorption	11.16	5	338	
Absorption	11.17	5	338	
Desorption	11.18	5.1	338	
Motionless air				10.4
Stirred air				38.5

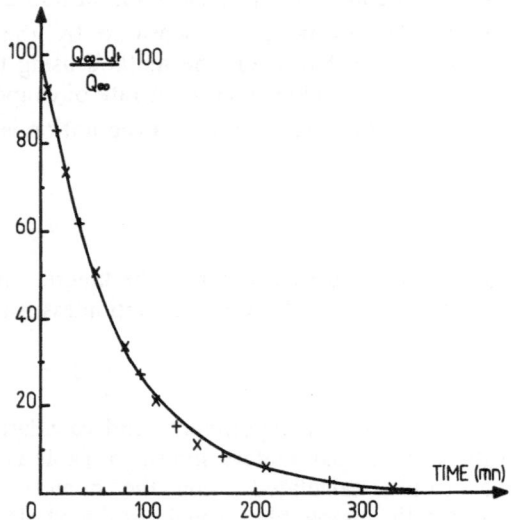

Fig. 11.3. Kinetics of drying of a 0.15 cm thick rubber sheet in motionless air, at 20 °C. $D = 5.1 \times 10^{-7}$ cm^2 s^{-1}, $F = 10.4 \times 10^{-5}$ (g cm^{-2} s^{-1}). (——, model; + experiments; ×, numerical expression.)

The value of diffusivity can be determined in many ways during the stages of absorption or desorption, and at various stages during the step of absorption: at the beginning, at the half-life point and at the end of the process. When the values of diffusivity obtained by all these means are the same, the diffusivity can be considered as constant.

Effect of Agitation of Air
The agitation of air plays an important role in the process of drying, as it affects the rate of evaporation.

Use of Thin Sheets
Thin rubber sheets are easily obtained by vulcanising the material in a mould. The purpose of using thin sheets is essentially for experiment time, this time being proportional to the square of the thickness. This is true not only for measuring the diffusivity, but also and especially for determining the amount of liquid absorbed (or desorbed) Q_∞ after a long time.

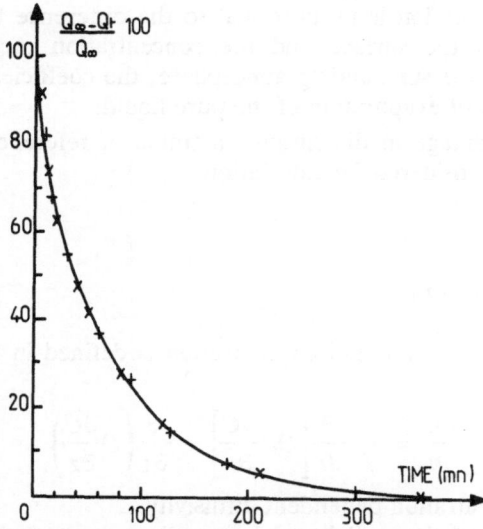

Fig. 11.4. Kinetics of drying of 0.15 cm thick rubber sheet in stirred air, at 20 °C. $D = 5.1 \times 10^{-7} \ cm^2 \ s^{-1}$ $F = 38.5 \times 10^{-5}$ $(g \ cm^{-2} \ s^{-1})$. (——, model; + experiments; ×, numerical expression.)

11.3 Cylinders of Finite Length

As experimental tests for determining the effects of liquids on rubbers are not only time-consuming but also destructive to the sample, it is of interest to build up models able to describe the process of drying.

The first objective in this section is to show that a numerical model with finite differences can describe the process of drying of rubber cylinders of finite length, when all the facts are known [6].

The second objective is to give an idea of the effect of the ratio of height and diameter of the cylinder on the process.

Some parameters are of interest for the simulation of the process by the model:

- Amount of liquid which can be evaporated after infinite time
- Diffusivity of the liquid
- Rate of evaporation of the liquid

11.3.1 Theory

Assumptions

The following assumptions are made:

1. The cylinder is made of an isotropic material.
2. The liquid transport is controlled by transient diffusion within the solid, and evaporation from the surface.
3. The diffusivity is the same whatever the directions, and it is concentration dependent, as found from experiments.

4. The rate of evaporation is proportional to the difference between the actual concentration on the surface and the concentration required to maintain equilibrium with the surrounding atmosphere, the coefficient of proportionality being the rate of evaporation of the pure liquid.

5. In spite of the change in dimensions, a frame of reference with the sample free of liquid is considered for calculation.

Mathematical Treatment

The rate of decrease of the liquid concentration is defined in two dimensions by Fick's law:

$$\frac{\partial C}{\partial t} = \frac{1}{r}\frac{\partial}{\partial r}\left[Dr\frac{\partial C}{\partial r}\right] + \frac{\partial}{\partial z}\left(D\frac{\partial C}{\partial z}\right) \tag{11.20}$$

where D is the concentration-dependent diffusivity.

The contributions of the radial and longitudinal transport are shown on the right-hand side.

The radial and longitudinal abscissae are r and z, where

$$0 \leqslant r \leqslant R$$

$$0 \leqslant z \leqslant H \tag{11.21}$$

H and R being the height and radius of the cylinder.

On the surface, the rate of evaporation remains constantly equal to the rate of liquid which is brought to the evaporating surface by diffusion:

$$-D\frac{\partial C}{\partial x} = \frac{F_0}{\rho}(C_s - C_{eq}) \qquad t > 0 \tag{11.3}$$

where C_s is the concentration on the surface and C_{eq} the concentration necessary to maintain equilibrium with the surrounding atmosphere.

Numerical Analysis

The cylinder is divided into various annuli of constant thickness Δr and constant length Δz, as shown in the concentration–time diagram (Fig. 11.5). The integers i and j are used to define the position of each annulus in the cylinder:

$$r = i\Delta r \qquad \text{with} \qquad R = N_r\Delta r$$

$$z = j\Delta z \qquad \text{with} \qquad H = 2N_z\Delta z \tag{11.22}$$

The matter balance during the increment of time Δt is calculated within an annulus.

Within the Cylinder
From the matter balance evaluated within the annulus of radius r and thickness Δr, at height j, the new concentration $CN_{i,j}$ is calculated as a function of the

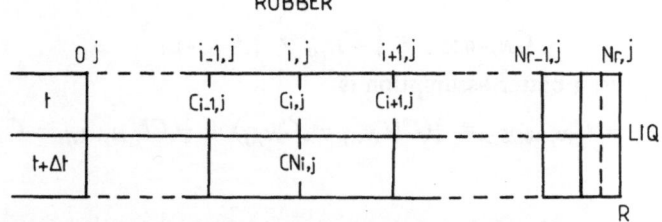

Fig. 11.5. Concentration–time diagram for numerical analysis for the cylinder of finite length.

previous concentration at the same place and adjacent places:

$$CN_{i,j} = C_{i,j} + \frac{1}{iM_r}[(i + 0.5)(C_{i+1,j} - C_{i,j}) - (i - 0.5)(C_{i,j} - C_{i-1,j})]$$

$$+ \frac{1}{M_h}[C_{i,j+1} - 2C_{i,j} + C_{i,j-1}] \qquad (11.23)$$

with the dimensionless numbers M_r and M_h

$$M_r = \frac{(\Delta r)^2}{D\Delta t} \qquad (11.24)$$

$$M_h = \frac{(\Delta h)^2}{D\Delta t} \qquad (11.24')$$

Centre of the Cylinder, r = 0
The matter balance must be rewritten by considering the small cylinder of radius $\Delta r/2$ and height Δh located at the centre of the rubber cylinder:

$$CN_{0,j} = C_{0,j} + \frac{4}{M_r}[C_{1,j} - C_{0,j}] + \frac{1}{M_h}[C_{0,j+1} - 2C_{0,j} + C_{0,j-1}] \quad (11.25)$$

Radial surface, i = N_r, j
The annulus of thickness $\Delta r/2$ and height Δh, next to the radial surface at position j, is considered. From the matter balance evaluated during the increment of time Δt within this annulus, the new concentration after time Δt on the radial surface can be obtained.
By assuming that

$$CN_{N_r-0.25,j} - C_{N_r-0.25,j} = CN_{N_r,j} - C_{N_r,j} \qquad (11.26)$$

the new concentration $CN_{N_r,j}$ can be written

$$CN_{N_r,j} = C_{N_r,j} + \left(\frac{2N_r - 1}{N_r - 0.25}\right)\frac{1}{M_r}(C_{N_r-1,j} - C_{N_r,j})$$

$$- \left(\frac{2N_r}{N_r - 0.25}\right)\frac{F_0\Delta t}{\rho\Delta r}(C_{N_r,j} - C_{ext})$$

$$+ \frac{1}{M_h}[C_{N_r-0.25,j-1} - 2C_{N_r-0.25,j} + C_{N_r-0.25,j+1}] \quad (11.27)$$

The concentration $C_{N_r-0.25,j}$ can be expressed by

$$C_{N_r-0.25,j} = C_{N_r,j} \qquad (11.28)$$

or better by

$$CN_{r,-0.25,j} = \tfrac{3}{4} C_{N_r,j} + \tfrac{1}{4} C_{N_r,-1,j} \tag{11.29}$$

In the same way, a better assumption is

$$CN_{N_r,-0.25,j} - C_{N_r,-0.25,j} = \tfrac{3}{4}(CN_{N_r,j} - C_{N_r,j}) + \tfrac{1}{4}(CN_{N_r,-1,j} - C_{N_r,-1,j}) \tag{11.30}$$

Plane Surfaces
An annulus of radius r, thickness Δr and height $\Delta h/2$ is considered. The matter balance is calculated by taking into account the radial and longitudinal diffusion as well as the evaporation from the plane surface. The new concentration can thus be written

$$CN_{i,N_h} = C_{i,N_h} + \frac{2}{M_h}[C_{i,N_h-1} - C_{i,N_h}] - \frac{2F_0\Delta t}{\rho\Delta h}[C_{i,N_h} - C_{ext}]$$

$$+ \frac{1}{M_r}[C_{i-1,N_h-0.25} - 2C_{i,N_h-0.25} + C_{i+1,N_h-0.25}]$$

$$- \frac{0.5}{iM_r}[C_{i+1,N_h-0.25} - C_{i-1,N_h-0.25}] \tag{11.31}$$

Edge
The annulus of radius $(N_r - 0.25)\Delta r$, of thickness $\Delta r/2$ and height $\Delta h/2$, located next to the edge, is considered. The matter balance is calculated during the time increment Δt by taking into account the radial and longitudinal diffusion as well as the evaporation from the surfaces next to the edge:

$$CN_{N_r,N_h} = C_{N_r,N_h} + \left(\frac{2N_r - 1}{N_r - 0.25}\right)\frac{1}{M_r}[C_{N_r-1,N_h-0.25} - C_{N_r,N_h-0.25}]$$

$$- \left(\frac{2N_r}{N_r - 0.25}\right)\frac{F_0\Delta t}{\rho\Delta r}[C_{N_r,N_h-0.25} - C_{ext}]$$

$$+ \frac{2}{M_h}[C_{N_r-0.25,N_h-1} - C_{N_r-0.25,N_h}]$$

$$- \frac{2F_0\Delta t}{\rho\Delta h}[C_{N_r-0.25,N_h} - C_{ext}] \tag{11.32}$$

Amount of Substance Remaining
The amount remaining at time t is obtained by integration of the concentration at this time with respect to space.

11.3.2   Experiment

Rubber Samples

Rubber cylinders of various dimensions were cured in a steel mould at 170 °C under a pressure of 50 bars. The cure conditions (Table 11.2), e.g., time and temperature, were calculated with the help of a numerical model. This model

Table 11.2. Cure conditions for rubber cylinders

Diameter (cm)	Thickness (cm)	Temperature (°C)	Time (min)
0.47	0.73		
3.8	0.92	170	25
2.2	0.49	170	15

takes into account the heat transferred through the mould and rubber by conduction and the heat evolved from the cure reaction [4, 5, 7]. The kinetics of the heat of reaction are determined from calorimeter curves obtained in scanning mode [8].

Operational Conditions for Desorption

The rubber samples were immersed in toluene at 20 °C, and weighed at intervals in order to obtain the kinetics of absorption.

The samples were then exposed to air at 20 °C. The samples were vertical, and the air either motionless or stirred. The weight was monitored in order to determine the kinetics of drying.

11.3.3 Results

After measuring the values of the parameters, the validity of the model was evaluated by comparing the theoretical and experimental kinetics of drying.

Determination of the Parameters

Diffusivity

The diffusivity was calculated from the experimental kinetics of drying determined with a thin sheet made of the same material as that of the cylinders.

By plotting $\ln((Q_\infty - Q_t)/Q_\infty)$ as a function of time, when $Q_t/Q_\infty > 0.6$, a straight line is obtained:

$$\ln\left(\frac{Q_\infty - Q_t}{Q_\infty}\right) = -\frac{\beta_1^2}{L^2}Dt + \ln\left(\frac{2S^2}{\beta_1^2(\beta_1^2 + S^2 + S)}\right) \quad (11.18)$$

Rate of Evaporation

The rate of evaporation was measured in the following two ways:

1. From the rate of desorption at the beginning of the drying step, when the process is controlled by evaporation.

2. From the constant rate of evaporation of the pure liquid determined under the same conditions.

Parameter Values
The values of the diffusivity and the rate of evaporation were

$$D = 12.5 \times 10^{-7} \text{ cm}^2 \text{ s}^{-1}$$

$$F_0 = 10.4 \times 10^{-5} \text{ g cm}^2 \text{ s}^{-1} \text{ (motionless air)}$$

The diffusivity is found to be almost constant towards the end of the drying process when $0.6 < Q_t/Q_\infty < 1$.

Validity of the Model

The validity of the model for the drying process of rubber cylinders of finite length was tested by comparing the theoretical and experimental kinetics of drying, for samples of

- Diameter 2.2 cm and height 0.49 cm (Fig. 11.6)
- Diameter 0.47 cm and height 0.92 cm (Fig. 11.7)

As shown in these figures, a good correlation can be seen between the theoretical and experimental curves, proving the validity of the model.

Of course, a shorter time of drying is obtained for the smaller sample.

11.3.4 Conclusions

The process of drying is very well described by the numerical model in the difficult case of cylinders of finite length.

The diffusivity was obtained from the experimental kinetics, with $Q_t/Q_\infty > 0.6$, for the second half of the process. The good correlation between the theoretical and experimental kinetics is proof of the constant value of the diffusivity throughout the whole process.

11.4   Tubes of Infinite Length

As already shown, volume change is a good measure of the resistance of a rubber to a liquid, a high degree of swelling indicating that the rubber is not suitable for use in this environment. Additionally, the degree of swelling is related to the state of cure of the rubber.

The action of the liquid on rubber generally results in absorption of liquid, but also in extraction of the soluble constituents from the rubber. As absorption is usually greater than extraction, the final result is an increase in volume [1]. So the swelling tests are considered sometimes as destructive tests. A numerical model able to describe the process of absorption and desorption is very useful, especially when the data such as the diffusivity, evaporation rate and amount of liquid absorbed, are determined by using a sample made of the same material.

The diffusion of liquid and relaxation processes in elastomers above the temperature of glass transition T_g are governed by considering the segmental

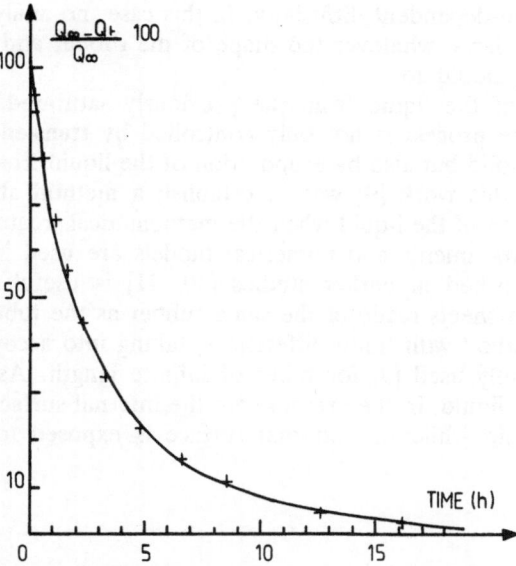

Fig. 11.6. Kinetics of drying for the cylinder of diameter 2.2 cm and height 0.49 cm. (——, model; +, experiments.)

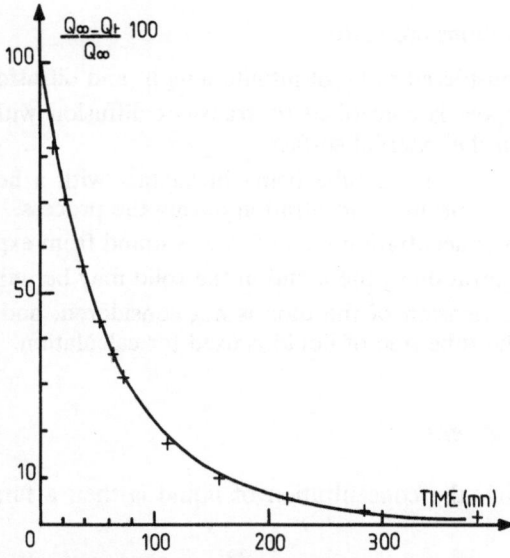

Fig. 11.7. Kinetics of drying for the cylinder of diameter 0.47 cm and height 0.92 cm. (——, model; +, experiments.)

mobility of the elastomer chains which is affected by the total free volume and its distribution in the elastomer system. Thus diffusivity depends on the relative mobility of the liquid penetrant and the elastomer chains, as well as the free volume. In elastomers, the chains adjust so quickly to the presence of the liquid that they do not cause diffusion anomalies, and the diffusion is generally Fickian with a concentration-dependent diffusivity. In this case, no analytical solution can be found for Fick's laws, whatever the shape of the rubber and the initial profile of concentration of the liquid.

The desorption of the liquid from the previously saturated rubber follows a similar pattern. The process is not only controlled by transient diffusion of the liquid through the solid but also by evaporation of the liquid from the surface.

The purpose of this work [9] was to establish a method able to predict the kinetics of desorption of the liquid when the mathematical treatment is inefficient. In this method, experiments and numerical models are used in turn. The short test technique described in earlier studies [10, 11] is useful to determine the diffusivity with thin sheets made of the same rubber as the tube. A model based on a numerical method with finite differences, taking into account all the known facts, was successfully used [9] for tubes of infinite length. As the main use of tube is to transport liquid, in the present case the internal surface of the tube is in contact with a liquid while the external surface is exposed to the surrounding atmosphere.

11.4.1 Theory

Assumptions

The following assumptions are made:

1. The tubes are considered to be of infinite length, and diffusion is radial only.
2. The liquid transport is controlled by transient diffusion within the solid and evaporation from the external surface.
3. The internal surface of the tube being in contact with a liquid, the internal surface is kept at constant concentration during the process.
4. The diffusivity is concentration dependent, as found from experiments.
5. The initial concentration of the liquid in the solid may be uniform or not.
6. The change in dimension of the tube is not considered, and a fixed frame of reference with the tube free of liquid is used for calculation.

Mathematical Treatment

Diffusion being radial, the concentration of liquid is then a function of radius r and time t only:

$$\frac{\partial C}{\partial t} = \frac{1}{r}\frac{\partial}{\partial r}\left(rD\frac{\partial C}{\partial r}\right) \tag{11.33}$$

with concentration-dependent diffusivity D.

The initial and boundary conditions are as follows:

$$t = 0, \quad R_i < r < R_e, \quad C_{r,t} \tag{11.34}$$

$$t > 0, \quad -D\frac{\partial C}{\partial r} = \frac{F_0}{\rho}(C_{R_e} - C_{eq}) \tag{11.35}$$

where R_i and R_e are the internal and external radii of the tube, C_{R_e} is the concentration of liquid on the external surface, and C_{eq} the concentration necessary to maintain equilibrium with the surrounding atmosphere.

As no analytical solution can be obtained, the problem is solved by using a numerical model.

Numerical Analysis

The circular cross-section located between the radii R_i and R_e is divided into N slices of constant thickness Δr, and each position is defined by the integer j (Fig. 11.8):

$$r = j\Delta r$$

$$R_i = N_i\Delta r$$

$$R_e = N_e\Delta r \qquad (11.36)$$

The concentration is calculated within the solid and on the internal and external surfaces.

Within the Solid, $N_i + 1 < j < N_e - 1$

From the matter balance calculated during the time increment Δt within the annulus of radius $j\Delta r$ by considering the radial diffusion, the new concentration after time Δt is obtained as a function of the previous concentrations at same and adjacent places through the function H (Chap. 6):

$$CN_j = C_j + \frac{\Delta t}{j\Delta r}[H_{j+0.5} - H_{j-0.5}] \qquad (11.37)$$

with the function H

$$H_{j+0.5} = (j + 0.5)D\frac{\partial C}{\partial r} \qquad \text{at } (j + 0.5) \qquad (11.38)$$

Internal Surface, $j = N_i$

As the internal surface is constantly in contact with the liquid, the liquid concentration on the surface is constant:

$$C_{N_i} = C_{eq} \qquad (11.39)$$

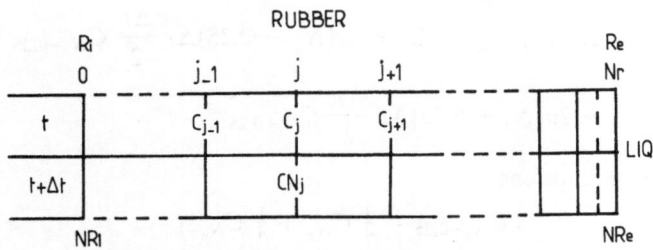

Fig. 11.8. Concentration–space diagram for a rubber tube.

External Surface, $j = N_e$
An annulus of thickness $\Delta x/2$ located next to the external surface is considered, with radius $(N_e - 0.25)\Delta r$. The matter balance during the increment of time Δt is calculated within this annulus, by considering the radial diffusion at position $(N_e - 0.5)\Delta r$ and the evaporation from the surface.
   By making the simple assumption

$$CN_{N_e-0.25} - C_{N_e-0.25} = CN_{N_e} - C_{N_e} \tag{11.40}$$

the new concentration after time Δt on the surface CN_{N_e} can thus be expressed in terms of the function H and the rate of evaporation:

$$CN_{N_e} = C_{N_e} - \frac{2\Delta t}{(N_e - 0.25)\Delta r}\left[H_{N_e-0.5} + N_e\frac{F_0}{\rho}(C_{N_e} - C_{eq})\right] \tag{11.41}$$

By making the better assumption

$$CN_{N_e-0.25} - C_{N_e-0.25} = \tfrac{3}{4}(CN_{N_e} - C_{N_e}) + \tfrac{1}{4}(CN_{N_e-1} - C_{N_e-1}) \tag{11.42}$$

the new concentration on the surface is then obtained as follows:

$$CN_{N_e} = C_{N_e} - \tfrac{1}{3}(CN_{N_e-1} - C_{N_e-1})$$
$$- \frac{8\Delta t}{3(N_e - 0.25)\Delta r}\left[H_{N_e-0.5} + \frac{N_e F_0}{\rho}(C_{N_e} - C_{eq})\right] \tag{11.41'}$$

with the function H

$$H_{N_e-0.5} = (N_e - 0.5)D\frac{\partial C}{\partial r} \qquad \text{at}(N_e - 0.5) \tag{11.38'}$$

   As the new concentration on the surface CN_{N_e} is expressed in terms of the new concentration at the adjacent position, CN_{N_e-1}, this new concentration must be calculated beforehand using Eq. (11.37).

Amount of Liquid Remaining
The amount of liquid remaining in the rubber tube itself can be calculated by integrating the concentration of liquid with respect to space from the internal to the external surfaces.
   The amount of liquid per unit length is given by

$$Q_t = 2\pi \int_{R_i}^{R_e} rC_r dr \tag{11.43}$$

With finite differences, this becomes

$$Q_t = 2\pi(\Delta r)^2 \sum_{N_i+1}^{N_e-1} jC_j + 2\pi(N_e - 0.25)\Delta r \frac{\Delta r}{2} C_{N_e-0.25}$$
$$+ 2\pi(N_i + 0.25)\Delta r \frac{\Delta r}{2} C_{N_i+0.25} \tag{11.44}$$

By making the assumptions

$$C_{N_e-0.25} = \tfrac{3}{4}C_{N_e} + \tfrac{1}{4}C_{N_e-1}$$
$$C_{N_i+0.25} = \tfrac{3}{4}C_{N_i} + \tfrac{1}{4}C_{N_i+1}$$

the amount of liquid remaining in the rubber becomes

$$Q_t = 2\pi(\Delta r)^2 \sum_{N_i+1}^{N_e-1} jC_j + \frac{\pi(\Delta r)^2}{4} [(N_e - 0.25)(3C_{N_e} + C_{N_e-1})$$
$$+ (N_i + 0.25)(3C_{N_i} + C_{N_i+1})] \tag{11.44'}$$

11.4.2 Experiment

Rubber Samples

Rubber tubes of various lengths were prepared by curing the elastomer in a steel mould operated by a power press under a pressure of 30 bars. Operational conditions for rubber cure were determined from earlier studies [4, 5] in order to obtain vulcanisates with a high value of the state of cure.

Test Conditions for Evaporation from the External Surface

Toluene was introduced into the rubber tube. Great care was taken in order to reduce the absorption of liquid on the plane faces, by protecting them from contact with the liquid, and using a long tube.

The sample with the liquid was weighed at intervals, after extracting the free liquid in contact with the internal surface.

Preparation of Rubber Sheets for Measuring the Parameters

Sheets made of rubber of the same composition as that of the tube were cured with selected conditions of temperature and time [4, 5, 9].

11.4.3 Results

After measuring the value of the parameters such as the diffusivity, rate of evaporation and amount of liquid absorbed, the model is tested by comparing the experimental and calculated kinetics of desorption.

Determination of the Parameters

Diffusivity
The diffusivity was determined during the process of absorption by immersing the rubber sheet in toluene, and following the kinetics by weighing the sample at intervals [2].

The three following methods were used:

1. *Short test technique*, $Q_t/Q_\infty < 0.5$. From the slope of the straight line obtained by plotting the amount of liquid Q_t which has entered the rubber as a function of the square root of time, the diffusivity is easily calculated:

$$\frac{Q_t}{Q_\infty} = \frac{2}{L}\left(\frac{Dt}{\pi}\right)^{0.5} \quad \text{when} \quad \frac{Q_t}{Q_\infty} < 0.5 \qquad (11.16)$$

$2L$ being the thickness of the sheet.

2. *Half-life of absorption.* The time necessary for Q_t to reach half the value of Q_∞ is determined from the kinetics of absorption, and the diffusivity is calculated by

$$D = 0.049 \left(\frac{4L^2}{t}\right) \qquad (11.15)$$

$2L$ being the thickness of the sheet.

3. *Use of final rate of absorption.* In the later stages of diffusion in the sheet, when $Q_t/Q_\infty > 0.7$, the first term in the series expressing the kinetics of absorption dominates:

$$\ln\left(\frac{Q_\infty - Q_t}{Q_\infty}\right) = -\frac{\pi^2 Dt}{.4L^2} + \ln\left(\frac{8}{\pi^2}\right) \qquad (11.17)$$

and the diffusivity is obtained using Eq. (11.17).

Rate of Evaporation

The rate of evaporation was determined by the following two techniques:

1. From the slope of the initial rate of evaporation from the rubber sheet,

$$\frac{F_0 C_0}{\rho} = \frac{dQ}{dt} \quad \text{for } t \to 0 \qquad (11.19)$$

2. By measuring the rate of evaporation of the pure liquid under the same conditions for the temperature and the surrounding atmosphere.

Parameter Values

The diffusivity and rate of evaporation were

$$D = 4 \times 10^{-8} \times \exp\left(3.52 \times \frac{C_{r,t}}{C_\infty}\right) \text{cm}^2 \text{ s}^{-1}$$

$$F_0 = 10.4 \times 10^{-5} \text{ g cm}^{-2} \text{ s}^{-1}$$

Validity of the Model

In this case, the liquid in contact with the internal surface enters the rubber by diffusion and then evaporates from the external surface of the tube.

The kinetics representing the uptake of liquid by the tube were obtained by experiments and calculation using the values of the parameters D and F_0 shown above (Fig. 11.9). A good correlation between these curves can be seen:

- In curve 1, where the rubber tube is immersed in the liquid, the liquid entering the tube through the internal and external surfaces.
- In curve 2, where the liquid is put in contact with the internal surface, the external surface of the tube being free of liquid.

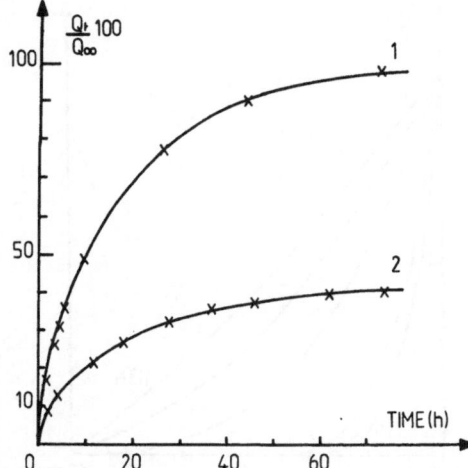

Fig. 11.9. Kinetics of absorption of liquid by rubber tube. 1: Tube of infinite length. 2: Kinetics of absorption by internal surface and evaporation from the external surface, for a tube of infinite length. (——, calculation; ×, experiments.)

The profiles of concentration developed through the thickness of the rubber tube are shown for various times in Fig. 11.10.

11.4.4 Conclusions

The following conclusions are worth noting:

1. The numerical model describes very well the process of absorption and desorption for the rubber tube. The liquid in contact with the internal surface diffuses through the tube wall and evaporates from the external surface.
2. For times shorter than 113 hours, the liquid transport occurs under transient conditions. At this time steady-state conditions are attained.
3. A comparison between the two kinetics of absorption shown in Fig. 11.9 may be instructive.
4. This case is of interest because it corresponds with the general use of rubber tube, where a liquid is in contact with the internal surface.

11.5 Annuli

As shown previously in this chapter, rubbers may absorb a large quantity of liquid. The change in volume of the rubber resulting from this absorption is a measure of the resistance of the rubber to the liquid, and its extent depends largely on the nature of the rubber and the liquid as well as on the crosslink density of the rubber.

Following the absorption of liquid (generally hydrocarbons) in the "oil ageing"

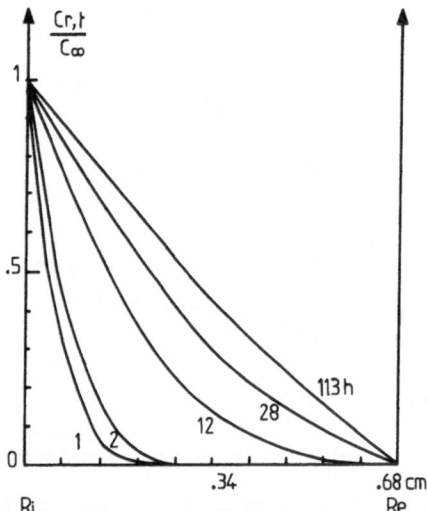

Fig. 11.10. Profiles of concentration of liquid developed through the thickness of the tube.

test, the process of desorption is also of great interest. In all applications, the rubber is in contact with the liquid, and from time to time it is in contact with air.

The problem of drying in the case of rubber annuli is rather complex in the sense that longitudinal and radial transfers take place simultaneously by diffusion, and that the liquid is evaporated either from the two radial surfaces, external and internal, and from two plane surfaces [12].

Of course, the diffusivity may be constant or concentration dependent, and the initial concentration of the liquid within the sample is very rarely uniform.

11.5.1 Theory

Assumptions

The following assumptions are made in order to set the problem:

1. The process of drying is controlled by diffusion within the solid and evaporation from the surface.
2. Two types of transfer take place: longitudinal and radial diffusion, under transient conditions.
3. The liquid evaporates from all the surfaces of the rubber annulus: both the internal and external circular surfaces, and the plane surfaces.
4. The diffusivity is constant or concentration dependent, as found from experiments.
5. The rate of evaporation is proportional to the difference between the liquid concentration on the surface and the concentration necessary to maintain equilibrium with the surrounding atmosphere.
6. The initial concentration of the liquid in the solid may be constant or not.

7. In spite of the change in dimensions of the sample during the drying process, a frame of reference is considered for calculation with the dimensions of the rubber annulus free of liquid.

Mathematical Treatment

The liquid is transferred through the annulus by transient diffusion in two dimensions: radial and longitudinal.

$$\frac{\partial C}{\partial t} = \frac{1}{r}\frac{\partial}{\partial r}\left(Dr\frac{\partial C}{\partial r}\right) + \left(D\frac{\partial C}{\partial z}\right) \tag{11.20}$$

where D is the diffusivity.
  The initial conditions are

$$t = 0 \qquad\qquad R_i \leqslant r \leqslant R_e$$
$$\text{and} \qquad 0 \leqslant z \leqslant H \qquad\qquad C_{in}$$

where R_i and R_e are the internal and external radii, H the height of the annulus and C_{in} is the initial concentration in the annulus, uniform or not.
  The boundary condition expresses the fact that the rate of evaporation from the surface is constantly equal to the rate at which the liquid is brought to the evaporating surface by diffusion.

$$t > 0 \qquad - D\frac{\partial C}{\partial r} = \frac{F_0}{\rho}(C_s - C_{eq}) \tag{11.3}$$

where C_s is the concentration of liquid on the surface and C_{eq} the concentration required to maintain equilibrium with the surrounding atmosphere.

Numerical Analysis

The rubber annulus is divided into a number of annuli of constant thickness Δr and constant height Δz, as shown in Fig. 11.11. The integers i and j are used to define the position of each small annulus:

$$r = i\Delta r \qquad z = j\Delta z \tag{11.22}$$

with the numbers N_i and N_e given by

$$R_i = N_i\Delta r$$
$$R_e = N_e\Delta r$$
$$H = 2N_z\Delta z \tag{11.45}$$

  The concentration of liquid is calculated at various places: within the solid, on the circular surfaces, on the plane surfaces and at the edges.

Within the Annulus, $N_i + 1 \leqslant i \leqslant N_e - 1$
The matter balance during the increment of time Δt is calculated within the annulus of radius $i\Delta r$, of dimensions Δr and Δz. The new concentration at

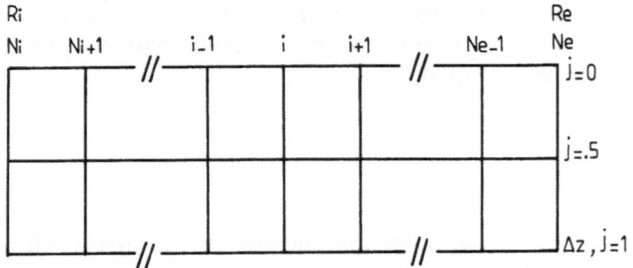

Fig. 11.11. Diagram for numerical analysis of the rubber annulus.

position i,j after time Δt is then obtained as a function of the previous concentrations:

$$CN_{i,j} = C_{i,j} + \frac{1}{iM_r}[(i + 0.5)(C_{i+1,j} - C_{i,j}) - (i - 0.5)(C_{i,j} - C_{i-1,j})]$$

$$+ \frac{1}{M_h}[C_{i,j+1} - 2C_{i,j} + C_{i,j-1}] \qquad (11.23)$$

with the dimensionless numbers

$$M_r = \frac{(\Delta r)^2}{D\Delta t} \qquad (11.24)$$

$$M_h = \frac{(\Delta h)^2}{D\Delta t} \qquad (11.24')$$

with constant diffusivity D.

External Circular Surface, $i = N_e$
An annulus of thickness $\Delta r/2$ located next to the external surface is considered. The matter balance during the time increment Δt is evaluated by considering the radial and longitudinal diffusion and the evaporation from the external radial surface.

The new concentration at the external surface after time Δt is calculated as follows. With the simple assumption

$$CN_{N_e-0.25} - C_{N_e-0.25} = CN_{N_e} - C_{N_e} \qquad (11.46)$$

the new concentration CN_{N_e} is

$$CN_{N_e,j} = C_{N_e,j} + \left(\frac{2N_e - 1}{N_e - 0.25}\right)\frac{1}{M_r}(C_{N_e-1,j} - C_{N_e,j})$$

$$- \left(\frac{2N_e}{N_e - 0.25}\right)\frac{F_0\Delta t}{\rho\Delta r}(C_{N_e,j} - C_{eq})$$

$$+ \frac{1}{M_h}[C_{N_e-0.25,j-1} - 2C_{N_e-0.25,j} + C_{N_e-0.25,j+1}]$$

$$(11.47)$$

The better assumption can also be made:

$$CN_{N_e-0.25,j} - C_{N_e-0.25,j} = \tfrac{3}{4}(CN_{N_e,j} - C_{N_e,j}) + \tfrac{1}{4}(CN_{N_e-1,j} - C_{N_e-1,j})$$

$$(11.48)$$

Internal Circular Surface, $i = N_i$

The new concentration after time Δt on the internal circular surface is obtained by calculating the matter balance during this time within the annulus of dimensions $\Delta r/2$ and Δh, located next to the internal surface.

With the simple assumption

$$CN_{N_i+0.25,j} - C_{N_i+0.25,j} = CN_{N_i,j} - C_{N_i,j} \qquad (11.49)$$

the new concentration $CN_{N_i,j}$ can be written

$$CN_{N_i,j} = C_{N_i,j} + \left(\frac{2N_i + 1}{N_i + 0.25} \right) \frac{1}{M_r}(C_{N_i+1,j} - C_{N_i,j})$$

$$- \left(\frac{2N_i}{N_i + 0.25} \right) \frac{F_0 \Delta t}{\rho \Delta r}(C_{N_i,j} - C_{eq})$$

$$+ \frac{1}{M_h}[C_{N_i+0.25,j-1} - 2C_{N_i+0.25,j} + C_{N_i+0.25,j+1}]$$

$$(11.50)$$

A better assumption is made with

$$CN_{N_i+0.25,j} - C_{N_i+0.25,j} = \tfrac{3}{4}(CN_{N_i,j} - C_{N_i,j}) + \tfrac{1}{4}(CN_{N_i+1,j} - C_{N_i+1,j}) \qquad (11.51)$$

but the expression of $CN_{N_i,j}$ is more complex.

Plane Surfaces, $j = N_h$

An annulus of radius r and dimensions Δr and $\Delta h/2$, located next to one of the plane surfaces, is considered. The matter balance during the increment Δt is determined by taking into account the radial and longitudinal diffusivity as well as the evaporation from the plane surface.

The new concentration after time Δt on the plane surface can be evaluated. By making the simple assumption

$$CN_{i,N_h-0.25} - C_{i,N_h-0.25} = CN_{i,N_h} - C_{i,N_h} \qquad (11.52)$$

the new concentration is written

$$CN_{i,N_h} = C_{i,N_h} + \frac{2}{M_h}[C_{i,N_h-1} - C_{i,N_h}] - \frac{2F_0 \Delta t}{\rho \Delta h}[C_{i,N_h} - C_{ext}]$$

$$+ \frac{1}{M_r}[C_{i-1,N_h-0.25} - 2C_{i,N_h-0.25} + C_{i+1,N_h-0.25}]$$

$$+ \frac{0.5}{iM_r}[C_{i+1,N_h-0.25} - C_{i-1,N_h-0.25}]$$

A better assumption can be made with

$$CN_{i,N_h-0.25} - C_{i,N_h-0.25} = \tfrac{3}{4}(CN_{i,N_h} - C_{i,N_h}) + \tfrac{1}{4}(CN_{i,N_h-1} - C_{i,N_h-1}) \qquad (11.53)$$

External Edge, $i = N_e$, $j = N_h$

An annulus of radius $(N_e - 0.25)\Delta r$ of dimensions $\Delta r/2$ and $\Delta h/2$, located next to the edge, is studied. The matter balance during the time increment Δt is calculated by considering the radial and longitudinal diffusion and the evaporation from the radial and plane surfaces.

The new concentration at the corner is thus obtained in terms of the previous concentrations at the same and adjacent places. By making the simple assumption

$$CN_{N_e-0.25,N_h-0.25} - C_{N_e-0.25,N_h-0.25} = CN_{N_e,N_h} - C_{N_e,N_h} \qquad (11.54)$$

the new concentration at the corner is given by

$$CN_{N_e,N_h} = C_{N_e,N_h} + \left(\frac{2N_e - 1}{N_e - 0.25}\right)\frac{1}{M_r}(C_{N_e-1,N_h-0.25} - C_{N_e,N_h-0.25})$$

$$- \left(\frac{2N_e}{N_e - 0.25}\right)\frac{F_0\Delta t}{\rho\Delta r}(C_{N_e,N_h-0.25} - C_{ext})$$

$$+ \frac{2}{M_h}(C_{N_e-0.25,N_h-1} - C_{N_e-0.25,N_h})$$

$$- \frac{2F_0\Delta t}{\rho\Delta r}(C_{N_e-0.25,N_h} - C_{ext}) \qquad (11.55)$$

A more complex expression is obtained by making the better assumption

$$CN_{N_e-0.25,N_h-0.25} - C_{N_e-0.25,N_h-0.25} = \tfrac{3}{4}(CN_{N_e,N_h} - C_{N_e,N_h})$$

$$+ \tfrac{1}{4}(CN_{N_e-1,N_h-1} - C_{N_e-1,N_h-1}) \qquad (11.56)$$

Internal Edge, $i = N_i$, $j = N_h$
An annulus of radius $(N_i + 0.25)\Delta r$ and dimensions $\Delta r/2$ and $\Delta h/2$, located at the edge, is examined. The matter balance during the time Δt is evaluated by considering the radial and longitudinal diffusion as well as the rate of evaporation from the plane and the internal surface.

By making the simple approximation

$$CN_{N_i+0.25,N_h-0.25} - C_{N_i+0.25,N_h-0.25} = CN_{N_i,N_h} - C_{N_i,N_h} \qquad (11.57)$$

the new concentration after time Δt at the internal corner becomes

$$CN_{N_i,N_h} = C_{N_i,N_h} + \left(\frac{2N_i + 1}{N_i + 0.25}\right)\frac{1}{M_r}(C_{N_i+1,N_h-0.25} - C_{N_i,N_h-0.25})$$

$$- \left(\frac{2N_i}{N_i + 0.25}\right)\frac{F_0\Delta t}{\rho\Delta r}(C_{N_i,N_h-0.25} - C_{ext})$$

$$+ \frac{2}{M_h}[C_{N_i+0.25,N_h-1} - C_{N_i+0.25,N_h}]$$

$$- \frac{F_0\Delta t}{\rho\Delta h}(C_{N_i+0.25,N_h} - C_{ext}) \qquad (11.58)$$

Amount of Matter Remaining
The amount of matter remaining in the annulus can be calculated at any time by integrating the concentrations obtained at this time with respect to space within the annulus.

11.5.2 Experiment

Rubber Samples

Rubber annuli of various sizes were cured in a heated mould with the conditions shown in Table 11.3, under a pressure of 30 bars.

The curing time of 15 min, determined by calculation using a numerical model which takes into account heat conduction and heat of reaction, is sufficient for the state of cure to reach a value higher than 90% [4, 5, 7].

Conditions of Desorption

The rubber samples were immersed in toluene at 20 °C. Great care was taken to ensure that the samples were exposed on all sides to the liquid by suspending them on wires. Experiments were done on various rubber samples to make sure that the extraction of the sample from the liquid during the short time necessary for weighing did not disturb the process of absorption.

The samples were then removed from the liquid and exposed to the surrounding atomsphere. The weight was constantly recorded for the kinetics of drying.

11.5.3 Results

The values of the parameters such as the diffusivity and rate of evaporation were determined, and the validity of the numerical model is tested using these values.

Determination of the Parameters

Diffusivity
The diffusivity was determined for absorption and desorption [12] using sheets of 0.15 cm thickness made of the same material as that of the rubber annuli. Two methods were used for the stage of absorption and one for desorption.

In the later stages of drying, the first term in the series in Eq. (11.7) needs to be considered

$$\ln\left(\frac{Q_\infty - Q_t}{Q_\infty}\right) = -\frac{\beta_1^2}{L^2}Dt + \ln\left(\frac{2S^2}{\beta_1^2(\beta_1^2 + S^2 + S)}\right)$$

The diffusivity can thus be calculated by iteration from the slope of the straight line obtained by using this equation in spite of the fact that β_1 depends on the value of the diffusivity.

Table 11.3. Preparation of rubber annuli

Sample	R_i (cm)	R_e (cm)	H (cm)	t (min)	T (°C)
1	0.985	1.66	3	30	160
2	0.985	1.66	0.7	30	160
3	0.985	1.66	0.7	15	160

Rate of Evaporation
The rate of evaporation was determined by

- Using the initial rate of drying, when the process is controlled by evaporation.
- Measuring the rate of evaporation of the pure liquid in steady conditions.

Of course, the value of the rate of evaporation depends largely on the agitation of the surrounding atmosphere. In the present case, the air is motionless [12].

Parameter Values
The parameters were found to be

$$D = 14 \times 10^{-7} \text{ cm}^2 \text{ s}^{-1}$$

$$F_0 = 10.4 \times 10^{-5} \text{ g cm}^2 \text{ s}^{-2}$$

$$\frac{Q_\infty}{Q_i} = 289\%$$

where Q_i is the weight of the rubber sample free of liquid.

Validity of the Model

The validity of the numerical model was tested by comparing the experimental kinetics of drying with the corresponding kinetics calculated with the model. Fig. 11.12 illustrates the validity of this method for a thin annulus when the calculation is made with a constant diffusivity. The experimental points represent uptake of toluene expressed in weight percent of the rubber free of liquid; the solid curve represents the uptake calculated with the model. Similar agreement between the calculated and experimental kinetics of drying can be seen in Fig. 11.13 for a rubber annulus having a greater thickness.

At the beginning of the process of drying, the amount of liquid transferred by

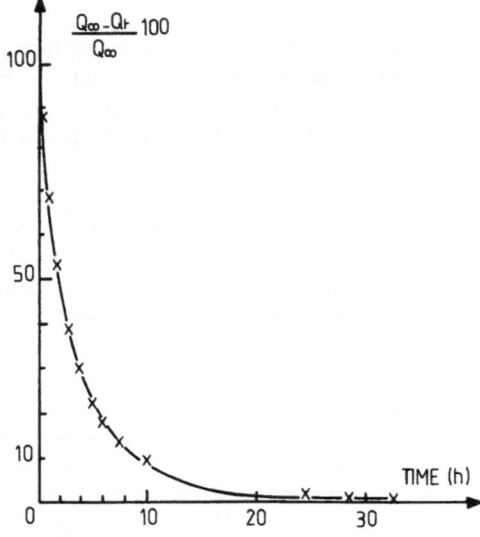

Fig. 11.12. Kinetics of drying of a flat annulus at 20 °C in motionless air. $R_i = 0.98$ cm; $R_e = 1.65$ cm; $h = 0.70$ cm. (——, calculation; ×, experiments.)

diffusion to the surface is less than that removed by evaporation, and the concentration of the liquid at the surface decreases rapidly. This fact can be clearly seen in Fig. 11.14 where the profiles of concentration are developed through the sample at various times.

11.5.4 Conclusions

The following conclusions can be drawn:

1. The drying process is described well by the numerical model in the difficult case of rubber annuli, in spite of the great change in dimensions of the rubber.
2. In the present case, the diffusivity was found to be constant.
3. The following assumptions must be made for calculation, as only concentrations with integer values are known:

$$C_{N_e-0.25,j} = \tfrac{3}{4}\, C_{N_e,j} + \tfrac{1}{4}\, C_{N_e-1,j}$$
$$C_{i,N_h-0.25} = \tfrac{3}{4}\, C_{i,N_h} + \tfrac{1}{4}\, C_{i,N_h-1}$$
$$C_{N_e-0.25,N_h-0.25} = \tfrac{3}{4}\, C_{N_e,N_h} + \tfrac{1}{4}\, C_{N_e-1,N_h-1}$$
$$C_{N_e-0.5,j} = \tfrac{1}{2}\, C_{N_e,j} + \tfrac{1}{2}\, C_{N_e-1,j} \qquad (11.59)$$

11.6 Conclusions on the Drying of Rubbers

As stated previously, testing of rubber samples often involves contact with liquids and it is of interest to know the kinetics of absorption and drying [13].

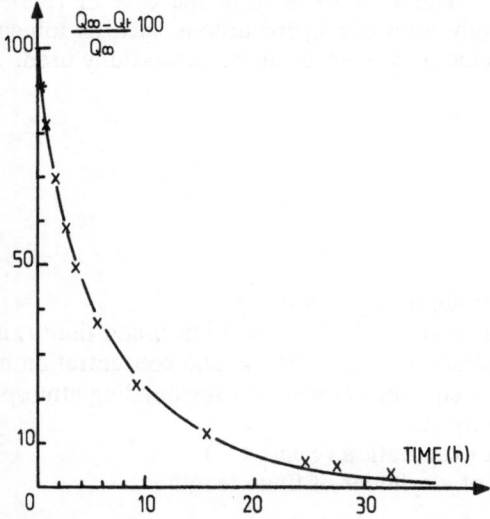

Fig. 11.13. Kinetics of drying of a long rubber annulus at 20 °C in motionless air. $R_i = 0.98$ cm; $R_e = 1.65$ cm; $h = 3.0$ cm. (——, calculation ×, experiments.)

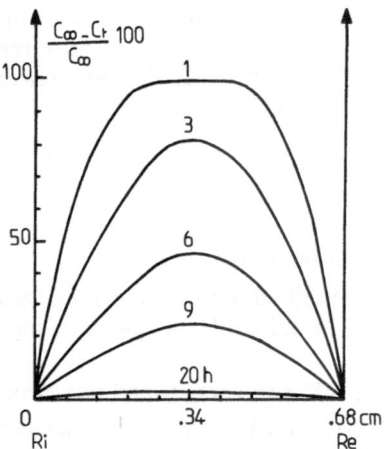

Fig. 11.14. Profiles of concentration of toluene within the rubber annulus at midheight in motionless air at 20 °C. $R_i = 0.98$ cm; $R_e = 1.85$ cm; $h = 0.70$ cm.

The general problem for testing the behaviour of a real sample in contact with a liquid is not easy in the sense that the test can be time-consuming if the sample is large and that the test is generally destructive since some additives are extracted. A method combining short tests and a numerical method is useful. The short tests are made with thin samples of the same material as the real sample, and the experiment time can be reduced as this time is roughly proportional to the square of the thickness. The numerical model must be able to predict the kinetics of drying for a real sample, whatever its size and shape.

Various examples were selected with rubber samples of different sizes and shapes. In each case, the numerical model was able to account for all the known facts regarding diffusion and evaporation, and it gave good correlation with the experimental results for the kinetics of drying.

In spite of the high degree of swelling in the case of rubbers in contact with hydrocarbons (especially aromatic hydrocarbons such as toluene), the numerical model with a fixed frame of reference can be successfully used.

Symbols

$C_{i,j}$	Concentration at position i, j
C_n, CN_n	Concentration of liquid at position n and time t, after time Δt
C_s, C_{eq}	Concentration on the surface, and concentration necessary to maintain equilibrium with the surrounding atmosphere
D	Diffusivity $(cm^2\ s^{-1})$
F_0	Rate of evaporation $(g\ cm^2\ s^{-1})$
H	Height of a cylinder of finite length
H_j	Function $H_j = jD\dfrac{\partial C}{\partial r}$ at position j

i	Integer characterising the radial position in a cylinder of finite length
j	Integer characterising the longitudinal position in a cylinder of finite length
$2L$	Thickness of the sheet
M	Dimensionless number $\left(\dfrac{(\Delta x)^2}{D\Delta t}\right)$
M_h	Dimensionless number $\left(\dfrac{(\Delta h)^2}{D\Delta t}\right)$
M_r	Dimensionless number $\left(\dfrac{(\Delta r)^2}{D\Delta t}\right)$
N_h	Number of slices in a cylinder of finite length
N_r	Number of annuli in a cylinder of finite length
N_i, N_e	Minimum and maximum value of N for a hollow cylinder
Q_t, Q_∞	Amount of liquid remaining in the sheet after time t, after infinite time, respectively
R	Radius
R_e, R_i	External and internal radii of a hollow cylinder
S	Dimensionless number $\left(\dfrac{LF_0}{D\rho}\right)$
t	time
β_n	Roots of $\beta\tan\beta = S$
$\Delta x, \Delta t$	Increments of space and of time, respectively
ρ	Density of the liquid (g cm^{-3})

References

1. Brown RP. Physical testing of rubbers. Applied Science Publishers, London, 1979, pp. 268–293
2. Khatir Y, Bouzon J, Vergnaud JM. Liquid sorption by rubber sheets and evaporation. Models and experiments. Polym Test 1986; 6: 253–265
3. Crank J. The mathematics of diffusion. 2nd edn. Clarendon Press, Oxford, 1975, pp. 60–61
4. Sadr A, Granger R, Vergnaud JM. Calculation of profiles of temperature and state of cure within the rubber mass in injection molding process. J Appl Polym Sci 1984; 29: 955–963
5. Abdul M, Vergnaud JM. Vulcanization progress in rubber sheets during cooling in motionless air after extraction from the mold. Thermochim Acta 1984; 76: 161–170
6. Khatir Y, Bouzon J, Vergnaud JM. Non-destructive testing of rubber for the sorption and desorption–evaporation of liquid by modelling (cylinders of finite length). J Polym Engng 1987; 7: 149–167
7. Accetta A, LeParlouer P, Vergnaud JM. Kinetic parameters of the overall reaction of scrap rubber vulcanization by 2% sulfur. Thermochim Acta 1982; 59: 149–156
8. Armand JY, Vergnaud JM. Comparative study of a reaction of low enthalpy (rubber cure) by DC and DSC. Thermochim Acta 1987; 116: 111–124
9. Khatir Y, Bouzon J, Vergnaud JM. Determination of processes of absorption and desorption of liquids with rubber tubings by using model and short tests. J Polym Engng 1987; 7: 275–299
10. Taverdet JL, Vergnaud JM. Study of transfer process of liquid into and plasticizer out of plasticized PVC by using short tests. J Appl Polym Sci 1984; 29: 3391–3400
11. Taverdet JL, Vernaud JM. Modelization of matter transfers between plasticized PVC and liquids in case of a maximum for liquid-time curves. J Appl Polym Sci 1986; 31: 111–122

12. Khatir Y, Bouzon J, Vergnaud JM. The kinetics of absorption and desorption of liquid by a rubber annulus using a model and short-term tests. Plast Rubber Process Applic 1988; 9: 53–58
13. Vergnaud JM. Liquid transport processes in polymeric materials. Modelling and Industrial Applications. Prentice Hall, 1990, pp. 168–210

Chapter 12
Drying of Plasticised PVC

The process of drying of plasticised PVC appears to be strange at first, but it can be explained by the formation of steep gradients of concentration which play a important role when the diffusivity is concentration dependent.

12.1 Introduction

PVC plays an important role in the plastics industry applied to packaging, and this polymer is the leading plastic material in most plastic-producing countries, except in the United States and Japan, where it is second to polyethylene. Most of the PVC is used in a plasticised form, and in packaging applications, plasticised PVC is in contact with liquid, food [1], blood [2, 3] or other solvents [4, 5]. Moreover, plasticised PVC contains the so-called plastics additives such as light and heat stabilisers, UV absorbers, antioxidants and lubricants, which may be physiologically objectionable. Of course, each of these additives is absolutely necessary for the processing and stability of the final plastics. The advantages in using plasticised PVC are that the material is very flexible, and it is inexpensive.

The drawback appears with the migration of these additives. When plasticised PVC is in contact with liquid food, blood or solvents, two types of matter transfer may take place: the liquid enters the plasticised PVC, enabling the plasticiser to leave the polymer, causing the following problems:

1. A decrease in mechanical properties of the polymer, and especially in flexibility, because of the loss of plasticiser.
2. Pollution of the liquid.

Many attempts have been made in order to discover new polymers exhibiting the same qualities as plasticised PVC without the drawbacks of matter transfer, but all of them without success [6]. Other ways have been explored to reduce matter transfer between PVC packaging and liquids, with irradiation [7], surface reticulation by electric discharge [8], UV treatment [9] and various chemical treatments [3], but no decisive results have followed.

Another method has been studied by preparing plasticised PVC sheets showing very low matter transfer when in contact with a liquid. The method consists of creating gradients of concentration of plasticiser next to the PVC surface with a low concentration on the surface [10]. A two-step method has been elaborated, by immersing the PVC sheet in a liquid for a short time (2–15 min) and drying the sheet very quickly. During the first step of immersion, two types of matter transfer take place, the liquid entering the PVC and enabling the plasticiser to leave the solid, as shown in earlier studies with various liquids [5, 11–15]. When the time of immersion is short, a small amount of matter is transferred, and two profiles of concentration are created next to the surface, with a high concentration of liquid and a low concentration of plasticiser on the PVC surfaces. Of course, the drying period is necessary to reduce the concentration of the liquid on the surface. But the process of drying this type of PVC is not simple.

In fact, the whole process of matter transfer is rather complex for the following reasons [16, 17]:

1. Two types of matter transfer take place, controlled by diffusion.
2. The diffusivity for each type of transfer is concentration dependent.
3. The two types of transfer are coupled in the sense that the diffusivity of each type depends on the concentration of both the plasticiser and the liquid.

Resulting from these facts, it appears that the rate of liquid transfer must be low next to the surface during drying. It is thus understandable that the operational conditions of drying must be very precisely defined, because the best final material must exhibit a steep gradient of plasticiser next to the surface and a small amount of liquid.

12.2   Drying Process

12.2.1   Theory

In the method of preparation of plasticised PVC with low matter transfer, the process of drying is the second stage. The study of the drying process necessitates a perfect knowledge of the state of the plasticised PVC, e.g. the profiles of concentration of the liquid and the plasticiser. These two stages are thus considered in succession.

Assumptions

In order to clarify this complex problem, the following assumptions are made [18–21].

For immersion:

1. Two types of matter transfer take place: the liquid entering the PVC, enabling the plasticiser to leave the polymer.
2. Both these types of transfer are controlled by transient diffusion.
3. In some studies, constant diffusivities were considered, as the time of absorption is short [18–20]. In other studies, concentration dependent diffusivities were preferred [21].

4. The concentration of liquid and plasticiser on the PVC surface reaches equilibrium as soon as the process starts.

For drying:

1. The process of drying is controlled by evaporation from the PVC surface and by diffusion through the solid.
2. The diffusivities are concentration dependent.
3. The rate of evaporation is proportional to the concentration of liquid on the PVC surface.
4. The plasticiser does not move during this stage.

Mathematical Treatment

The transfer of both the liquid and the plasticiser is described by Fick's law in one dimension with concentration-dependent diffusivity:

$$\frac{\partial C}{\partial t} = \frac{\partial}{\partial x}\left(D\,\frac{\partial C}{\partial x}\right) \tag{12.1}$$

Initial and boundary conditions for the immersion stage are

$$
\begin{array}{llll}
t = 0 & 0 < x < L & C = 0 & \text{(liquid)} \\
 & & C_{in} & \text{(plasticiser)} \quad (12.2) \\
t > 0 & x = 0 \text{ and } x = L & C = C_{eq} & \text{(liquid, plasticiser)} \quad (12.3)
\end{array}
$$

Initial and boundary conditions for the drying stage are:

$$
\begin{array}{llll}
t = 0 & 0 < x < L & C_x^l & \text{(liquid)} \\
 & & C_x^p & \text{(plasticiser)} \quad (12.4) \\
t > 0 & x = 0 \text{ and } x = L
\end{array}
$$

$$-D\,\frac{\partial C}{\partial x} = \frac{F_0}{\rho}\,(C_s - C_{ext}) \tag{12.5}$$

Of course, the profiles of concentration of liquid and plasticiser obtained at the end of the absorption stage are identical to the corresponding profiles at the beginning of the process of drying.

As no analytical solution can be found, a numerical method is used.

Numerical Analysis

A numerical method with finite differences is used. The thickness of PVC sheet is divided into finite slices of constant thickness Δx, by concentration–time reference planes, n being an integer which characterises the positions (Fig. 12.1).

Within the Solid During Immersion and Drying

The balance of matter transferred in the slice n during the increment of time Δt is evaluated for each liquid. The new concentration after time Δt of each liquid is

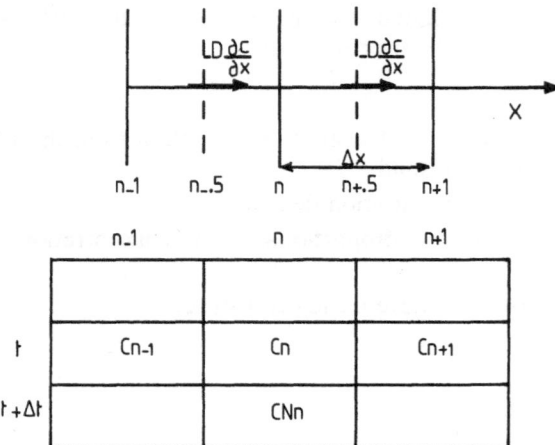

Fig. 12.1. Space–time diagram for numerical analysis. Matter balance within the slice at position n.

thus determined as a function of the previous concentrations obtained at the previous time:

$$CN_n = \frac{1}{M_n} [C_{n-1} + (M_n - 2)C_n + C_{n+1}] \qquad (12.6)$$

with the dimensionless number M_n

$$M_n = \frac{(\Delta x)^2}{D_n \Delta t} \qquad (12.7)$$

while each diffusivity is expressed in terms of the concentrations of the plasticiser and liquid, C^p and C^l:

$$D_n = D_0 \exp \frac{-A}{C_n^p + C_n^l} \qquad (12.8)$$

As the profiles of concentration are very steep next to the surface, a better equation is written for calculating each concentration:

$$CN_n = C_n + \frac{\Delta t}{(\Delta x)^2} [D_{n-0.5}(C_{n-1} - C_n) + D_{n+0.5}(C_{n+1} - C_n)] \qquad (12.9)$$

Of course, Eqs (12.6) or (12.9) are used for calculating the concentrations of the liquid and the plasticiser in succession, after each increment of time.

On the Surface During Immersion
The concentrations of liquid and plasticiser on the surface reach the values at equilibrium, as soon as the process starts.

$$C_s = C_{eq} \qquad (12.10)$$

On the Surface During Drying
A slice of thickness $\Delta x/2$ next to the surface is considered (Fig. 12.2), and the liquid balance evaluated by considering the diffusion within the solid and evaporation from the surface. The new concentration on the surface CN_N is obtained as follows:

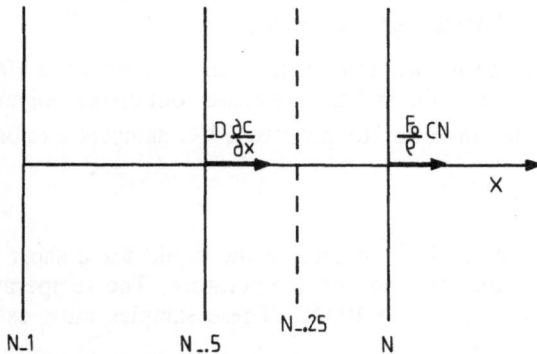

Fig. 12.2. Slice of thickness $\Delta x/2$ next to the surface with diffusion and evaporation.

$$CN_N = C_N - \tfrac{1}{3}[CN_{N-1} - C_{N-1}] + \tfrac{8}{3}\left[\frac{\Delta t}{(\Delta x)^2} D_{N-0.5}(C_{N-1} - C_N) - \frac{F_0\Delta t}{\rho\Delta x} C_N\right]$$

$$(12.11)$$

with the assumption

$$CN_{N-0.25} - C_{N-0.25} = \tfrac{3}{4}(CN_N - C_N) + \tfrac{1}{4}(CN_{N-1} - C_{N-1}) \quad (12.12)$$

Of course the new concentration at position $N - 1$ must be calculated beforehand by using Eqs (12.6) or (12.9).

Amount of Matter Remaining

The amount of liquid and plasticiser is determined for each time, by integrating the corresponding concentration with respect to space.

12.2.2 Experiment

Materials

Sheets of plasticised PVC of thickness 0.06 cm were used, with the following composition:

- Plasticiser 30% by weight
- Additives 2% by weight

The plasticiser was diethylhexylphthalate (DEHP).

The liquid was n-hexane (this liquid having been chosen by the Food and Drug Administration for simulating fatty liquids).

Apparatus and Methods

Immersion

PVC samples of $3 \times 3 \times 0.06$ cm weighing about 0.75 g were soaked in 50 ml of n-hexane at 20 °C. The samples were weighed at intervals, while the concentration of plasticiser in the liquid was measured by UV spectrometry (UV-1100, Hitachi).

Two types of experiments were performed:

1. For a long time, up to equilibrium, in order to obtain the kinetics of transfer of the liquid into the PVC and the plasticiser out of the polymer.
2. For a short time, in order to prepare PVC samples exhibiting low matter transfer.

Drying

The PVC samples previously immersed in the liquid for a short time were dried under specific conditions of time and temperature. The temperature was chosen within a wide range from 20 to 100 °C. These samples must exhibit low matter transfer.

Testing

After drying, the PVC samples were then tested by again putting them in contact with the liquid. During this long test, the kinetics of matter transfer were determined by weighing the samples and measuring the plasticiser concentration in the liquid at intervals.

12.2.3 Results for Immersion

In spite of the fact that this work is devoted to the principles of drying, it is of interest in the case of drying of plasticised PVC to consider the immersion stage, because the initial profiles of concentration of liquid in the drying stage are the same as those obtained at the end of the immersion stage.

Two PVC samples with the same dimensions ($3 \times 3 \times 0.06$ cm with a weight of 0.68 g) were immersed in n-hexane (50 ml) in two flasks.

Determination of the Kinetics of Matter Transfer

Kinetics of Absorption

In the first case, the PVC sample was immersed in the liquid over a long period of time until equilibrium of matter transfer is reached. The kinetics of the two types of transfer are shown in Fig. 12.3, and the following facts must be emphasised:

1. At the beginning of the process, for times shorter than 1 h, the rate of transfer is higher for the liquid than for the plasticiser.
2. Both types of transfer are controlled by diffusion as indicated by the square-root dependence on time of the amount of matter transferred (Fig. 12.4).
3. The amount of liquid transferred passes through a maximum. This fact has been observed many times with various liquids: benzyl alcohol [12], ethanol [13] acetic acid [14], and a model was able to describe the process in this difficult case [17].

Diffusivity

The diffusivities for the plasticiser and n-hexane can be determined from the slopes of the straight lines drawn in Fig. 12.4 by using the following equation:

$$\frac{Q_t}{Q_\infty} = \frac{4}{L}\left(\frac{Dt}{\pi}\right)^{0.5} \tag{12.13}$$

where L is the thickness of the PVC sheet.

Amount of Matter Transferred at Equilibrium

The value of the matter transferred at equilibrium is easy to obtain for the plasticiser as shown in Fig. 12.3. However, the problem is more complicated for the liquid (Fig. 12.3). In fact, three values can be considered for the amount of liquid transferred at equilibrium:

1.   The values actually attained after long contact.
2.   The value reached at the maximum.
3.   The value extrapolated from the values measured at the beginning of the process, for times shorter than 4 h.

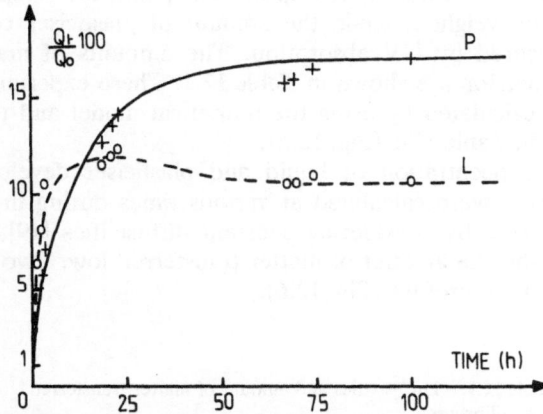

Fig. 12.3. Kinetics of transfer of liquid (entering the PVC) and plasticiser (leaving the PVC) during immersion of the PVC in liquid. (P, plasticiser; L, n-hexane; +, O, experiment; ——, ---, calculation.)

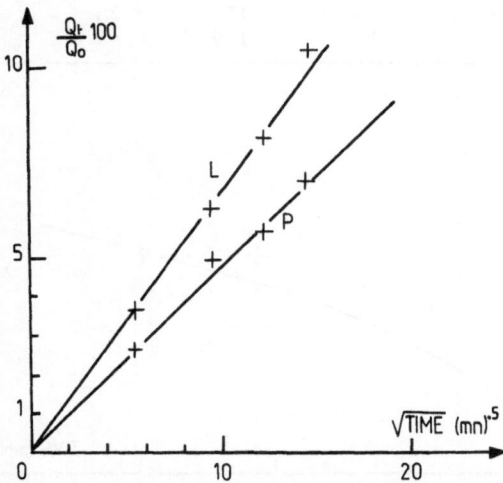

Fig. 12.4. Amount of matter transferred as a function of the square-root of time, during immersion of the PVC in liquid. (P, plasticiser; L, n-hexane.)

As in the present case the time of immersion selected for preparing the PVC samples is 15 min, the third value of the amount of liquid transferred at equilibrium is chosen.

The values of diffusivities and amounts of matter transferred at equilibrium are given in Table 12.1. The amounts of matter transferred after 15 min of immersion are also given. These amounts of matter transferred are expressed in grams for 100 g of the initial PVC.

Preparation of Samples

New PVC samples with the same composition and dimensions ($3 \times 3 \times 0.06$ cm) as those taken for determining the kinetics of transfer, were immersed in n-hexane (100 ml) for 15 min in order to prepare the samples to be studied for the evaporation. At the end of this short step of immersion, the sample was removed from the liquid and weighed, while the amount of plasticiser contained in the n-hexane was measured by UV absorption. The amounts of matter transferred after this short immersion are shown in Table 12.1. These experimental values are the same as those calculated by using the numerical model and the data such as diffusivity and Q_∞ in Table 12.1 (Fig. 12.5).

The profiles of concentration of liquid and plasticiser developed within the thickness of the sheet were calculated at various times during immersion. These calculations were made by considering constant diffusivities [19], as the time of transfer was short and the amount of matter transferred low. Two conclusions are worth noting from these profiles (Fig. 12.6):

Table 12.1. Diffusivities and amount of matter transferred at equilibrium

Matter	$D \times 10^8$ (cm^2 s^{-1})	Q_∞ (%)	Q_{15} (%)
Plasticiser	0.7	17	2.1
Liquid	5	16.9	4.2

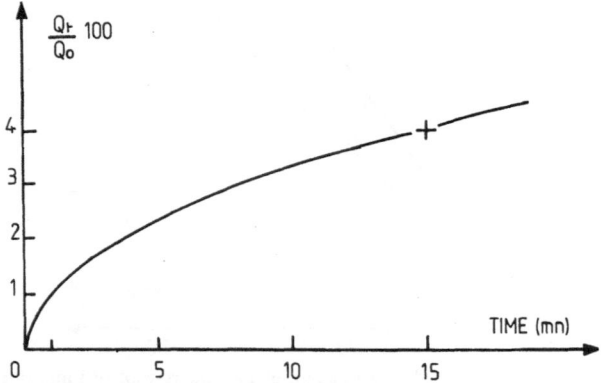

Fig. 12.5. Kinetics of n-hexane transferred in PVC during immersion (——, calculaton; +, experiment.)

1. A steep gradient of concentration is observed for the plasticiser, with a low concentration on the surface of PVC sheet. As proved previously by experiments and calculation [22], the plasticiser concentration on the PVC surface is small, but not zero. In fact, this concentration depends on the ratio of the volumes of liquid and PVC. These profiles of concentration of plasticiser are calculated by assuming that equilibrium is reached on the PVC surface as soon as the process starts.

2. Another steep gradient of concentration for the liquid is obtained by using the model with a high concentration on the PVC surface. The value of this concentration of liquid on the PVC surface is assumed to be that at equilibrium.

12.2.4 Results for Drying

Experimental Kinetics of Drying

After the immersion in liquid for 15 min, the plasticised PVC sheet was exposed to air and dried. The drying process was performed in motionless air firstly at 22 °C up to equilibrium (or rather until a low rate of evaporation was obtained). After this treatment, the sample was dried at 85 °C up to equilibrium, and then at 120 °C. The loss in weight, following the drying, is shown as a function of time in Fig. 12.7. The following observations are worth noting:

1. It is very difficult, if not impossible, to dry the PVC sample at room temperature. This observation is quite different from the results obtained with ethylene-vinyl acetate copolymers which can be evaporated to dryness in the same conditions.

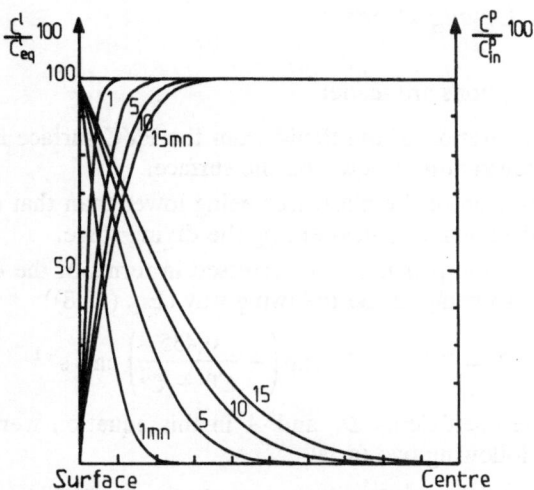

Fig. 12.6. Profiles of concentration of plasticiser and liquid developed through the PVC thickness during immersion. C_{in}^p, initial concentration of plasticiser in the sheet. C_{eq}^l, concentration of liquid at equilibrium.

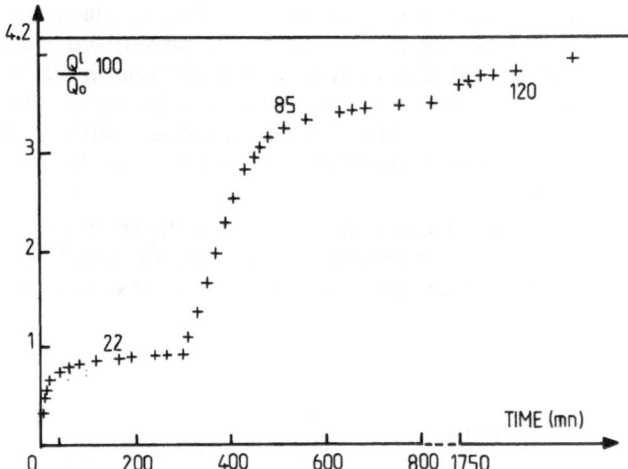

Fig. 12.7. Kinetics of drying of the PVC sheet (previously immersed in n-hexane for 15 min) at temperatures of 22, 85 and 120 °C. (O_0, initial weight of the PVC sheet.)

2. The drying process at 85 °C is far from being complete. After a drying time of 25 h at 85 °C, more than 10% of the liquid remains in the sample.

3. A third stage of drying at 120 °C is necessary to dry the PVC sample up to 95%.

4. The kinetics of drying at 22 °C are peculiar. The rate of drying is quite high at the beginning, but becomes very low after only 40 min.

5. Drying at 85 °C during the second stage is not very efficient, since the rate at the beginning of this stage is lower than the rate obtained at the beginning of drying at 22 °C.

Modelling of the Drying Process

Assumptions
The following assumptions are made:

1. The rate of evaporation of the liquid from the PVC surface is proportional to the actual concentration of liquid on the surface.

2. The rate of diffusion of the plasticiser being lower than that of the liquid, the plasticiser is taken as motionless during the drying stage.

3. The diffusivity of the plasticiser is expressed in terms of the concentrations of the plasticiser and liquid in the following way (Eq. (12.8)):

$$D_1 = 2.7 \times 10^{-7} \exp\left(-\frac{0.435}{C^p + C^l}\right) \text{cm}^2\text{s}^{-1} \qquad (12.8')$$

The values of the coefficients D_0 and A in this equation were determined by accounting for the following two facts:

1. The liquid and plasticiser play the same role for the diffusion of the liquid, as shown previously [17, 23].

2. The value of the diffusivity in Eq. (12.8′) is the same as that obtained for the

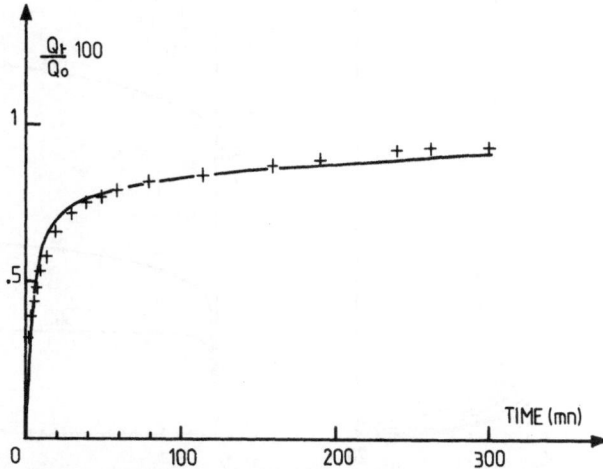

Fig. 12.8. Validity of the model of drying. Kinetics of drying for the PVC sheet at 22 °C. (+, experiment; ——, calculation.)

liquid entering the PVC at the beginning of the immersion stage, when the initial concentration of plasticiser is equal to the sum of the concentrations of liquid and plasticiser.

The Model

A major difficulty for calculating the profile of concentration of the liquid with the help of the model was concerned with the number of slices through the PVC thickness [19]. After several attempts, the following way was found to be the most convenient:

1. Twenty slices of equal thickness are considered through half the thickness of the sheet (only half the thickness need be studied because of the symmetry).
2. The two slices located next to the surface are divided into 30 or 40 thinner small slices, the results being similar with 30 and 40 slices.

This particular way of dividing the PVC thickness into slices was chosen for the reason that the profile of the concentration of liquid is especially steep next to the surface.

The validity of the model was tested by comparing the kinetics of drying the sheet at 22 °C obtained by calculation and by experiments (Fig. 12.8). Good superimposition of the kinetics curves is seen for drying at 22 °C. The profiles of concentration of liquid developed through the PVC thickness during the process of drying were plotted by considering half the thickness of the sheet (Fig. 12.9) and half the slice located on the surface (Fig. 12.10). The following conclusions are worth noting:

1. The drying process is described well by the model, good correlation being obtained for the experimental and theoretical kinetics of drying.
2. The concentration of liquid on the surface decreases very quickly because of evaporation.
3. Because of liquid transfer by diffusion, the value of the concentration of liquid on the surface is never zero at 22 °C.

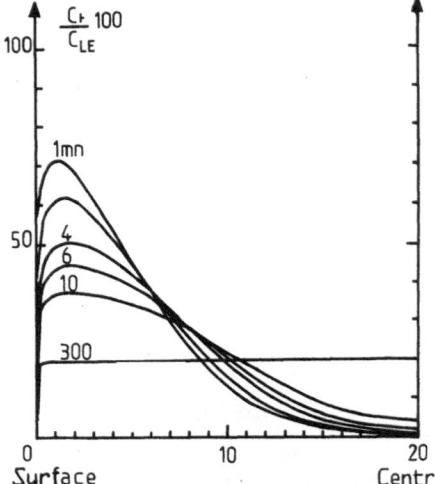

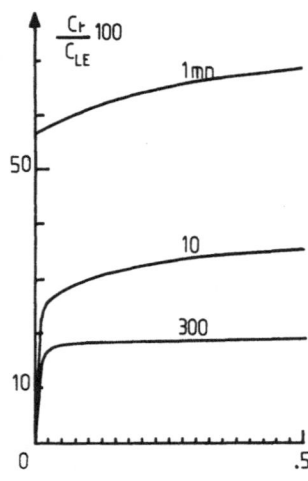

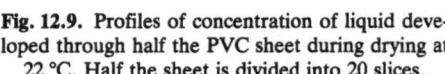

Fig. 12.9. Profiles of concentration of liquid developed through half the PVC sheet during drying at
22 °C. Half the sheet is divided into 20 slices.

Fig. 12.10. Profiles of concentration of liquid developed through the PVC sheet next to the surface. Half the slice next to the surface is divided
into 20 small slices.

4. Slight transfer of liquid towards the centre of the sheet can be observed.

5. A very steep gradient of concentration of liquid is developed next to the
    surface. This fact has made it necessary to divide the thickness into many
    slices near the surface.

6. This high gradient of concentration of liquid with a low concentration of
    liquid next to the surface, associated with the steep gradient of concentration
    of plasticiser with a low concentration of plasticiser on the surface, are
    responsible for the very slow rate of drying of the sheet. As the diffusivity is
    concentration-dependent, its value becomes very low next at the surface. This
    fact explains why the liquid is entrapped in the sheet.

12.3 Conclusions

Generally the process of drying of polymers is not very complex, being controlled
by diffusion of the liquid through the polymer and by evaporation from the
surface. Therefore if no change in the polymer occurs during absorption or
drying, the polymer can be dried to completion at temperatures lower than the
boiling point of the liquid.

In the case of plasticised PVC the process is more complex. The liquid cannot
evaporate at temperatures lower than the boiling point of the liquid, and for
n-hexane, very high temperatures, between 85 and 120 °C, are necessary to
evaporate a large part of the liquid contained in the sheet.

This complex process is explained not only with the help of experiments but
also by using a numerical model. The following facts are considered: during the
liquid absorption by the PVC, steep gradients of concentration are developed for
the liquid and plasticiser with a high concentration of liquid and a low concentra-

tion of plasticiser on the surface. These gradients are especially steep when the time of absorption is short. The stage of drying is controlled by diffusion of the liquid through the solid and evaporation from the surface. Moreover, the diffusivity of the liquid, being concentration-dependent, is very low next to the surface when the concentration of liquid is decreased. The liquid which has previously entered the PVC is thus trapped in the polymer.

The drying conditions are very important in order to obtain a plasticised PVC with very low matter transfer [18–21].

Symbols

A	Constant in Eq. (12.8)
C_{ext}	Concentration necessary to maintain equilibrium with the surrounding atmosphere (zero in this case)
C_{in}	Initial concentration
C_N, C_s	Concentration at the surface
C_n^l, C_n^p	Concentration of liquid and of plasticiser at position n, respectively
CN_n	New concentration at position n after time Δt
D_n	Diffusivity at position n ($cm^2\ s^{-1}$)
D_0	Constant in Eq. (12.8)
F_0	Rate of evaporation of the pure liquid ($g\ cm^2\ s^{-1}$)
L	Thickness of the sheet
M_n	Dimensionless number at position n
N	Number of slices
Δx, Δt	Increments of space and of time, respectively
ρ	Density of the liquid ($g\ cm^{-3}$)
Q_t, Q_∞	Amount of liquid evaporated after time t, and after infinite time, respectively

References

1. Haesen G, Schwarze A. Migration phenomena in food packaging. Commission of the European Communities, 1978
2. Blais P. DEHP in blood bags and medical plastics: their limitations. Can Res 1984; June/July: 13–25
3. Igarashi H, Kioi S, Gejko F, Arakawa M. Physiologic approach to dialysis-induced hypoxemia: effects of dialyser material and dialysate composition. Nephron 1985; 41: 62–96
4. Kampouris EM. The migration of plasticizers into petroleum oils. Europ Polym J 1975; 11: 705–710
5. Vergnaud JM. Scientific aspects of plasticizer migration from plasticized PVC into liquids. Polym Plast Technol Engng 1983; 20: 1–22
6. Garreau M. Contribution à l'amélioration de la biocompatibilité de circuits extracorporels pour hémodialyse. Thesis, Compiegne, France, 1989
7. Kessler M. Bio-compatibilité à long terme en hémodialyse. Nephrologie 1985; 6: 231–234
8. Ishikawa Y, Honda K, Sasakawa S, Katada K, Kabayashi H. Prevention of leakage of di-(2-ethyl-hexyl)phthalate from blood bags by glow discharge treatment and its effect on aggregability of stored platelets. Vox Sang 1983; 45: 68–76

9. Papaspyrides CD. Flexible poly (vinyl chloride) sheets: 1-interrelation between UV-irradiation and plasticizer migration into alcohols. Polymer 1986; 27: 1967–1970
10. Taverdet JL, Vergnaud JM. Preparation of plasticized PVC samples with very low matter transfers. Europ Polym J 1986; 12: 959–962
11. Messadi D, Hivert M, Vergnaud JM. A new approach for study of plasticizer migration from PVC into methanol. J Appl Polym Sci 1981; 26: 667–677
12. Messadi D, Vergnaud JM. Simultaneous diffusion of benzyl alcohol into plasticized PVC and of plasticizer from polymer into liquids. J. Appl Polym Sci 1981; 26: 2315–2324
13. Messadi D, Vergnaud JM. Plasticizer transfer from plasticized PVC into ethanol-water mixture. J Appl Polym Sci 1982; 27: 3945–3955
14. Grigorakakis A, Taverdet JL, Vergnaud JM. Transfer of liquids between plasticized PVC and acetic acid. Europ Polym J 1985; 11: 967–972
15. Messadi D, Taverdet JL, Vergnaud JM. Plasticizer migration from plasticized PVC into liquids. Effect of several parameters on the transfer, I and EC Prod Res Dev 1983; 22: 142–146
16. Taverdet JL and Vergnaud JM. Study of transfer process of liquid into and plasticizer out of plasticized PVC by using short tests. J Appl Polym Sci 1984; 29: 3391–3400
17. Taverdet JL, Vergnaud JM. Modelization of matter transfers between plasticized PVC and liquids in case of a maximum for liquid-time curves. J Appl Polym Sci 1986; 31: 111–122
18. Aboutaybi A, Bouzon J, Vergnaud JM. Modelling the process of drying of plasticized PVC previously immersed in n-hexane. Europ Polym J 1989; 25: 1013–1018
19. Aboutaybi A, Bouzon J, Vergnaud JM. Drying at room temperature of plasticized PVC previously immersed in liquids. Modelling and experiments. Europ Polym J 1990; 26: 285–291
20. Aboutaybi A, Bouzon J, Vergnaud JM. Study of the preparation of plasticized PVC with a surface with barrier properties. J Vinyl Technol 1990; 12: 58–64
21. Aboutaybi A, Bouzon J, Vergnaud JM. Conditions of preparation of plasticized PVC with low matter transfers. In: Proceedings of conference on food and beverages, Crete, July 1989
22. Taverdet JL, Vergnaud JM. Surface analysis of plasticized PVC packagings by attenuated total reflectance. In: Instrumental analysis of foods, vol. 1. Academic Press, New York, 1983, pp 367–377
23. Khatir Y, Taverdet JL, Vergnaud JM. Theoretical study of preparation of plasticized PVC samples exhibiting low matter transfers. J Polym Engng 1988; 8: 111–131

Chapter 13

Drying of Wood

13.1 Introduction

Wood is continually exchanging moisture and this phenomenon is observed not only for living trees but also for pieces of wood in use, whatever the surrounding atmosphere. This surrounding atmosphere can be air or water, and in all these cases, the problem of drying the pieces of wood is of interest.

The process of drying is very complex for pieces of wood, for the following reasons: the origin of the driving force of the water in the wood, the principal directions of transport, the value of the moisture content relative to the fibre saturation point and the presence of various forms of water located in the wood.

Forms of Water and Fibre Saturation Point

The movement of moisture through wood follows a very complex process since moisture in wood exists in three basic forms:

- Bound water within the cell wall
- Water vapour in the voids of the wood
- Free water in liquid form in the voids of the wood

The fibre saturation point (FSP) is the moisture content attained when the relative humidity (RH) of the surrounding air tends to 100%.

Below the FSP, when the RH is less than 100%, only bound water and water vapour exist in the wood. Equilibrium is reached between the content of bound water and the RH of air. This equilibrium moisture content (EMC) increases with the value of RH until the cell wall becomes saturated, this saturation being attained when RH tends to 100%. The bound water, also called hygroscopic water, is found in the cell wall and is believed to be hydrogen bound to the hydroxyl groups, primarily cellulose and hemicellulose, and to a lesser extent to the hydroxyl groups of liquids. The content of bound water is thus limited by the number of sorption sites available in the cell and by the number of molecules of water which can be held on a sorption site.

Beyond the FSP, additional water is in the free liquid form. This case occurs when the wood is in contact with liquid water. Free water in wood is in liquid form in the lumens or voids of the wood. The amount of free water which can be held is limited by the porosity of the wood. As there is no hydrogen bonding, free water is held only by weak capillary forces and cannot cause normal swelling or shrinking, because the cell wall is already saturated by the much more tightly bound hygroscopic water.

As the bound water content increases, wood swells, the rate of the bound water transport increases, and mechanical strength decreases. These changes are usually gradual and continuous until the cell wall is saturated. Beyond this saturation, little change in these properties is observed. However, the presence of free water in wood, beyond the FSP drastically increases its decay susceptibility as well as the possibility of fungal attacks [1, 2].

Principal Directions of Transport of Moisture in Wood

The rate of transport of water within the wood depends largely on the direction taken in the wood. Three principal directions of transport exist (Fig. 13.1): the longitudinal direction along the vertical axis of the living tree, and within the circular cross-section of the tree the radial direction passing through the centre, and the tangential direction which is perpendicular to the radial one.

Generally, wood in the form of beams or boards is cut so that their length is along the longitudinal direction.

Driving Force of Water Within the Wood

Two schools of thought can be recognised in the matter of bound water transport in wood. One school, representing the vast majority of wood scientists, believes that the transport is described by diffusion obeying Fick's laws, and the gradient of moisture concentration is considered as the driving force. Many attempts have been made to measure the diffusivity either for the bound water diffusion [3] or for the combined bound water and water vapour diffusion [4–8]. The other school of thought assumes that bound water diffusion takes place in response to a vapour pressure gradient and not to a gradient of moisture concentration [9].

Although the value of diffusivity depends on the technique of determination

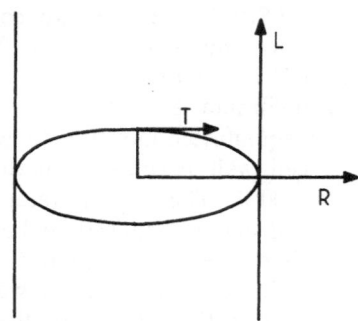

Fig. 13.1. Cross-section of a tree showing the principal directions of moisture transport.

[6, 10], some results of interest have been found about the dependence of diffusivity with the principal direction in the wood and with the moisture concentration. The transverse diffusivity increases with the moisture content, while the longitudinal diffusivity decreases with increasing moisture content [7, 11, 12].

Exponential relationships have been found to describe the dependency of diffusivity of moisture content during absorption [5, 8, 12, 13]. The effect of moisture content on the radial diffusion is very pronounced, but the tangential diffusion remains constant with moisture content [5].

Most studies have been limited to the movement of moisture below the FSP. In this case, where no free liquid exists, the movement of moisture in wood is said to be somewhat less complicated since the movement of liquid water is eliminated [6]. However, the problem of moisture transport above the FSP is also of interest, especially in the following two cases:

1. When the wood is in contact with a surrounding atmosphere of RH around 100%, and water condenses on the wood surface. This condensation occurs very often during the night in form of dew at lower temperature.
2. When the wood is immersed in water.

Problems of Drying

The first purpose of this investigation is to study the process of desorption under transient conditions, whatever the initial moisture content in the wood with regard to the FSP, and whatever the directions of transport. Cases are considered for one dimension [14–17], two dimensions [18] and three dimensions [19].

A second purpose is to build up numerical models able to describe the process of drying when the moisture content is below or above the FSP, by taking all the facts into account: the principal diffusivities along the principal directions, the dependency of diffusivities with the moisture content, the rate of evaporation. The rate of evaporation is taken to be proportional to the difference between the actual moisture content on the surface and the moisture content which is at eqilibrium with the surrounding atmosphere.

13.2   Thin Sheets

It is perhaps unusual to dry a thin sheet of wood. This case appears when one dimension is so small with regard to the other two that the thin plane can be considered as infinite, the effect of the edges becoming negligible.

The case of a thin sheet is very important, especially in woods for which three principal diffusivities exist along the principal axes of diffusion. As the diffusivities do not have the same value according to the direction, it is not possible to determine each diffusivity from the experiments performed with a three-dimensional sample. Each diffusivity must be determined from experiments carried out for one dimension, by eliminating transfer in the other two directions.

13.2.1 Theory

Some assumptions are made in order to clarify the problem, as the diffusivity may be either constant or concentration dependent, while the rate of evaporation is finite.

Assumptions

1. The moisture is transferred in one dimension.
2. The moisture transport is expressed by transient diffusion.
3. The diffusivity is
   - Concentration dependent when the moisture content is below the FSP.
   - Constant when the moisture content is beyond the FSP.
   - Generally the same during aborption and desorption.
4. The rate of evaporation from the surface is proportional to the difference of the moisture concentration on the surface and that which is at equilibrium with the surrounding atmosphere, the coefficient of proportionality being the rate of evaporation of pure water under the same conditions.
5. For mathematical treatment leading to analytical solutions, the diffusivity is constant and the initial concentration in the solid is uniform.
6. For numerical treatment, the diffusivity can be constant or not, and the initial concentration through the thickness of the sheet can be irregular.

Mathematical Treatment

The transfer of moisture through the thickness of the sample is governed by transient diffusion in one dimension:

$$\frac{\partial MC}{\partial t} = \frac{\partial}{\partial x}\left(D\,\frac{\partial MC}{\partial x}\right)$$ (13.1)

which becomes when the diffusivity D is constant

$$\frac{\partial MC}{\partial t} = D\,\frac{\partial^2 MC}{\partial x^2}$$ (13.1')

where MC is the moisture content (expressed in grams per gram of dried wood).
   The initial and boundary conditions are for desorption:

$$t = 0, \quad -L < x < L, \quad MC = MC_{\text{in}} \text{ (solid)}$$ (13.2)

$$t > 0, \quad -D\frac{\partial MC}{\partial x} = F_0(MC_L - MC_{\text{ext}}) \text{ (surface)}$$ (13.3)

where MC_L is the actual moisture content on the surface, MC_{ext} the moisture content required to maintain equilibrium with the surrounding atmosphere and F_0 the evaporation rate of pure liquid in the same conditions of temperature and agitation of air.
   The profiles of moisture content developed through the thickness are given by [20]

$$\frac{MC_{\text{ext}} - MC_{x,t}}{MC_{\text{ext}} - MC_{\text{in}}} = \sum_{n=1}^{\infty} \frac{2S \cos \dfrac{\beta_n x}{L}}{(\beta_n^2 + S^2 + S) \cos \beta_n} \exp\left(-\frac{\beta_n^2 D}{L^2} t\right) \quad (13.4)$$

where the β_ns are the positive roots of

$$\beta \tan \beta = S \quad (13.5)$$

with the dimensionless number S

$$S = \frac{F_0 L}{D} \quad (13.6)$$

Roots of Eq. (13.5) are given in Table 2.2 for various values of S.

The total amount of diffusing substance Q_t leaving the sheet up to time t is expressed as a fraction of Q_∞, the corresponding quantity after infinite time:

$$\frac{Q_\infty - Q_t}{Q_\infty} = \sum_{n=1}^{\infty} \frac{2S^2}{(\beta_n^2 + S^2 + S)\beta_n^2} \exp\left(-\frac{\beta_n^2 D}{L^2} t\right) \quad (13.7)$$

Numerical Analysis

The thickness of the sheet is divided into NS slices of constant thickness Δx by concentration reference planes (Fig. 13.2).

The concentration is determined within the solid and on the surface, by considering a constant and a concentration-dependent diffusivity.

Within the Solid

Constant Diffusivity. The matter balance within the increment of time Δt is calculated within the slice of thickness Δx at position n, n being an integer (Fig. 13.3).

The new moisture concentration after time Δt is thus obtained in terms of the previous concentration by:

$$MCN_n = \frac{1}{M} [MC_{n-1} + (M - 2)MC_n + MC_{n+1}] \quad (13.8)$$

where MCN_n is the new moisture concentration after time Δt at position n, MC_n

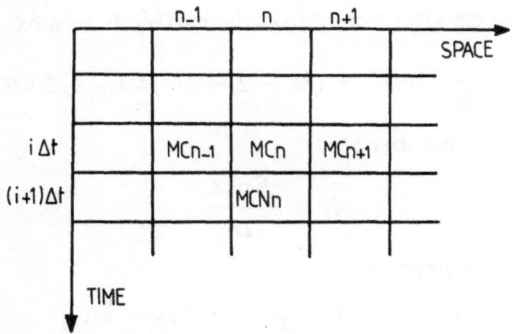

Fig. 13.2. Space–time diagram for numerical analysis for one-dimensional transport.

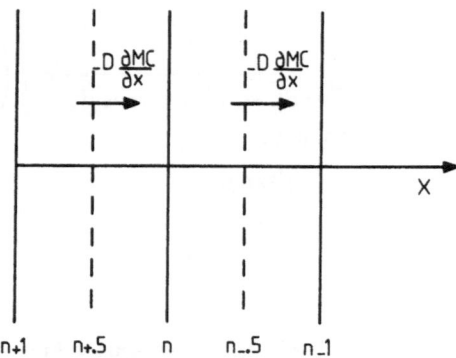

Fig. 13.3. Matter transfer by diffusion within the slice of thickness Δx at position n within the thickness of the sheet.

is the moisture concentration at previous time t and the dimensionless number M is

$$M = \frac{(\Delta x)^2}{D\Delta t} \tag{13.9}$$

Concentration-Dependent Diffusivity. The matter balance is determined in the same way as for constant diffusivity (Fig. 13.3).

The new moisture concentration after time Δt at position n is thus obtained in terms of the previous moisture concentration at the same place and of the function G:

$$MCN_n = MC_n + \frac{\Delta t}{\Delta x}(G_{n-0.5} + G_{n+0.5}) \tag{13.10}$$

with the function G

$$G_{n-0.5} = D_{n-0.5}\frac{\partial C}{\partial x} \quad \text{at } (n - 0.5) \tag{13.11}$$

On the Surface
Constant Diffusivity. The matter balance during the increment of time Δt is evaluated within the slice of thickness $\Delta x/2$ located next to the surface, by considering the diffusion of moisture through the solid and evaporation from the surface (Fig. 13.4).

The new concentration after time Δt on the surface is thus expressed by

$$MCN_0 = \frac{1}{M}[2MC_1 + (M - 2 - 2N)MC_0 + 2NMC_{ext}] \tag{13.12}$$

where the dimensionless number is

$$N = \frac{F_0\Delta x}{D} \tag{13.13}$$

by making the simple assumption

$$MCN_{0.25} - MC_{0.25} = MCN_0 - MC_0 \tag{13.14}$$

By making the better assumption

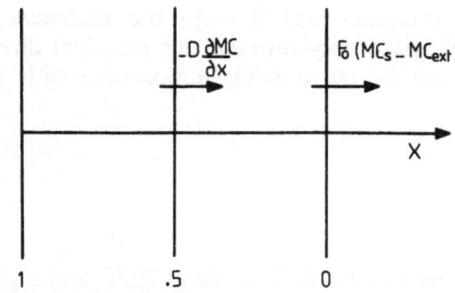

Fig. 13.4. Slice of thickness $\Delta x/2$ next to the surface with diffusion and evaporation of moisture.

$$MCN_{0.25} - MC_{0.25} = \tfrac{3}{4}(MCN_0 - MC_0) + \tfrac{1}{4}(MCN_1 - MC_1) \quad (13.15)$$

the new moisture concentration on the surface is thus given by

$$MCN_0 = MC_0 - \tfrac{1}{3}(MCN_1 - MC_1) + \frac{8}{3M}[MC_1 - (N+1)MC_0 + NMC_{ext}]$$

$$(13.16)$$

where MCN_1 is the new concentration after time Δt at position 1.

Of course, it is necessary to determine the new concentration at position 1 before calculating the new concentration on the surface, by using Eqn. 13.16. The value of MCN_1 is obtained with the help of Eq. (13.8)

Concentration-Dependent Diffusivity. From the matter balance evaluated during the increment of time Δt within the half-slice next to the surface (Fig. 13.4), the new concentration can be determined in terms of the previous concentration.

By making the assumption

$$MCN_{0.25} - MC_{0.25} = \tfrac{3}{4}(MCN_0 - MC_0) + \tfrac{1}{4}(MCN_1 - MC_1) \quad (13.15)$$

the new concentration on the surface is expressed by

$$MCN_0 = MC_0 - \tfrac{1}{3}(MCN_1 - MC_1) - \frac{8\Delta t}{3\Delta x}[G_{0.5} + F_0(MC_0 - MC_{ext})]$$

$$(13.17)$$

Of course, the new moisture concentration at position 1 must be determined beforehand by using Eq. (13.10), as MCN_0 is function of MCN_1 in Eq. (13.17).

Moisture Content Remaining

The amount of moisture remaining in the sheet at any time can be obtained by integrating the moisture concentration with respect to space.

13.2.2 Experiment

Samples

Rectangular blocks of Scots pine sapwood (*Pinus sylvestris*) of sides 2×2 cm and with thicknesses of 1 cm and 0.5 cm were used. The samples were cut in such a way that a principal direction was through the thickness of the block. The moisture diffusion took place only through this principal direction, by protecting the other four faces from the surrounding atmosphere with an impermeable film of silicone elastomer.

Methods

Absorption
After equilibration of the wood sample in air at 20 °C and a given RH, the sample was either put in contact with air at a higher RH, or in water at 20 °C. In the first case when the surrounding atmosphere is air with an RH lower than 100%, the moisture content in the wood is below the FSP. When the sample is immersed in water, the moisture content becomes higher than the FSP.

Drying
After the absorption stage, the sample was exposed to air at a given temperature and RH. The kinetics were followed by weighing the sample at intervals.

The reference weight of the sample was obtained by drying it at 103 °C to constant weight, the moisture content being zero.

13.2.3   Results for Moisture Content Below FSP

In this case, the value of the moisture content in the wood is quite low, below the value obtained at the FSP. The wood samples were exposed to air with a given RH, and then dried in air with a lower RH.

Data of interest for calculation, namely the diffusivity, the moisture content and the rate of evaporation were determined, and the validity of the model established for drying of the samples.

Determination of the Parameters

Moisture Content at Equilibrium
The moisture content at equilibrium (EMC) was determined for various RH of the atmosphere at constant temperature (30 °C). Resulting from these measurements, the isotherm of absorption was obtained by plotting the EMC versus the RH (Fig. 13.5).

Within the 55%–80% range considered for the RH of air, the EMC can be expressed by a linear functon:

$$EMC = 0.088\,RH + 6.15 \qquad (13.18)$$

Diffusivity
The diffusivity was determined using the short test technique performed under transient conditions [14, 21]. Successive absorption of moisture was conducted with the same sample, by increasing the RH of the surrounding atmosphere in a discontinuous way, each step corresponding with an RH increase of 5%. Between each step, the time of contact was chosen so that the absorption equilibrium could

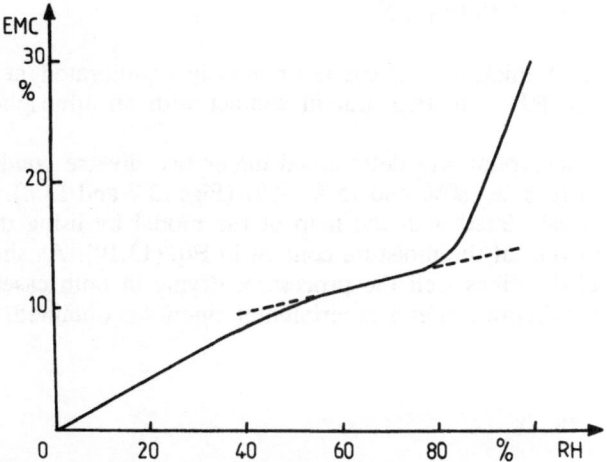

Fig. 13.5. Equilibrium moisture content (EMC) versus relative humidity at 30 °C.

be attained. The successive absorption of moisture, step by step versus time is shown for moisture transport along the longitudinal axis at 30 °C [14].

The transverse diffusivity was determined using the same method [15], and the variation of this transverse diffusivity with the moisture content *MC* is illustrated in Fig. 13.6. From the straight line obtained in this figure, the diffusivity is expressed in terms of the moisture content by the relationship:

$$D_T = 1.5 \times 10^{-6} \exp(0.023 \, MC) \, cm^2 \, s^{-1} \qquad (13.19)$$

Rate of Evaporation

The rate of evaporation was determined by evaporating pure water from a flask of constant area. In these studies [14, 15] the rate of evaporation of the moisture from the wood surface was assumed to be very high with regard to the rate of diffusion. Equilibrium on the wood surface was thus reached as soon as the process starts.

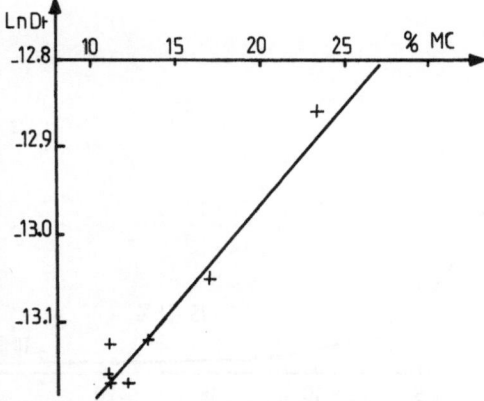

Fig. 13.6. Logarithm of transverse diffusivity versus moisture content, below the FSP. (——, calculation; +, experiment.)

Validity of the Model Below FSP

A piece of wood of thickness 0.5 cm is previously equilibrated at 30 °C in an atmosphere of high RH, and then put in contact with an atmosphere of lower RH.

The kinetics of desorption are determined under two diverse conditons for the RH, within the range 87%–80% and 75%–55% (Figs 13.7 and 13.8). The kinetics of drying are also calculated with the help of the model by using the diffusivity expressed as a function of the moisture content in Eq. (13.19). As shown in these figures, the model describes well the process of drying in both cases, as a good correlation between theoretical and experimental kinetics is obtained.

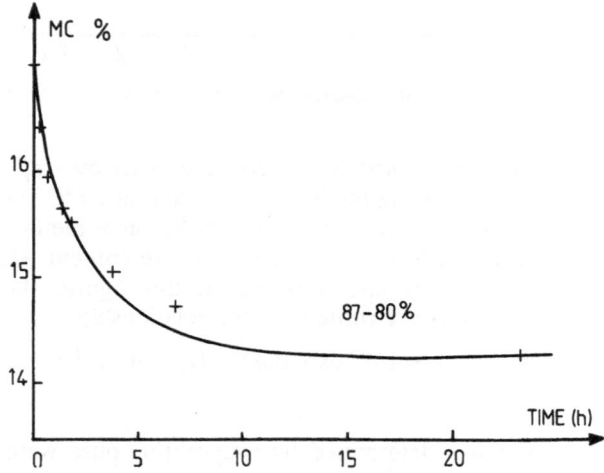

Fig. 13.7. Kinetics of drying at 30 °C of a sample 0.5 cm thick along the transverse direction, 87%–80% RH. (——, calculation; +, experiment.)

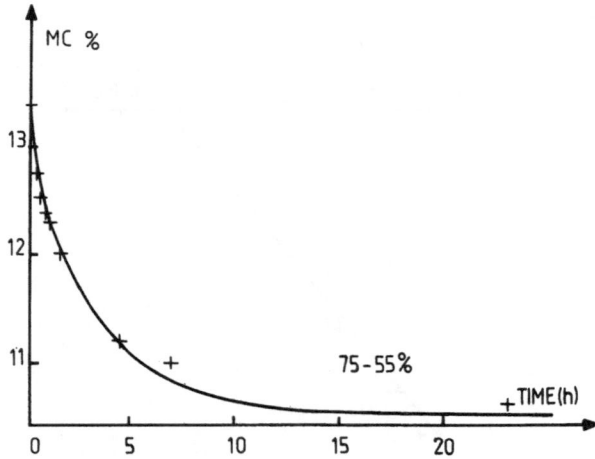

Fig. 13.8. Kinetics of drying at 30 °C of a sample 0.5 cm thick along the transverse direction, 75%–55% RH. (——, calculation; +, experiment.)

13.2.4 Results for Moisture Content Above FSP

The wood samples were prepared in such a way that the moisture could be transferred only along one principal direction. Two principal directions were considered: tangential and longitudinal.

- For the tangential diffusion, the $2 \times 2 \times 0.5$ cm samples were used, 0.5 cm being the thickness through which the moisture was transported.
- For the longitudinal diffusion, the $2 \times 2 \times 1$ cm samples were used, 1 cm being the thickness through which the moisture was transported.

In each case, the sample was immersed in water at 20 °C, and the kinetics of absorption followed by weighing the sample at intervals.

The diffusivity, the moisture content and the rate of evaporation were determined by experiments. The validity of the model was then tested by calculating the kinetics using the model and these parameters.

Determination of the Parameters

Tangential Diffusivity

The kinetics of matter absorption through the tangential direction are rather complex when conducted for times as long as 500 h [16]. However, the kinetics for times shorter than 5 h are very simple to describe, with a quasi equilibrium attained at around 4 h (Fig. 13.9). This time of 4 h has been selected for two reasons: the curve of absorption is simple; this is the average time for the condensation of water vapour during a night.

The diffusivity was calculated from the slope of the straight line obtained by plotting the amount of water absorbed as a function of the square root of time using Eq. (13.20).

$$\frac{Q_t}{Q_\infty} = \frac{4}{L} \left(\frac{Dt}{\pi} \right)^{0.5} \tag{13.20}$$

The diffusivity for the drying process was assumed to be the same as that obtained during the absorption stage.

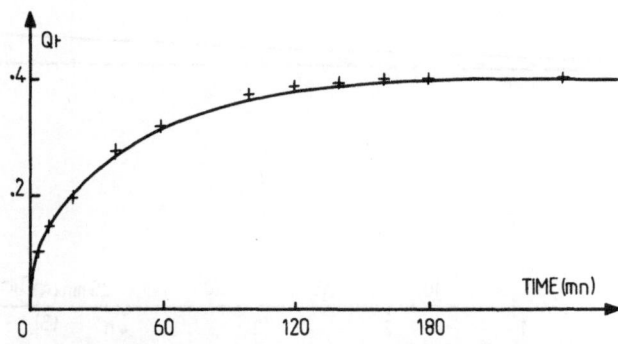

Fig. 13.9. Kinetics of absorption of water by wood along the tangential direction, below and above the FSP, at 20 °C. (——, calculation; +, experiment.)

Longitudinal Diffusivity
The kinetics of absorption of water through the longitudinal direction are rather complex when the time of absorption is long, but become very simple when the time of absorption is short, 30 min for instance [17]. As shown in Fig. 13.10, simple diffusional kinetics are obtained with a short time of absorption.

Rate of Evaporation
The rate of evaporation of pure water at 20 °C was determined by weighing at intervals the amount of water contained in a flask of constant area.

Parameter Values
The values of the parameters were found to be

• Tangential diffusivity $D_T = 10^{-5}$ cm^2 s^{-1}, $MC_{in} = 41\%$
• Longitudinal diffusivity $D_L = 2.2 \times 10^{-3}$ cm^2 s^{-1}, $MC_{in} = 76.8\%$
• Rate of evaporation $F_0 = 2.8 \times 10^{-5}$ cm s^{-1}

The value of the diffusivity is constant either for the tangential or the longitudinal direction. Moreover, the longitudinal diffusivity is considerably greater than the tangential diffusivity.

Validity of the Model Above FSP

The validity of the model based on the assumption of transient diffusion, as well as the accuracy of the parameters, is tested by comparing the experimental kinetics of drying with the corresponding kinetics calculated with the help of the model or with the analytical solutions.

Tangential Diffusion [16]
The validity of the model of the drying process was tested for tangential diffusion, in the following three cases [16]:

• Where a pseudo-equilibrium of absorption was previously attained (Fig. 13.11).

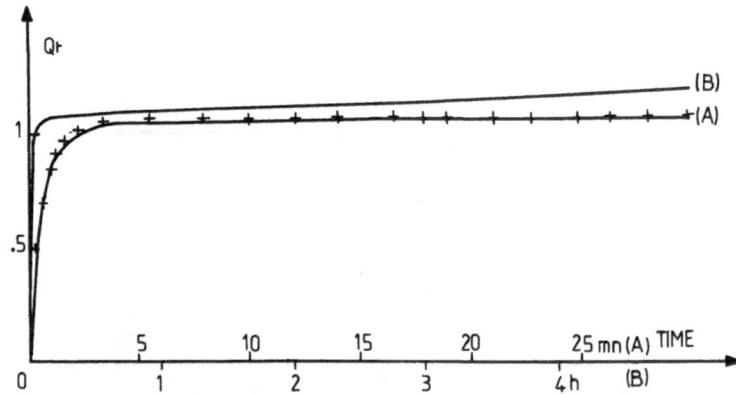

Fig. 13.10. Kinetics of absorption of water by wood along the longitudinal direction, below and above the FSP, at 20 °C. (——, calculation; +, experiment.)

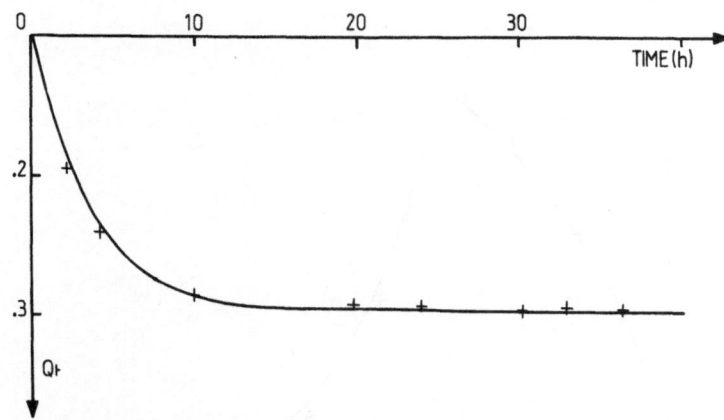

Fig. 13.11 Validity of the model in one dimension. Kinetics of drying along the tangential direction at 20 °C; pseudo-equilibrium of absorption was previously attained. (+, experiments; ——, model.)

- When the pseudo-equilibrium of absorption was not reached, either with a time of absorption of 20 min (Fig. 13.12), or 50 min (Fig. 13.13).

As shown in these three cases, a good correlation is obtained between the experimental kinetics of drying and the corresponding kinetics obtained by calculation using either the analytical solution (Eq. (13.7)) or the numerical model.

The profiles of moisture content developed through the thickness (0.5 cm along the tangential direction) can be calculated by using the numerical model, whatever the initial profiles of moisture content, or by using the analytical solution (Eq. (13.4)) when the profile of moisture is initially uniform.

Three types of profiles of moisture content were calculated with the numerical model during the stage of drying, when the time of absorption was 4 h (Fig. 13.14), 50 min (Fig. 13.15) or 20 min (Fig. 13.16).

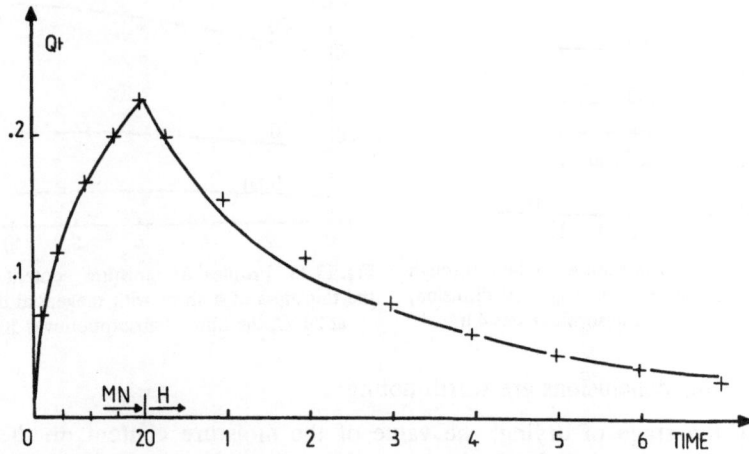

Fig. 13.12 Validity of the model in one dimension. Kinetics of drying along the tangential direction at 20 °C; the time of absorption was 20 min. (+, experiments; ——, model.)

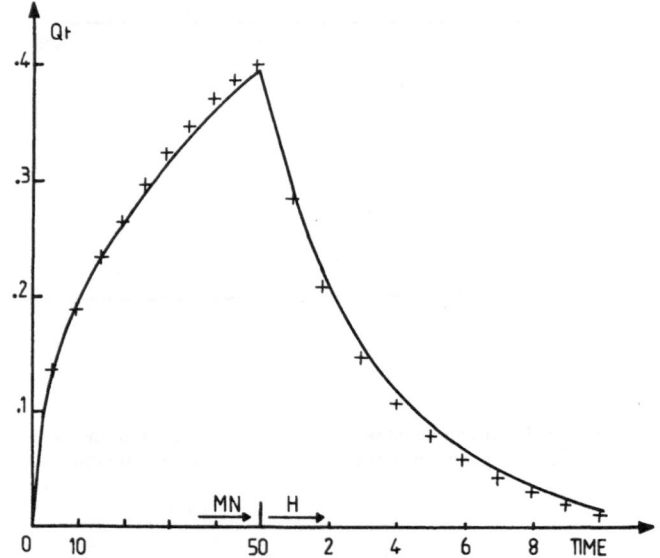

Fig. 13.13. Validity of the model in one dimension. Kinetics of drying along the tangential direction at 20 °C; the time of absorption was 50 min. (+, experiments; —, model.)

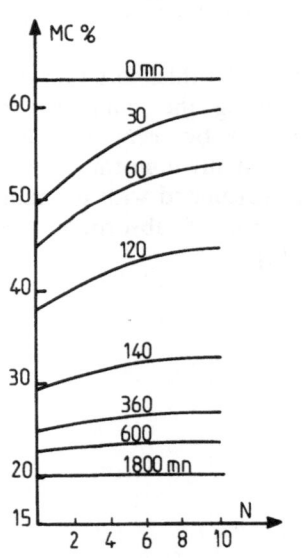

Fig. 13.14. Profiles of moisture content through the thickness of a sheet with tangential diffusion, at 20 °C, the time of absorption was 4 h.

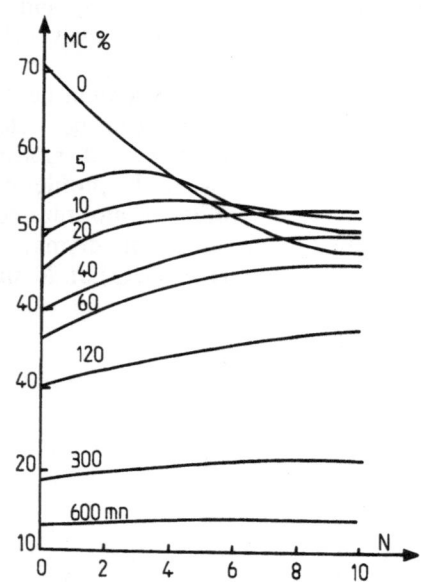

Fig. 13.15. Profiles of moisture content through the thickness of a sheet with tangential diffusion, at 20 °C, the time of absorption was 50 min.

The following conclusions are worth noting:

1. During the stage of drying, the value of the moisture content on the wood surface decreases very slowly, because of the rather low rate of evaporation.
2. Because of the steep profiles of moisture content with a high value on the

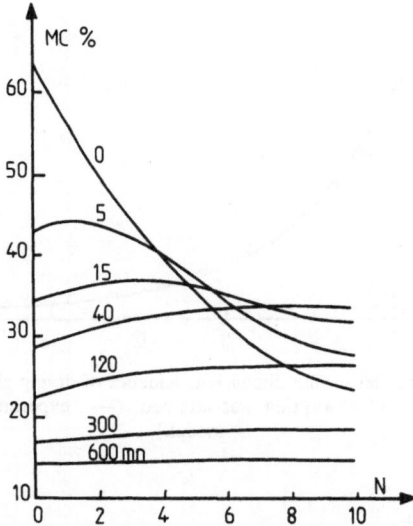

Fig. 13.16. Profiles of moisture content through the thickness of a sheet with tangential diffusion, at 20 °C, the time of absorption was 20 min.

surfaces, the moisture not only evaporates form the surface but also diffuses within the wood toward lower values of the moisture content. As a result of these transfers, the moisture content gradients become less and less steep, and after approximately 15 min (Fig. 13.15) and 5 min (Fig. 13.16), an inversion is observed, with the highest value at the centre of the sample.

3. No experiment was performed in order to determine these profiles of moisture content through the thickness of the sample. Such procedures are not only tedious but also destructive to the sample, as shown in previous studies [22, 23]. However, proof of the validity of these profiles exists, since the kinetics of desorption are obtained by integrating these profiles with respect to space.

4. Of course, a flat gradient of moisture content is obtained when the absorption has reached equilibrium (Fig. 13.14).

Longitudinal Diffusion [17]
The model was tested for two cases:

- Where pseudo-equilibrium of absorption was attained (Fig. 13.17) after 30 min of longitudinal diffusion, and desorption performed at 20 °C.
- Where equilibrium of absorption was far from being attained, with a time of absorption of 30 s (Fig. 13.18) or 60 s (Fig. 13.19), and desorption was performed at 30 °C.

Good correlation was obtained in these cases between the experimental and theoretical kinetics.

The profiles of moisture content developed through the thickness of the sample were calculated using the numerical model for all cases and the analytical solution where the initial moisture content was uniform. Profiles are shown for the latter case in (Fig. 13.20), and where the initial moisture content was not uniform, after a time of absorption of 30 s (Fig. 13.21) and 60 s (Fig. 13.22).

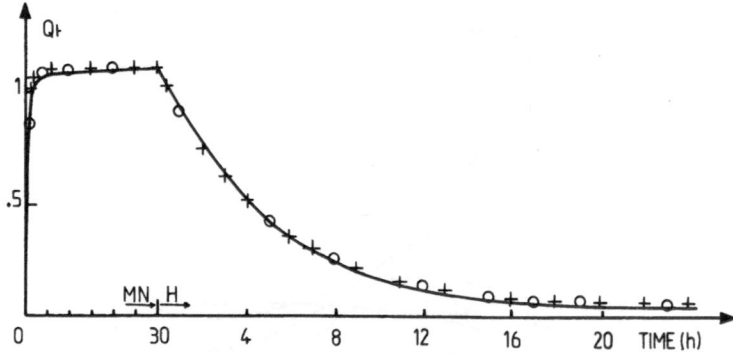

Fig. 13.17. Validity of the model in one dimension. Kinetics of drying along the longitudinal direction, at 20 °C; pseudo-equilibrium of absorption was attained. (——, experiment; + analytical solution; O, model.)

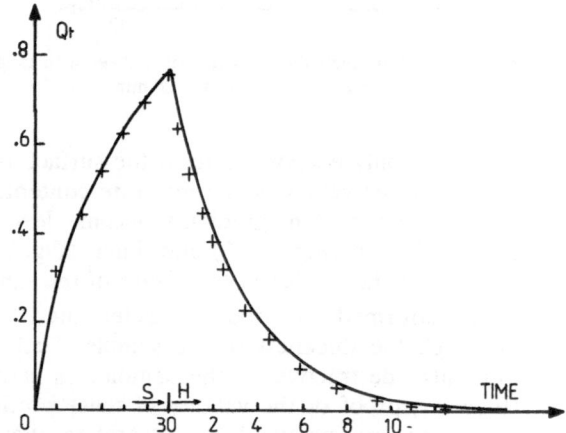

Fig. 13.18. Validity of the model in one dimension. Kinetics of drying along the longitudinal direction, at 30 °C, after a time of absorption in water of 30 s (——, experiment; + numerical model; $F_0 = 4.8 \times 10^{-5}$ cm s^{-1}.)

The following results are observed:

1. After 30 min of absorption in water with longitudinal diffusion, a pseudo-equilibrium is attained, and a flat gradient of moisture content is thus observed (Fig. 13.20) through the sample.

2. Flat gradients of moisture content are developed during the step of drying in this case (Fig. 13.20), because of the great longitudinal diffusivity and the rather low rate of evaporation. The liquid is brought to the surface by diffusion as soon as this liquid is evaporated.

3. For times of absorption lower than 30 min, as shown in Fig. 13.21 and Fig. 13.22, the moisture content is not uniform at the end of the absorption stage. Thus a steep gradient of moisture content exists at the beginning of the drying stage.

4. During the first few seconds of the desorption process (Figs 13.21 and 13.22), the liquid evaporates from the surface, and it diffuses also to the centre of the

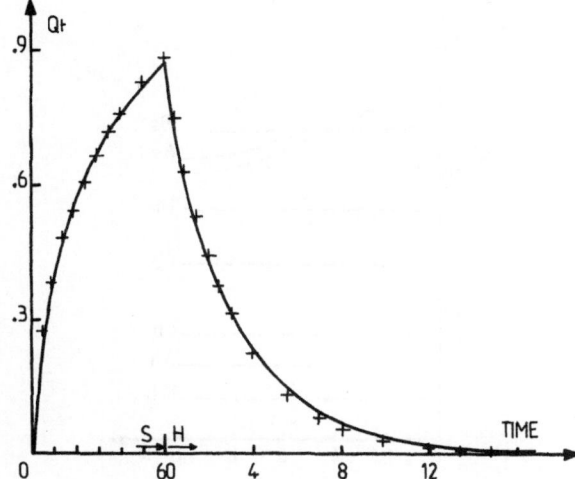

Fig. 13.19. Validity of the model in one dimension. Kinetics of drying along the longitudinal direction, at 30 °C, the time of absorption was 60 s (——, experiment; +, numerical model; $F_0 = 4.8 \times 10^{-5}\ cm\,s^{-1}$.)

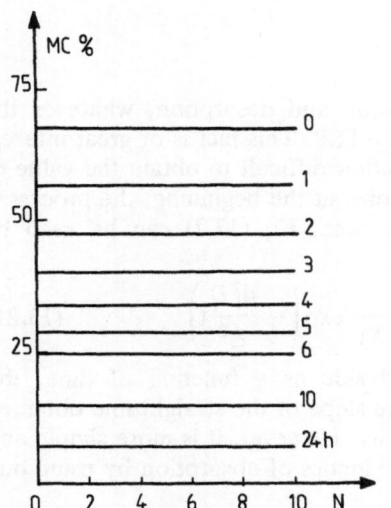

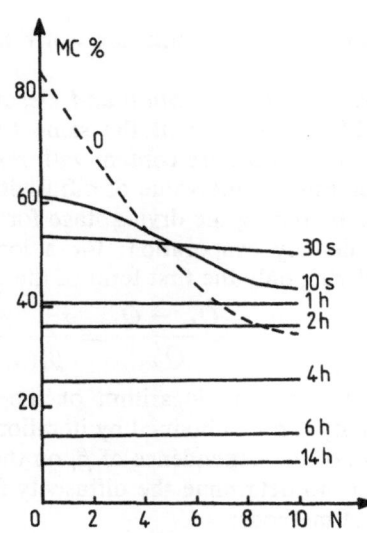

Fig. 13.20. Profiles of moisture content through the thickness of a sheet for longitudinal diffusion at 20 °C; the time of absorption was 30 min.

Fig. 13.21. Profiles of moisture content through the thickness of a sheet for longitudinal diffusion at 30 °C; the time of absorption was 30 min.

sheet, because of the high gradient of moisture with a high moisture content on the surface.

5. Following evaporation from the surface and diffusion within the sample, the profiles of moisture content become less and less steep. After around half an hour a flat gradient is obtained.

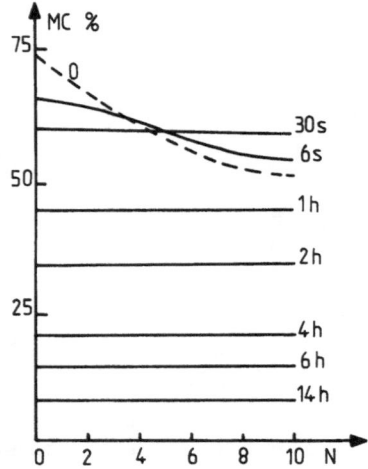

Fig. 13.22. Profiles of moisture content through the thickness of a sheet for longitudinal diffusion at 30 °C; the time of absorption was 60 s.

13.2.5 Conclusions

The following conclusions are worth noting.

Diffusivity for Absorption and Desorption
The diffusivity is about the same for absorption and desorption, whatever the value of the moisture content with regard to the FSP. This fact is of great interest for determining the value of diffusivity. It is rather difficult to obtain the value of diffusivity during the drying stage for two reasons: at the beginning, the process is controlled by evaporation; for a long drying time, Eq. (13.7) can be used by considering only the first term of the series:

$$\frac{Q_\infty - Q_t}{Q_\infty} = \frac{2S^2}{\beta_1^2(\beta_1^2 + S^2 + S)} \exp\left(-\frac{\beta_1^2 D}{L^2}t\right) \qquad (13.21)$$

By plotting the logarithm of the left-hand side as a function of time, the diffusivity can be obtained by iteration from the slope of the straight line obtained in spite of the dependence of β_1 on the diffusivity. However, it is more simple and accurate to determine the diffusivity from the kinetics of absorption by using one of three methods:

- Short experiments and Eq. (13.20);
- The half-life of absorption;
- A long time of absorption.

Process of Drying
The process of drying is controlled by diffusion of moisture within the solid and evaporation from the surface. The values of diffusivity vary largely depending on the principal direction, the longitudinal diffusivity being considerably greater than the other two principal diffusivities. It is thus very useful to make measurements with thin sheets of wood and one-dimensional transfer in order to determine the value of each principal diffusivity.

13.3 Diffusion–Evaporation in Two Dimensions

The problem considered is that of a two-dimensional transfer of moisture obtained using parallelepiped boards, where

1. One direction of the board is very long with regard to the other two, and the objective is to determine the profiles of moisture content developed in the cross-section perpendicular to this very long direction (Fig. 13.23), as well as the kinetics of transfer. Of course, these calculations can be made at the places where the moisture transport through this long direction is negligible.
2. The moisture transfer is only in two directions, the third being stopped by some means.

The moisture content can be below or above the fibre saturation point.

The process of drying is controlled by diffusion in two dimensions within the solid and by evaporation form the surfaces.

13.3.1 Theory

The two-dimensional transfer of moisture is studied by considering diffusion within the solid with two principal diffusivities and the rate of evaporation.

Assumptions

The following assumptions are made to simplify the problem:

1. Two-dimensional transfer is considered through the cross-section of the sample (rectangular or square).
2. Moisture transfer is controlled by transient diffusion through the wood, with a principal direction of diffusion on each side.

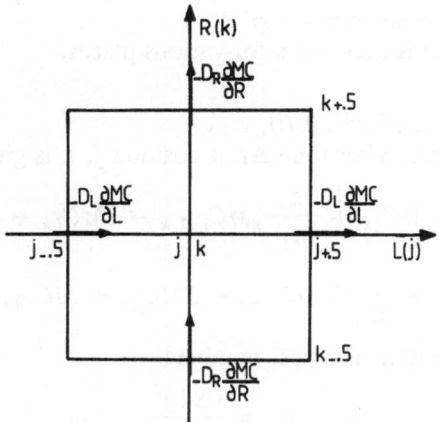

Fig. 13.23. Scheme for two-dimensional transfer within the sample.

3. Each principal diffusivity is constant [16, 17] or concentration-dependent [18] as found from experiments.
4. During the desorption stage, the rate of evaporation is constantly equal to the rate at which the moisture is brought to the evaporating surface by diffusion.
5. The rate of evaporation is also proportional to the difference between the moisture content on the surface and the moisture content necessary to maintain equilibrium with the surrounding atmosphere.
6. Changes in the dimensions of the sample are neglected.
7. The decrease in temperature resulting from the enthalpy of evaporation from the surface is negligible.

Mathematical Treatment

The transport of moisture is governed by Fick's law in two dimensions with two principal directions of diffusion (longitudinal and radial):

$$\frac{\partial MC}{\partial t} = D_L \frac{\partial^2 MC}{\partial L^2} + D_R \frac{\partial^2 MC}{\partial R^2} \qquad (13.22)$$

D_L and D_R are the longitudinal and radial diffusivities, which are constant, and L and R are the longitudinal and radial directions.

The rate of evaporation from the surface is given by

$$- D_X \frac{\partial MC}{\partial X} = F_0(MC_s - MC_{ext}) \qquad (13.23)$$

where X represents each of the principal directions, MC_s is the moisture content at the surface and MC_{ext} the moisture content necessary to maintain equilibrium with the surrounding atmosphere.

Numerical Analysis

The problem is solved using a mathematical model. Only the results are given here, as calculations are made in Chap. 8.

The moisture content is thus given for various places.

Within the Sample (Fig. 13.23), L(j), R(k)
The new moisture content after time Δt at position j, k is given by

$$MCN_{j,k} = MC_{j,k} + \frac{1}{M_L} [MC_{j-1,k} - 2MC_{j,k} + MC_{j+1,k}]$$

$$+ \frac{1}{M_R} [MC_{j,k-1} - 2MC_{j,k} + MC_{j,k+1}] \qquad (13.24)$$

where M_L and M_R are dimensionless numbers

$$M_L = \frac{(\Delta L)^2}{D_L \Delta t}$$

$$M_R = \frac{(\Delta R)^2}{D_R \Delta t} \qquad (13.25)$$

and D_L and D_R are the constant principal diffusivities.

On the Longitudinal Surface (Fig. 13.24)
By assuming that

$$MCN_{N_l-0.5,k} - MC_{N_l-0.5,k} = MCN_{N_l,k} - MC_{N_l,k} \qquad (13.26)$$

the new concentration on the face N_l after time Δt is expressed by

$$MCN_{N_l,k} = MC_{N_l,k} + \frac{2}{M_L}(MC_{N_l-1,k} - MC_{N_l,k})$$

$$- \frac{2F_0 \Delta t}{\Delta L}(MC_{N_l,k} - MC_{ext})$$

$$+ \frac{1}{M_R}[MC_{N_l-0.25,k-1} - 2MC_{N_l-0.25,k} + MC_{N_l-0.25,k+1}] \qquad (13.27)$$

On the Radial Surface
The new concentration on the radial surface after time Δt is given in the same way by

$$MC_{j,N_r} = MC_{j,N_r} + \frac{2}{M_R}(MC_{j,N_r-1} - MC_{j,N_r}) - \frac{2F_0 \Delta t}{\Delta R}(MC_{j,N_r} - MC_{ext})$$

$$+ \frac{1}{M_L}[MC_{j-1,N_r-0.25} - 2MC_{j,N_r-0.25} + MC_{j+1,N_r-0.25}] \qquad (13.28)$$

At the Corner (Fig. 13.25)
By making the simple assumption

$$MCN_{N_l-0.25,N_r-0.25} - MC_{N_l-0.25,N_r-0.25} = MCN_{N_l,N_r} - MC_{N_l,N_r} \qquad (13.29)$$

the new concentration at the corner after time Δt

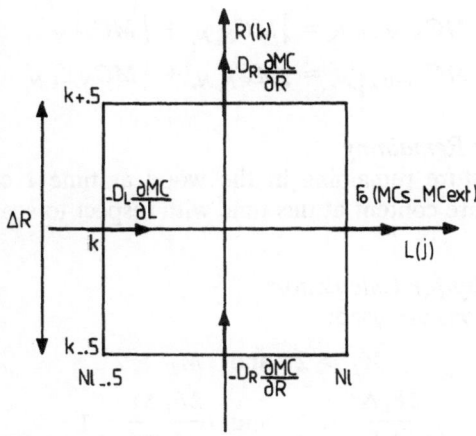

Fig. 13.24. Scheme for two-dimensional transfer on the surface, with diffusion and evaporation.

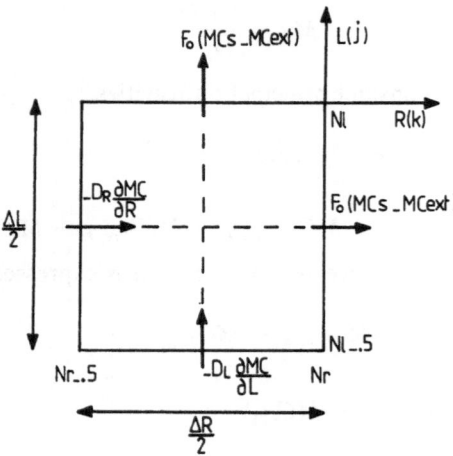

Fig. 13.25. Scheme for two-dimensional transfer at the corner, with diffusion and evaporation.

$$MCN_{N_b,N_r} = MC_{N_b,N_r} - \frac{2}{M_L} \left(MC_{N_b,N_r-0.25} - MC_{N_l-1,N_r-0.25} \right)$$

$$- \frac{2F_0\Delta t}{\Delta L} \left(MC_{N_b,N_r-0.25} - MC_{\text{ext}} \right)$$

$$- \frac{2F_0\Delta t}{\Delta R} \left(MC_{N_l-0.25,N_r} \right) - MC_{\text{ext}} \right)$$

$$- \frac{2}{M_R} \left(MC_{N_l-0.25,N_r} - MC_{N_l-0.25,N_r-1} \right) \qquad (13.30)$$

Other Assumptions for Calculation
Some other assumptions must be made for calculation, as only the places corresponding to an integer are considered:

$$MC_{N_b,N_r-0.25} = \tfrac{3}{4} MC_{N_b,N_r} + \tfrac{1}{4} MC_{N_b,N_r-1} \qquad (13.31)$$

$$MC_{N_l-0.25,N_r} = \tfrac{3}{4} MC_{N_b,N_r} + \tfrac{1}{4} MC_{N_l-1,N_r} \qquad (13.31')$$

Amount of Moisture Remaining
The amount of moisture remaining in the wood at time t can be obtained by integrating the moisture content at this time with respect to space.

Conditions of Stability for Calculation
The following conditions are used:

$$M_L > 2 \quad \text{and} \quad M_R > 2 \qquad (13.32)$$

$$\frac{2F_0\Delta t}{\Delta L} < 1 \quad \text{and} \quad \frac{2F_0\Delta t}{\Delta R} < 1 \qquad (13.33)$$

13.3.2 Experiment

Materials

Samples were cut from Scots pine sapwood (*Pinus sylvestris*) and prepared as follows:

1. Cubes of dimensions $2 \times 2 \times 2$ cm were used for determining diffusivity. The longitudinal and radial diffusivities were measured separately by covering four surfaces with a silicone film to isolate them from the surrounding atmosphere (water in the case of absorption, air during the desorption). Other samples were used for measuring the transverse diffusivity.

2. A sheet of dimensions $10 \times 10 \times 1$ cm was used to study the two-dimensional transport of moisture, one direction being along the longitudinal axis, the other along the radial axis. Two faces of dimensions 10×1 cm were covered by a silicone film in order to prevent tangential transport. The sheet was cut in such a way that radial transport could take place through the surfaces of dimensions 10×10 cm and longitudinal transport through two 10×1 cm surfaces (Fig. 13.26).

3. Experiments were also made on a block with dimensions $2 \times 2 \times 6$ cm. The block was cut so that the length was along the longitudinal direction. The other two directions were the transverse directions. Longitudinal transport was avoided with the help of silicone film (Fig. 13.27).

Methods

Great care was taken in preventing diffusion along the third principal direction in order to maintain two-dimensional transport.

Absorption
After equilibration in air at 20 °C and 50% RH, each sample was immersed in water at 20 °C. The kinetics of absorption were followed by weighing each sample at intervals. The values of the diffusivities were tested by comparing the experimental and theoretical kinetics.

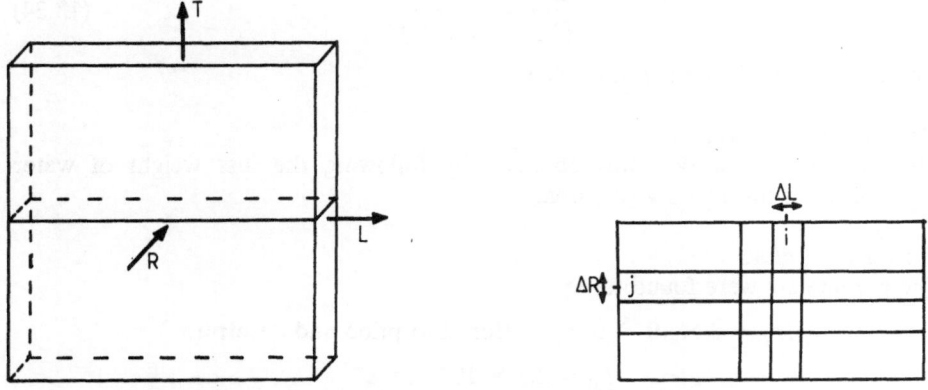

Fig. 13.26. Sheet of dimensions $10 \times 10 \times 1$ cm, with longitudinal–radial transfer.

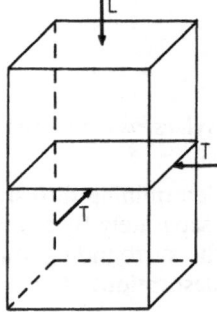

Fig. 13.27. Block of dimensions $2 \times 2 \times 6$ cm, with transverse–transverse transfer.

Desorption
After removal from the water, each sample was exposed to surrounding atmosphere at 20 °C and 50% RH. The kinetics of drying were followed by weighing each sample at intervals. The reference weight of each sample was determined after drying each sample at 103 °C to constant weight.

13.3.3 Results for Transverse–Transverse Drying

The values of the transverse diffusivity and the rate of evaporation were determined first. Evidence for the validity of the model is thus provided by comparing the theoretical and experimental kinetics of drying.

Determination of the Parameters

Transverse Diffusivity
The diffusivity was calculated from experiments performed with a sheet through which water diffused along the transverse direction, by protecting the other surfaces from the contact with the surrounding atmosphere. From the straight lines obtained by plotting the amount of water absorbed as a function of the square root of time, the diffusivity was obtained using

$$\frac{Q_t}{Q_\infty} = \frac{4}{L} \left(\frac{Dt}{\pi} \right)^{0.5} \qquad (13.34)$$

where L is the thickness of the sheet.

Rate of Evaporation
The rate of evaporation was obtained by following the loss weight of water exposed to air through a given area.

Parameter Values
The parameters were found to be

$$D_T = 3 \times 10^{-5} \text{ cm}^2 \text{ s}^{-1} \text{ (for absorption and desorption)}$$

$$F_0 = 2.5 \times 10^{-5} \text{ cm s}^{-1}$$

Some conclusions can be drawn from these values.

1. The diffusivity has the same value for absorption and desorption.
2. The moisture content is above the FSP during a large part of the process. At the beginning of desorption, the moisture content is around 40%.
3. The diffusivity is constant. This is quite different from the results obtained below the FSP where the diffusivity is concentration-dependent [14, 15, 18, 19].

Validity of the Model

The validity of the model was evaluated by comparing the kinetics of drying obtained by experiments and calculation, using a sheet previously immersed in water for 8 h. A rather good superimposition is seen between the experimental and theoretical kinetics, giving evidence for the validity of the model (Fig. 13.28).
 The following conclusions are drawn:

1. The numerical model describes the whole process of drying when the moisture content is above and beyond the FSP.
2. The constant transverse diffusivity is proved to give theoretical kinetics corresponding well to experiments.

Profiles of Moisture Content

It is of interest to represent the profiles of moisture content developed through the cross-section of the sample at various times. As already said, it is difficult to measure these values, as experiments are destructive [22, 23]. But circumstantial evidence for these profiles exists, as the kinetics of transfer are determined by integrating the profiles of moisture content with respect to space.

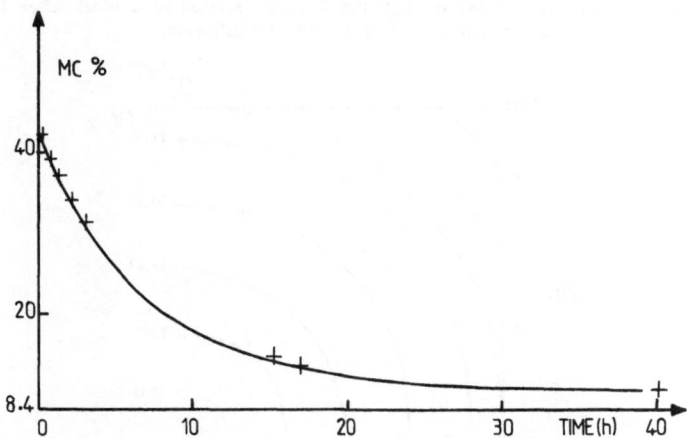

Fig. 13.28. Kinetics of drying in two dimensions above and below the FSP, for a beam of dimensions $2 \times 2 \times 6$ cm with transverse diffusion, at 20 °C. (——, calculation; +, experiment.)

Two profiles of moisture content are shown:

- After 1 h 30 min of desorption (Fig. 13.29)
- After 10 h of desorption (Fig. 13.30)

Some conclusions are worth noting:

1. These profiles of moisture content are able to illustrate the process very easily, as shown in a previous paper when the moisture content was below the FSP [19].
2. Of course, these profiles are symmetrical for two reasons: the cross-section is square, and the diffusivity is the same for the two perpendicular directions considered.
3. The gradient of moisture content can be estimated by the distance between the lines of iso-moisture content.
4. During the stage of drying, the process being controlled by diffusion and evaporation, the value of the moisture content is not the same all over the surface at any time. The lowest value is observed at the corner, where the effect of evaporation is greatest.

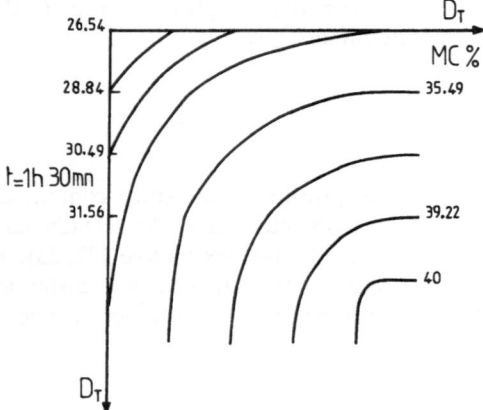

Fig. 13.29. Profiles of moisture content through the 2 × 2 cm section of a block after 1 h 30 min of desorption at 20 °C, for transverse diffusion.

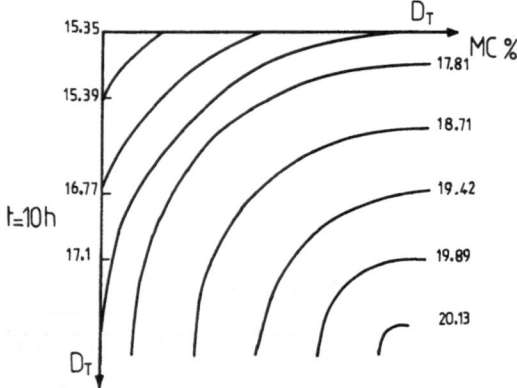

Fig. 13.30. Profiles of moisture content through the 2 × 2 cm section of a block after 10 h of desorption at 20 °C, for transverse diffusion.

13.3.4 Results for Longitudinal–Radial Drying

The following is considered:

1. Determination of the parameters necessary for simulating the process.
2. Testing the validity of the model in the case of a plane sheet with longitudinal and radial diffusivities.
3. Affording a further insight into the nature of the process of drying, by calculating the profiles of moisture content developed within the cross-section of the sample.

Determination of the Parameters

Longitudinal and Radial Diffusivities
The diffusivities were determined from the experimental curves of absorption by using wood sheets cut in such a way that water was transported through their thickness either by longitudinal or by radial diffusion. From the straight lines obtained by plotting the amount of water absorbed as a function of the square root of time, the diffusivities were calculated when the EMC was known, using the following equation:

$$\frac{Q_t}{Q_\infty} = \frac{4}{L} \left(\frac{Dt}{\pi}\right)^{0.5} \tag{13.34}$$

where L is the thickness of the sheet.

Of course, these small samples must be representative of the larger sheet. In fact, all these samples were cut from the same tree at about the same places, and they followed the same history.

Constant diffusivities were found for the longitudinal and the radial transport. This result, obtained when the moisture content was above the FSP, is different from the values determined when the moisture content was below the FSP [14, 15, 18, 19].

Moisture Content at Equilibrium
The moisture content at equilibrium is of interest in the following two extreme cases: firstly when the sample is saturated with water at the end of the process of absorption, and secondly when the sample is dried.

Rate of Evaporation
The rate of evaporation was determined either with pure water under steady conditions or from the former stage of evaporation from the wood sample [24–28].

Parameter Values
The parameters were found to be

- Longitudinal diffusivity $D_L = 2.2 \times 10^{-3} \text{ cm}^2 \text{ s}^{-1}$
- Radial diffusivity $D_R = 3 \times 10^{-5} \text{ cm}^2 \text{ s}^{-1}$
- Rate of evaporation $F_0 = 2.5 \times 10^{-5} \text{ cm s}^{-1}$

The diffusivities are the same for absorption and desorption.

Validity of the Model

The validity of the model was tested by comparing the kinetics of desorption obtained by experiments and calculation. After immersion of the wood sample in water at 20 °C for 8 h, the sheet was exposed to air at 20 °C. As shown in Fig. 13.31, a good correlation can be seen between the theoretical and experimental kinetics of drying.

The following conclusions can be drawn [28]:

1. The model describes the stage of desorption during the whole process, when the moisture content is above and below the FSP.
2. Constant values are found for the longitudinal and the radial diffusivity, in contrast with the results obtained when the moisture content is below the FSP [19].
3. The diffusivities for each principal direction are the same during the stages of absorption and desorption.
4. The longitudinal diffusivity is greater than the radial diffusivity.

Profiles of Moisture Content

The profiles of moisture content can be calculated using the numerical model for any time. Two of these profiles are shown during desorption, after 1 h 30 min (Fig. 13.32) and after 10 h (Fig. 13.33).

Some conclusions are worth noting:

1. These profiles of moisture content are able to give a further insight to the process.
2. They can give a good idea of the contribution of each transport (longitudinal and radial) at any place and any time. For instance, in places where these lines are parallel to each other and also parallel to one surface of the sample, one of these two transfers is dominant. This fact can be observed especially for the longitudinal diffusivity.
3. The gradient of moisture content can be evaluated easily.

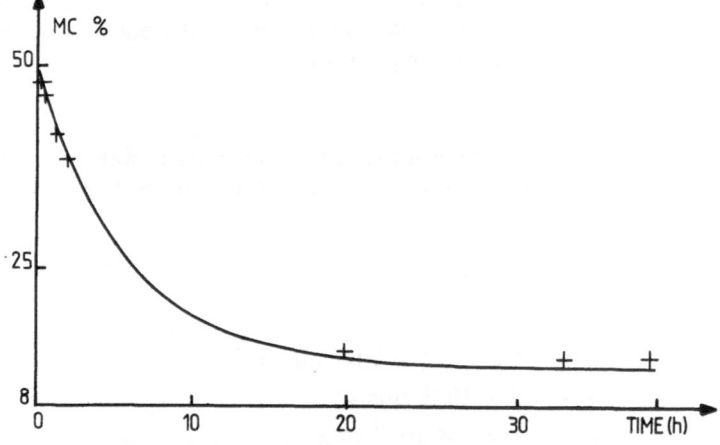

Fig. 13.31 Kinetics of drying in two dimensions, for a sheet of dimensions 10 × 10 × 1 cm at 20 °C, with longitudinal–radial transfer. (——, calculation; +, experiment.)

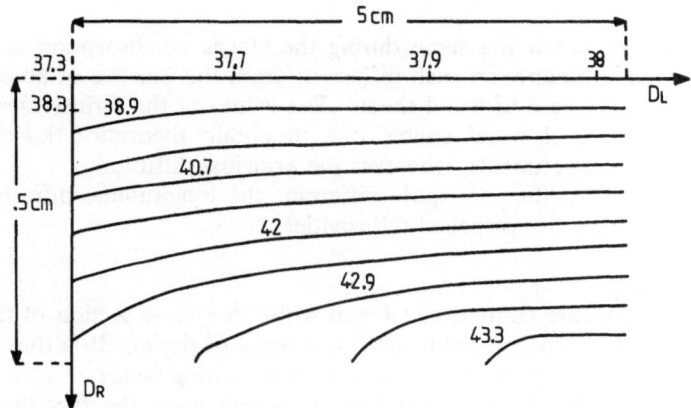

Fig. 13.32. Profiles of moisture content through the 10 × 1 cm section of a sheet after 1 h 30 min of desorption at 20 °C, for longitudinal–radial transfer.

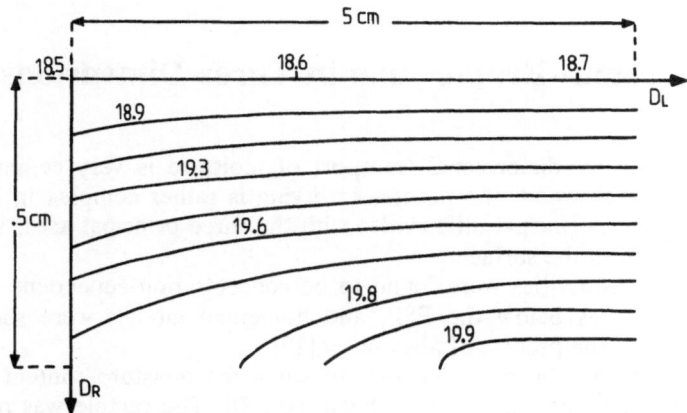

Fig. 13.33. Profiles of moisture content through the 10 × 1 cm section of a sheet after 10 h of desorption at 20 °C, for longitudinal–radial transfer.

4. During the stage of drying, the lines of equal moisture content and the surface perpendicular to the longitudinal direction cut each other, while this fact does not appear so obviously for the other surface.

5. During the process of drying, the value of the moisture content on each surface (longitudinal or radial) varies not only with time but also with position. For instance, the moisture content is lowest next to the corner.

6. The rate of evaporation is the limiting factor for the process of drying on the surface perpendicular to the longitudinal direction. On the other surface perpendicular to the radial direction, the radial diffusion seems to be the limiting factor.

13.3.5 Conclusions

Some general conclusions for the process of drying wood samples in two dimensions can be drawn:

Diffusivity

As diffusivities are about the same during the stages of absorption and desorption, it is simpler to determine each diffusivity from the kinetics of absorption, as already shown in the case of wood sheets. The values of the diffusivities obtained by using thin sheets of wood enable one to obtain theoretical kinetics which correlate well with experiments, whatever the principal diffusion.

The principal diffusivities are quite different, the longitudinal diffusivity being greater than the other two principal diffusivities.

Process of Drying

The profiles of moisture content developed within the cross-section of the sample are able to provide a fuller insight into the process of drying. It is thus very easy to distinguish which parameter plays the role of a limiting factor.

In Figs 13.29 and 13.30, some emphasis is placed upon the fact that the two perpendicular axes of diffusion which are considered are transverse, by calling them T and T.

13.4   Diffusion–Evaporation in Three Dimensions

The problem of three-dimensional transport of moisture is very common for a solid. In the case of wood, the process of drying is rather complex in the sense that there are three principal diffusivities with the three principal axes, as well as the evaporation from the surface.

The principal diffusivities were found to be concentration-dependent when the moisture content was below the FSP, and numerical models were successfully built for describing the process of absorption [19].

Further studies have been carried out for when the moisture content is above the FSP, by considering a sheet and a beam [29, 30]. The sample was previously immersed in water over a period of time and a pseudo-equilibrium attained; the sample was then exposed to air and dried. The main diffusivities determined from experiments were constant over the whole process of drying.

13.4.1   Theory

The three-dimensional transport of moisture takes place along the three principal axes of diffusion.

Assumptions

The following assumptions are made:

1. Transport is controlled by transient diffusion within the solid and evaporation from the surface.
2. Three-dimensional transport takes place through the sample with three principal axes of transport and three principal diffusivities.

3. Each principal diffusivity may be constant [16, 17, 29, 30] or concentration-dependent [18], as found from experiments.
4. The rate of evaporation from the surface is constantly equal to the rate at which the liquid is brought by diffusion to the evaporating surface.
5. The rate of evaporation is proportional to the difference between the moisture content on the surface and the moisture content necessary to maintain equilibrium with the surrounding atmosphere.
6. The dimensions of the sample are constant.
7. The temperature is uniform throughout the solid, and decrease in temperature on the surface resulting from endothermic evaporation is neglected.

Mathematical Treatment

The transport of water throughout the wood is expressed by Fick's law in three dimensions with three principal directions of diffusion:

$$\frac{\partial MC}{\partial t} = D_L \frac{\partial^2 MC}{\partial L^2} + D_R \frac{\partial^2 MC}{\partial R^2} + D_T \frac{\partial^2 MC}{\partial T^2} \tag{13.35}$$

with constant diffusivities, D_L being the longitudinal diffusivity, and D_R and D_T the radial and tangential diffusivities.

The rate of evaporation from the surface is

$$-D_X \frac{\partial MC}{\partial X} = F_0(MC_s - MC_{ext}) \tag{13.36}$$

where X is the principal axis of diffusion considered, MC_s is the moisture content on the surface and MC_{ext} the moisture content necessary to maintain equilibrium with the surrounding atmosphere.

Numerical Analysis

As no analytical solution can be obtained, the problem is solved using a numerical model with finite differences. As calculation for the moisture content is made in Chap. 8, the results only are given here.

The moisture content is given for various places.

Within the Solid, i, j, k (Fig. 13.34)
The new moisture content after time Δt at the position defined by the integers (i, j, k) is expressed in terms of the previous moisture content at the same and adjacent places:

$$MCN_{i,j,k} = MC_{i,j,k} + \frac{1}{M_L}[MC_{i-1,j,k} - 2MC_{i,j,k} + MC_{i+1,j,k}]$$

$$+ \frac{1}{M_R}[MC_{i,j-1,k} - 2MC_{i,j,k} + MC_{i,j+1,k}]$$

$$+ \frac{1}{M_T}[MC_{i,j,k-1} - 2MC_{i,j,k} + MC_{i,j,k+1}] \tag{13.37}$$

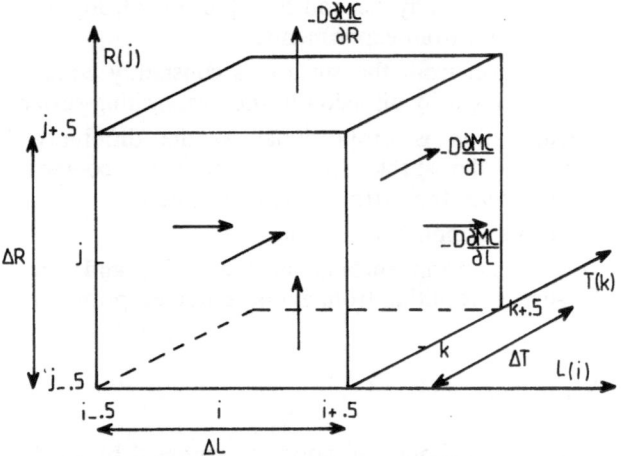

Fig. 13.34. Scheme for three dimensional transfer within the wood.

with the dimensionless numbers

$$M_L = \frac{(\Delta L)^2}{D_L \Delta t}$$

$$M_R = \frac{(\Delta R)^2}{D_R \Delta t}$$

$$M_T = \frac{(\Delta T)^2}{D_T \Delta t} \tag{13.38}$$

Surfaces $i = N_L$, j, k (Fig. 13.35)
The moisture is calculated on the surface $i = N_L$ shown in Fig. 13.35. The new concentration after time Δt at this surface is given by

$$MCN_{N_b j,k} = MC_{N_b j,k} + \frac{2}{M_L}(MC_{N_l-1 j,k} - MC_{N_b j,k}) - \frac{2\Delta t F_0}{\Delta L}(MC_{N_b j,k} - MC_{\text{ext}})$$

$$+ \frac{1}{M_T}[MC_{N_b j,k-1} - 2MC_{N_b j,k} + MC_{N_b j,k+1}]$$

$$+ \frac{1}{M_R}[MC_{N_b j-1,k} - 2MC_{N_b j,k} + MC_{N_b j+1,k}] \tag{13.39}$$

with the assumptions

$$MCN_{N_l-0.25,j,k} - MC_{N_l-0.25,j,k} = MCN_{N_b j,k} - MC_{N_b j,k} \tag{13.40}$$

$$MC_{N_l-0.25,j,k} = MC_{N_b j,k} \tag{13.41}$$

A better assumption is made with

$$MCN_{N_l-0.25,j,k} - MC_{N_l-0.25,j,k} = \tfrac{3}{4}(MCN_{N_b j,k} - MC_{N_b j,k})$$
$$+ \tfrac{1}{4}(MCN_{N_l-1,j,k} - MC_{N_l-1,j,k}) \tag{13.42}$$

The moisture content is calculated on the other surface by permuting the indexes, as shown in Chap. 8.

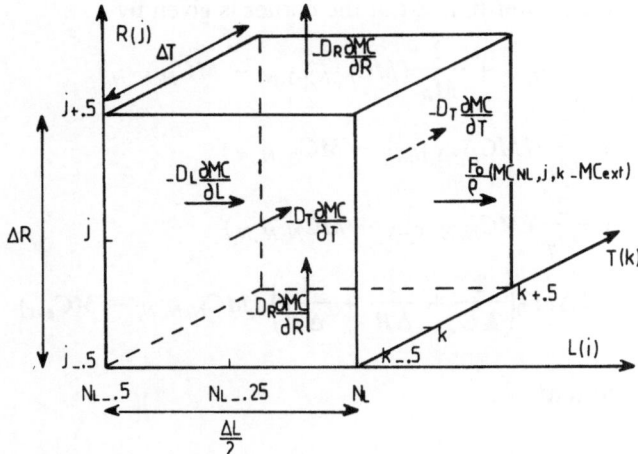

Fig. 13.35. Scheme for three dimensional transfer on the surface with diffusion and evaporation.

Edges $i = N_l$, $j = N_r$, k
With the simple assumption

$$MCN_{N_l-0.25,N_r-0.25,k} - MC_{N_l-0.25,N_r-0.25,k} = MCN_{N_b,N_n,k} - MC_{N_b,N_n,k}$$

$$MC_{N_l-0.25,N_r-0.25,k} = MC_{N_l-0.25,N_n,k} = MC_{N_b,N_n,k} \qquad (13.43)$$

the new moisture content after time Δt is expressed in terms of the previous values by

$$MCN_{N_b,N_n,k} = MC_{N_b,N_n,k} + \frac{2}{M_L}(MC_{N_l-1,N_n,k} - MC_{N_b,N_n,k})$$

$$- 2\Delta t F_0\left(\frac{1}{\Delta L} + \frac{1}{\Delta R}\right)(MC_{N_b,N,k} - MC_{\text{ext}})$$

$$+ \frac{2}{M_R}(MC_{N_b,N_r-1,k} - MC_{N_b,N_n,k})$$

$$+ \frac{1}{M_T}[MC_{N_b,N_n,k-1} - 2MC_{N_b,N_n,k} + MC_{N_b,N_n,k+1}] \qquad (13.44)$$

The moisture content at the edges is obtained by permuting the indexes, as shown in Chap. 8.

Corners, N_l, N_r and N_t
With the following assumptions:

$$MCN_{N_l-0.25,N_r-0.25,N_t-0.25} - MC_{N_l-0.25,N_r-0.25,N_t-0.25} = MCN_{N_b,N_n,N_t} - MC_{N_b,N,N_t}$$
$$(13.45)$$

and

$$MC_{N_l-0.25,N_t N_t-0.25} = MC_{N_l N_r-0.25,N_t-0.25} = MC_{N_l-0.25,N_r-0.25,N_t} = MC_{N_b,N,N_t}$$
$$(13.46)$$

the moisture content after time Δt at the corner is given by

$$MCN_{N_b,N_r,N_t} = MC_{N_b,N_r,N_t} + \frac{2}{M_R}(MC_{N_b,N_r-1,N_t} - MC_{N_b,N_r,N_t})$$

$$+ \frac{2}{M_L}(MC_{N_l-1,N_r,N_t} - MC_{N_b,N_r,N_t})$$

$$+ \frac{2}{M_T}(MC_{N_b,N_r,N_t-1} - MC_{N_b,N_r,N_t})$$

$$- 2\Delta t F_0\left(\frac{1}{\Delta L} + \frac{1}{\Delta R} + \frac{1}{\Delta T}\right)(MC_{N_b,N_r,N_t} - MC_{\text{ext}}) \qquad (13.47)$$

13.4.2 Experiment

Materials

Samples were cut from Scots pine sapwood (*Pinus sylvestris*) as follows:

1. Sheets of dimensions $10 \times 10 \times 1$ cm, cut in such a way that the directions of the sheets were the principal directions of diffusion (Fig. 13.36).
2. Blocks of dimensions $2 \times 2 \times 6$ cm, cut in such a way that the length (6 cm) was along the longitudinal direction and the other two directions were transverse (Fig. 13.37).
3. Cubes of dimensions $2 \times 2 \times 2$ cm, used for determining the diffusivity, as shown for drying in one and two dimensions.

Methods

The samples were immersed in water for a selected period of time, and then

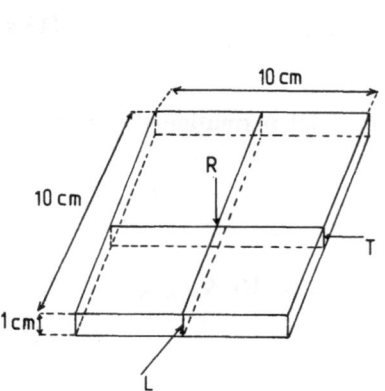

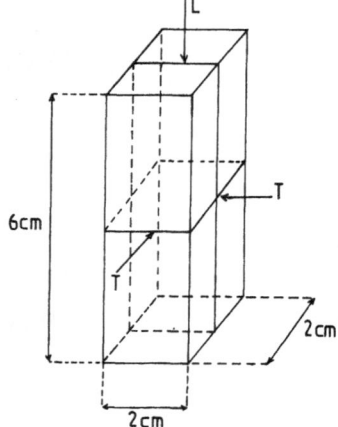

Fig. 13.36. Sheet of dimensions $10 \times 10 \times 1$ cm with three principal directions of diffusion: longitudinal, radial and tangential.

Fig. 13.37. Block of dimensions $2 \times 2 \times 6$ cm with three principal directions of diffusion: longitudinal, transverse and transverse.

exposed to air for drying. As all six faces of the samples were in contact either with water or with the surrounding atmosphere, transport was in three dimensions.

Absorption
After equilibrating the samples in air at 20 °C and 50% RH, they were immersed in water at 20 °C. The kinetics of absorption were followed by weighing the samples at intervals. The values of the three principal diffusivities were tested by comparing the theoretical kinetics with experiments.

Desorption
After removal from the water, the samples were exposed to air at 20 °C with 50% RH. The kinetics of drying were followed by weighing the samples at intervals.

The reference weight for each sample was obtained by drying to constant weight at 103 °C.

13.4.3 Results for a Sheet

The values of the three principal diffusivities and of the rate of evaporation were determined using the small cubes. The validity of the model was then tested by comparing the theoretical and experimental kinetics of drying. A fuller insight to the process was then given by calculating the profiles of moisture content developed through two midplanes of the sheet.

Determination of the Parameters

Principal Diffusivities
Each of the three principal diffusivities is determined in turn, separately, by using the cubic samples of $2 \times 2 \times 2$ cm where only two opposite faces are in contact with the surrounding atmosphere. Each principal diffusivity is obtained during the step of absorption, by plotting the amount of water absorbed as a function of the square root of time, and by using Eq. 13.34.

$$\frac{Q_t}{Q_\infty} = \frac{4}{L} \left(\frac{Dt}{\pi} \right)^{0.5} \tag{13.34}$$

where L is the thickness of the sample.

Rate of Evaporation
The rate of evaporation was determined from the kinetics of loss in weight of water exposed to air through a given area.

Parameter Values
The parameters were found to be

- Longitudinal diffusivity $D_L = 1 \times 10^{-3}$ cm^2 s^{-1}
- Radial diffusivity $D_R = 3 \times 10^{-5}$ cm^2 s^{-1}
- Tangential diffusivity $D_T = 1 \times 10^{-5}$ cm^2 s^{-1}
- Rate of evaporation $F_0 = 2.5 \times 10^{-5}$ cm s^{-1}

Two results are of interest:

1. The radial and tangential diffusivities are about the same, while the longitudinal diffusivity is about one hundred times greater.
2. The principal diffusivities are constant during the whole process of absorption.

Validity of the Model

The sheet was immersed in water for 4 h, this period of time being chosen because a pseudo-equilbrium is attained under these conditions. The validity of the model and the accuracy of the parameters were tested by comparing the experimental and theoretical kinetics of absorption.

The model was also evaluated for the stage of drying by comparing the kinetics obtained either by experiments or by calculation. The values of the principal diffusivities are those obtained in the stage of absorption. A good correlation is observed between these kinetics of drying in Fig. 13.38.

The following conclusions can be drawn [29]:

1. The process of drying a sheet can be described by diffusion in three dimensions through the solid and evaporation from the surface.
2. The principal diffusivities determined during the stage of absorption can successfully be used for the following stage of drying.
3. As stated above the longitudinal diffusivity is considerably higher than the other two principal diffusivities.
4. The stage of drying is longer than the stage of absorption. For instance, the time for which $Q_t/Q_\infty = 0.5$ is of around 4 h for desorption while it is only 8 min for absorption.
5. After around 30 h, the sheet is completely dried under the operational conditions of temperature and relative humidity.

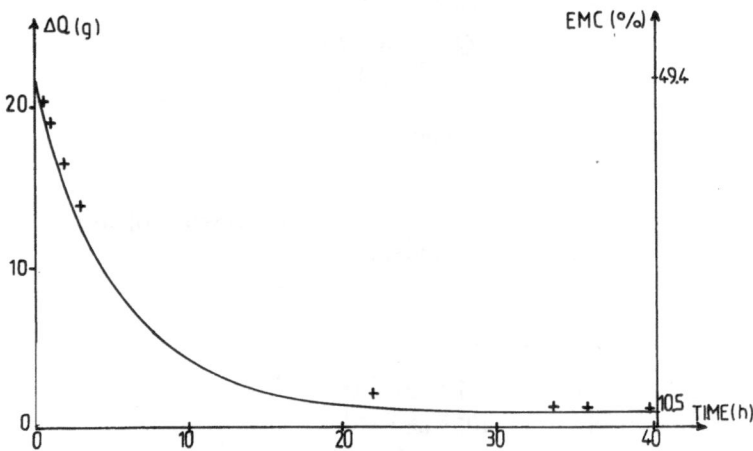

Fig. 13.38. Kinetics of drying for the sheet with transport of water in three directions (20 °C, 50% RH). (——, calculation; +, experiment.)

Profiles of Moisture Content

The profiles of moisture content which develop through the sample are able to provide a fuller insight into the process of drying. As it is difficult to represent these profiles in three dimensions, two midplanes are considered within the sheet:

- At the midlength of the transverse direction, with longitudinal and radial transport.
- At the midlength of the longitudinal direction with tangential and radial transport.

Longitudinal and Radial Transport

The profiles of moisture content developed within the midplane at the midlength of the transverse direction are shown after 2 h (Fig. 13.39), 5 h (Fig. 13.40) and 10 h (Fig. 13.41) of drying.

Tangential and Radial Transport

The profiles of moisture content developed within the midplane at the midlength of the longitudinal direction are shown after 2 h (Fig. 13.42), 5 h (Fig. 13.43) and 10 h (Fig. 13.44) of drying.

The following results are worth noting:

1. The lines with same moisture content are shown at various times in two

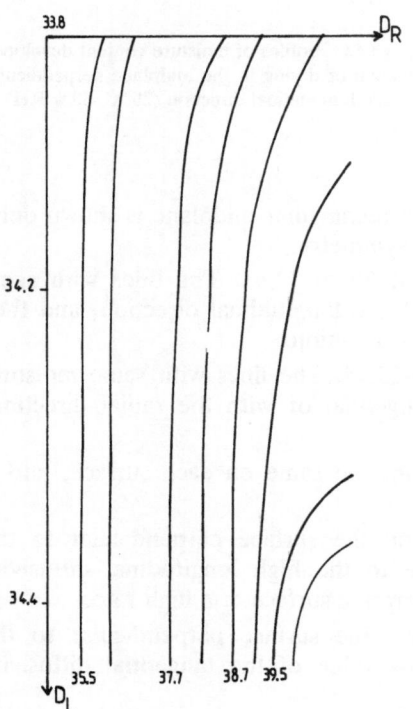
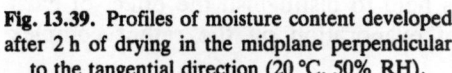

Fig. 13.39. Profiles of moisture content developed after 2 h of drying in the midplane perpendicular to the tangential direction (20 °C, 50% RH).

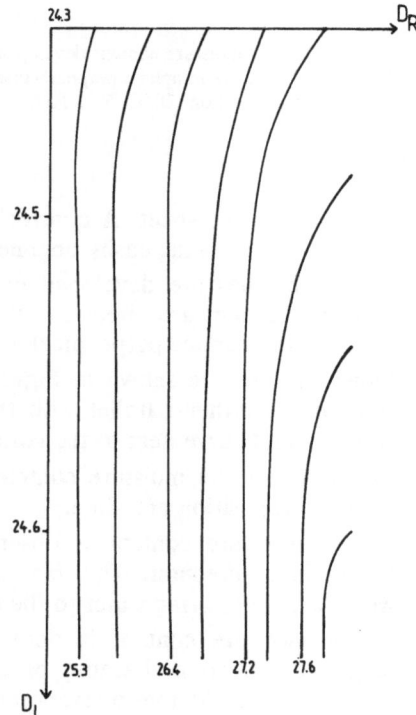

Fig. 13.40. Profiles of moisture content developed after 5 h of drying in the midplane perpendicular to the tangential direction (20 °C, 50% RH).

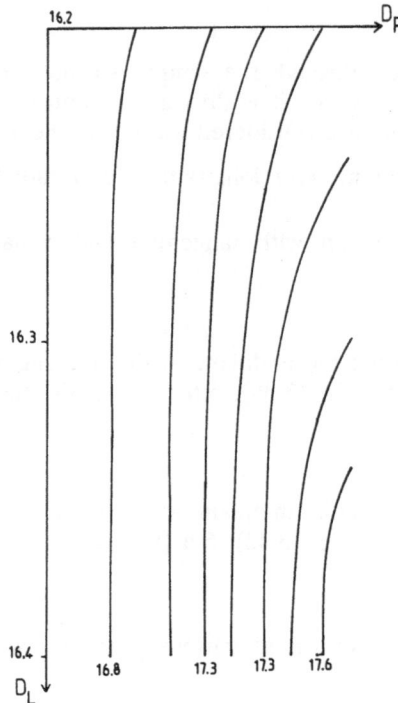

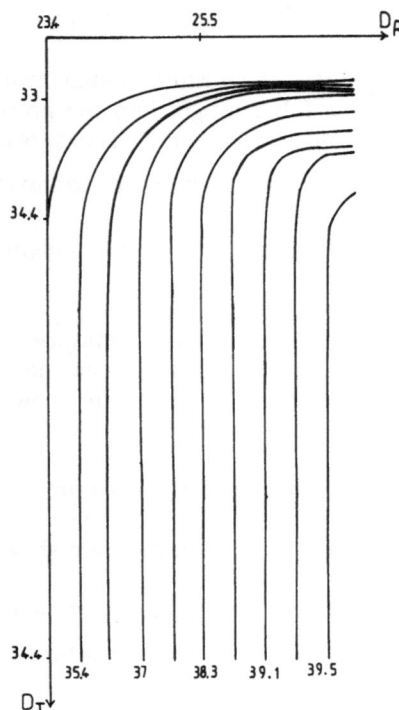

Fig. 13.41. Profiles of moisture content developed after 10 h of drying in the midplane perpendicular to the tangential direction (20 °C, 50% RH).

Fig. 13.42. Profiles of moisture content developed after 2 h of drying in the midplane perpendicular to the longitudinal direction (20 °C, 50% RH).

midplanes of the sheet. A quarter of the rectangular midplane is shown only, the other three being easily obtained by symmetry.

2. Typical profiles are developed in Figs 13.39 to 13.41. The lines with same moisture content are about parallel with the longitudinal direction, and they intersect the surface perpendicular to this direction.

3. Other profiles are shown in Figs 13.42–13.44. The lines with same moisture content are either parallel with the tangential or with the radial direction, with a round curve next to the corner.

4. The value of the moisture content is not the same on each surface, and it varies with position and time.

5. Higher moisture content is observed on the surface perpendicular to the longitudinal direction. This fact is due to the high longitudinal diffusivity which is able to bring water to the evaporating surface at a high rate.

6. Lower moisture content is observed on the surface perpendicular to the tangential direction, because of the low value of the tangential diffusivity compared with the rate of evaporation.

7. These profiles of moisture content thus help to distinguish the effect of each principal diffusivity and of the rate of evaporation on the rather complex process of drying.

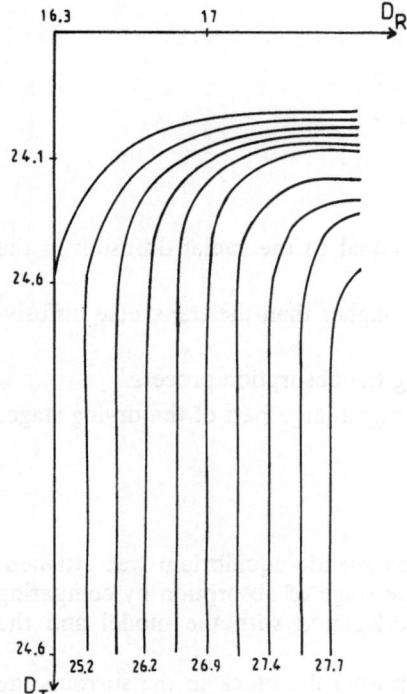

Fig. 13.43. Profiles of moisture content developed after 5 h of drying in the midplane perpendicular to the longitudinal direction (20 °C, 50% RH).

Fig. 13.44. Profiles of moisture content developed after 10 h of drying in the midplane perpendicular to the longitudinal direction (20 °C, 50% RH).

13.4.4 Results for a Block

After determining the values of the principal diffusivities and the rate of evaporation, the numerical model was tested by comparing the experimental and theoretical kinetics of drying. Some profiles of moisture content were then calculated in various planes selected through the block. The block was cut in such a way that longitudinal and transverse diffusion could take place (Fig. 13.37).

Determination of the Parameters

Principal Diffusivities
The blocks were cut from adjacent places in the same tree, and thus each principal diffusivity is assumed to be the same. The diffusivities were determined from the kinetics of absorption, as the absorption process is simpler. Each principal diffusivity was measured separately, by protecting four surfaces from absorption.

Rate of Evaporation
The rate of evaporation was obtained from the kinetics of evaporation of water exposed to the surrounding atmosphere under the same conditions as for the drying process for the block.

Parameter Values

The parameters were found to be

- Longitudinal diffusivity $D_L = 2.2 \times 10^{-3}$ cm^2 s^{-1}
- Transverse diffusivity $D_T = 3 \times 10^{-5}$ cm^2 s^{-1}
- Rate of evaporation $F_0 = 2.5 \times 10^{-5}$ cm s^{-1}

The following conclusions are drawn:

1. The transverse diffusivity in the block is equal to the radial diffusion in the sheet.
2. The longitudinal diffusivity is considerably higher than the transverse diffusivity.
3. Each principal diffusivity is constant during the absorption process.
4. The moisture content is above the FSP during a large part of the drying stage.

Validity of the Model

The block was immersed in water for 6 h, and pseudo-equilibrium was attained. The validity of the model was tested during the stage of absorption by comparing the experimental kinetics with the kinetics calculated with the model and the values of diffusivities [30].

The stage of drying was then studied by exposing the block to the surrounding air at 20 °C and RH of 50%. The validity of the model was evaluated by comparing the experimental and theoretical kinetics of drying. A good correlation between these curves is seen in Fig. 13.45, proving the validity of the model as well as the accuracy of the values of parameters.

The following conclusions can be drawn [30]:

1. The stage of drying for a block with longitudinal and transverse transport can be represented by transient diffusion through the solid and evaporation from the surface.

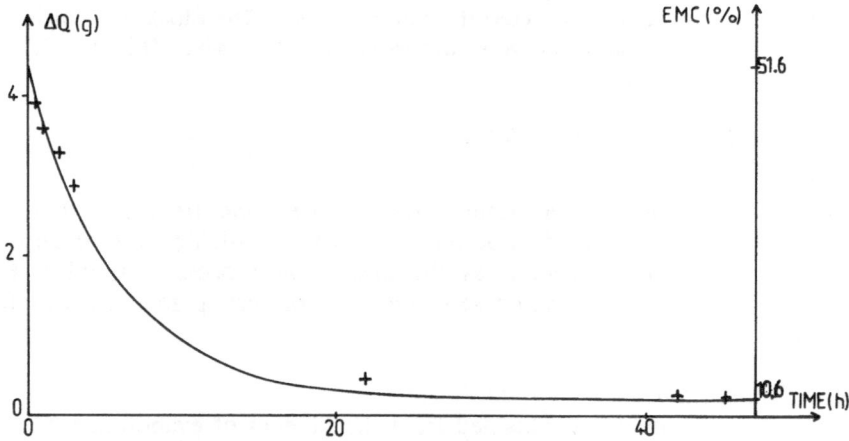

Fig. 13.45. Kinetics of drying for the beam with transport of water in three dimensions: longitudinal and transverse diffusion (20 °C, 50% RH).

2. The principal diffusivities obtained from the kinetics of absorption can be successfully used for the drying stage.

3. The principal diffusivities (longitudinal and transverse) are constant throughout the whole process of drying.

4. A comparison between the kinetics of absorption and evaporation can be instructive. For instance, the time for which $Q_t/Q_\infty = 0.5$ is less than 10 min in the stage of absorption, while it reaches 4 h during the stage of drying.

5. Up to 25 h is necessary to dry the beam perfectly under the operational conditions of temperature and RH.

Profiles of Moisture Content

The rather complex process of drying with diffusion and evaporation becomes more understandable by examining the profiles of moisture content developed through the solid. Two midplanes within the beam were selected (Fig. 13.37):

- At midlength, perpendicular to the longitudinal direction
- At mid-thickness, parallel with the longitudinal direction

Transverse–Transverse Transport
The profiles of moisture content developed within the midplane perpendicular to the longitudinal direction are shown after 2 h (Fig. 13.46), 5 h (Fig. 13.47) and 10 h (Fig. 13.48) of drying.

Longitudinal–Transverse Transport
The midplane parallel to the longitudinal direction and at midthickness of the block was considered, and the profiles of moisture calculated after 2 hours (Fig. 13.49), 5 h (Fig. 13.50) and 10 h (Fig. 13.51) of drying.

Relevant conclusions are:

1. The lines with same moisture content are able to give a fuller insight into the

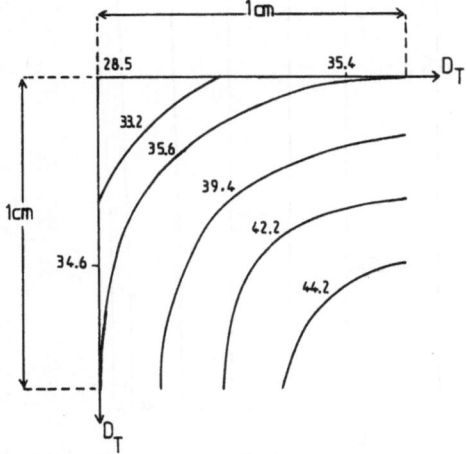

Fig. 13.46. Profiles of moisture content developed through the midplane perpendicular to the longitudinal direction, after 2 h of drying in air (20 °C, 50% RH).

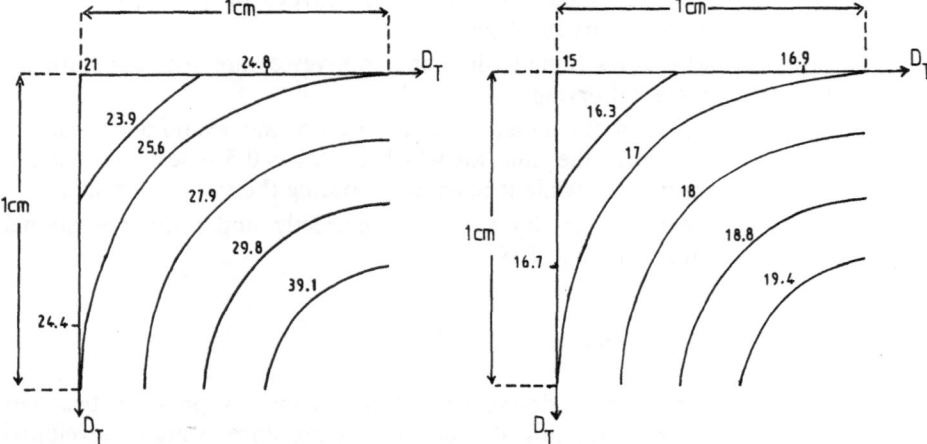

Fig. 13.47. Profiles of moisture content developed through the midplane perpendicular to the longitudinal direction, after 5 h of drying in air (20 °C, 50% RH).

Fig. 13.48. Profiles of moisture content developed through the midplane perpendicular to the longitudinal direction, after 10 h of drying in air (20 °C, 50% RH).

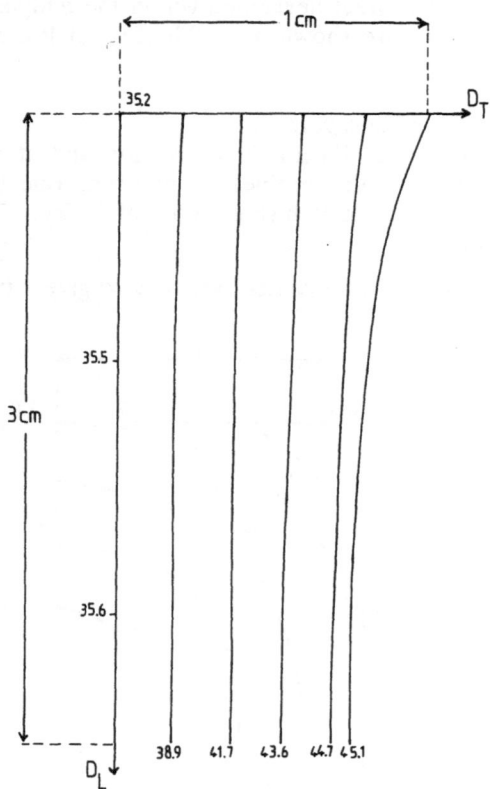

Fig. 13.49. Profiles of moisture content developed through the midplane parallel to the longitudinal direction, after 2 h of drying in air (20 °C, 50% RH).

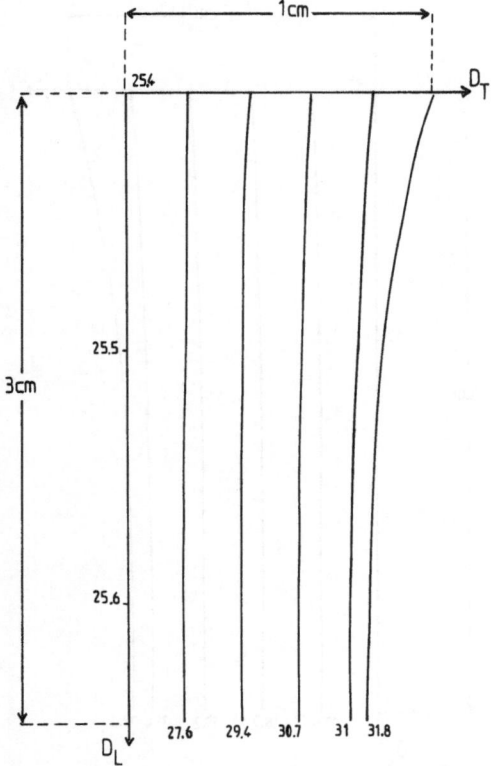

Fig. 13.50. Profiles of moisture content developed through the midplane parallel to the longitudinal direction, after 5 h of drying in air (20 °C, 50% RH).

nature of the process. Only a quarter of the rectangular and square midplane is described, as the other three can be obtained by symmetry.

2. Characteristic profiles are developed within the square midplane, following the two transverse transports perpendicular to each other (Figs 13.46–13.48). The lines with same moisture content are nearly circular in shape, with the highest value next to the centre.

3. Quite different profiles are shown within the rectangular midplane parallel to the longitudinal direction. In this case, the lines with same moisture content are almost parallel to the longitudinal direction.

4. The value of the moisture content on each surface is not the same, and it varies not only with time but also with position.

5. Higher values of the moisture content are obtained on the surface perpendicular to the longitudinal direction. In the same way as for the sheet, the longitudinal diffusion is so high with regard to the evaporation rate that water is brought to the surface very quickly as soon as it evaporates.

6. Lower values of the moisture content are shown on the surfaces perpendicular to the transverse directions. This is due to the low value of the transverse diffusivity which is not able to bring water to these surfaces very fast.

7. As for the sheet, the effect of each diffusional transport on the process of

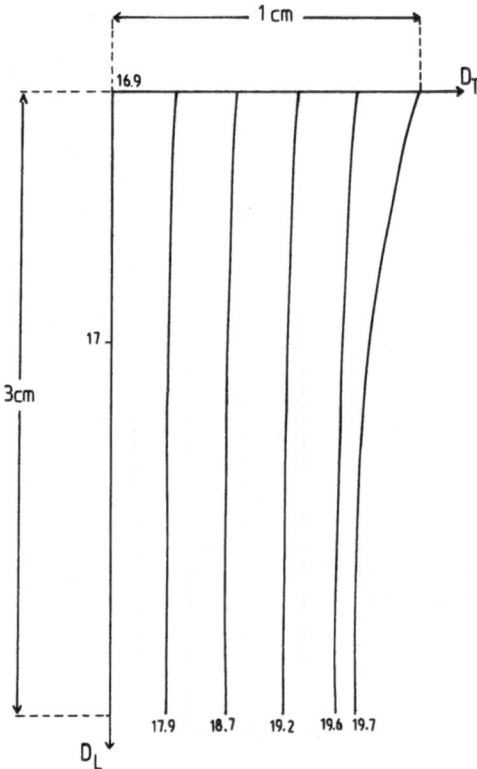

Fig. 13.51. Profiles of moisture content developed through the midplane parallel to the longitudinal direction, after 10 h of drying in air (20 °C, 50% RH).

drying can be differentiated, the behaviour of the longitudinal and the transverse diffusions being quite different.

13.4.5 Conclusions

The following general conclusions can be drawn on the process of drying of wood samples in three dimensions.

Process of Drying
The process of drying of wood samples can be described by diffusion of water within the solid and evaporation from the surface, whatever the shape of the samples. Two common shapes have been successfully studied: a sheet and a block.

Three principal directions of diffusion are considered with three principal diffusivities, as the samples are usually cut along these principal directions.

Diffusivities
As the three principal diffusivities are quite different from each other, it is impossible to determine the value of these diffusivities from experiments performed using a three-dimensional solid.

The method using small cubic samples representative of the large sample, is proved to be efficient. Not only is the method reliable for determining each principal diffusivity, but it is also of help to establish the dependency of each diffusivity on the moisture content.

Moisture Content
The moisture content is a relevant factor. When the moisture content is below the FSP, each principal diffusivity is concentration dependent, while these diffusivities can be considered as constant when the moisture content is above the FSP.

Numerical Model and Results
The numerical model with finite differences, taking into account not only the rate of evaporation of the liquid from the surface, but also the transport by diffusion along three principal axes within the solid, is able to describe the process of drying.

The kinetics of drying can thus be calculated for various shapes of the samples cut along the three principal axes of diffusion, when the principal diffusivities and the rate of evaporation are known.

Moreover, the profiles of moisture content developed at various places in the sample, and especially within the midplanes perpendicular to each principal direction, are able to afford a fuller insight to the process of drying. For instance the effect of each principal diffusion participating in the whole process can be clearly established. Various data which are difficult to obtain by experiments, can be found by calculation.

Symbols

D	Principal diffusivity ($cm^2\ s^{-1}$)
$D_L,\ D_R,\ D_T$	Longitudinal, radial and tangential diffusivity
F_0	Rate of evaporation ($cm\ s^{-1}$)
G_n	Function $\left(D_n \dfrac{\partial C}{\partial x}\right)$
$i,\ j,\ k$	Integers defining position
$L, 2L$	Sheet thicknesses
$M_T,\ M_L,$ $M_R,\ M_t$	Dimensionless numbers
$MC,\ MCN$	Moisture content at time t and at time $t + \Delta t$, respectively
MC_{ext}	Moisture content necessary to maintain equilibrium with surrounding atmosphere
$MC_{i,j,k}$	Moisture content at position $i,\ j,\ k$
MC_{in}	Initial moisture content
$MC_{N_l N_r N_t}$	Moisture content at the corner ($N_l,\ N_r,\ N_t$)
N	Dimensionless number
NS	Number of slices
$Q_t,\ Q_\infty$	Amount of water evaporated after time t and after infinite time, respectively

S Dimensionless number $\left(\dfrac{F_0 L}{D}\right)$

β Positive roots of $\beta \tan \beta = S$ (Table 2.2)
$\Delta L, \Delta R, \Delta T$ Increments of space
$\Delta x, \Delta t$ Increments of space and of time, respectively

References

1. Griffin DM. Water potential and wood decay fungi, Ann Rev Phytopathol 1977; 15: 319–329
2. Dirol D. Evaluation d'une methode d'essais mycologiques sur vermiculite. Holzforschung 1985; 39: 11–16
3. Stamm AJ. Bound-water diffusion into wood in the fiber direction. Forest Prod J 1959; 9: 27–32
4. Stamm AJ. Combined bound-water and water vapor diffusion into Sitka Spruce. Forest Prod J 1960; 10: 644–648
5. Skaar C. Moisture movement in beach below the fiber saturation point. Forest Prod J 1958; 8: 352–357
6. Comstock GL. Moisture diffusion, coefficients in wood as calculated from absorption desorption, and steady state data. Forest Prod J 1963; 13: 97–103
7. Choong ET. Diffusion coefficients of softwoods by steady state and theoretical methods. Forest Prod J 1965; 15: 21–27
8. Moschler WW, Martin RE. Diffusion equation solutions in experimental wood drying. Wood Sci 1968; 1: 47–57
9. Bramhall G. Fick's law and bound-water diffusion. Wood Sci 1976; 8: 153–160
10. Ashworth JC. The mathematical simulation of the batch drying of softwood timber. PhD Thesis, University Canterbury, 1977
11. Siau JF. Flow in wood. Syracuse University Press, New York, 1971
12. Avramidis ST, Siau JF. An investigation of the external and internal resistance to moisture diffusion in wood. Wood Sci Technol 1987; 21: 249–256
13. Rosen HN. Exponential dependency of the moisture diffusion coefficient on moisture content. Wood Sci 1976; 8: 174–179
14. Droin A, Taverdet JL, Vergnaud JM. Modeling the kinetics of moisture absorption by wood. Wood Sci Technol 1988; 22: 11–20
15. Droin-Josserand A, Taverdet JL, Vergnaud JM. Modelling the absorption and desorption of moisture by wood in an atmosphere of constant and programmed relative humidity. Wood Sci Technol 1988; 22: 299–310
16. El Kouali M, Vergnaud JM. Modelling the process of absorption and desorption of water above and below the fiber saturation point. Wood Sci Technol 1991; 25: 327–339
17. Mounji H, El Kouali M, Vergnaud JM. Absorption and desorption of water along the longitudinal axis above and below the fiber saturation point (short time of absorption) Holzforschung 1991; 45: 141–146
18. Droin-Josserand A, Taverdet JL, Vergnaud JM. Modelling of moisture absorption within a section of parallelepipedic sample of wood by considering longitudinal and transversal diffusion. Holzforschung 1989; 43: 297–302
19. Droin-Josserand A, Taverdet JL, Vergnaud JM. Modelling the process of moisture absorption in three dimensions by wood samples of various shapes: cubic, parallelepipedic. Wood Sci Technol 1989; 23: 259–271
20. Crank J. The mathematics of diffusion, 2nd edn. Clarendon Press, Oxford, 1976, pp 60–62
21. Taverdet JL, Vergnaud JM. Study of transfer process of liquid into and plasticizer out of plasticized PVC by using short test. J Appl Polym Sci 1984; 29: 3391–3400
22. Messadi D, Hivert M, Vergnaud JM. A new approach to the study of plasticizer migration from PVC into methanol. J Appl Polym Sci 1981; 26: 667–677
23. Messadi D, Vergnaud JM. Plasticizer transfer from plasticized PVC into ethanol-water mixtures. J Appl Polym Sci 1982; 27: 3945–3955
24. Mounji H, Bouzon J, Vergnaud JM. Modelling the process of absorption and desorption of water in two-dimension (transverse) in a square wood beam. Wood Sci Technol 1992; 26: 23–27

25. Blandin HP, David JC, Illien JP, Malizewicz M, Vergnaud JM. Modelling the drying process of coatings with various layers. J Coat Technol 1987; 59: 27–32
26. Blandin HP, David JC, Illien JP, Malizewicz M, Vergnaud JM. Modelling the drying process of coatings: effect of the thickness, temperature and concentration of solvent, Progr Organic Coat 1987; 15: 163–172
27. Khatir Y, Bouzon J, Vergnaud JM. Liquid sorption by rubber sheets and evaporation; models and experiments. Polym Test 1986; 6: 253–267
28. El Kouali M, Bouzon J, Vergnaud JM. Modelling the process of absorption and desorption of water in a sheet beyond the fiber saturation point by considering a two-dimensional transfer. Holzforschung (in press)
29. Mounji H, El Kouali M, Bouzon J, Vergnaud JM. Modelling of the drying process of wood in 3-dimensions. Dry Technol 1991; 9: 1295–1314
30. Mounji H, Bouzon J, Vergnaud JM. Process of absorption and desorption of water with a wood beam in three-dimensions beyond the fiber saturation point. Unpublished results

Chapter 14

Drying of Thermosetting Coatings

14.1 Introduction

Epoxy resins are often used for high performance coatings, as well as for the polymeric matrix in composites. Many studies have been undertaken on the composition of epoxy resins, and on the process of cure. Because of the low thermal conductivity of the resin associated with the high exothermicity of the cure reactions, the process gives rise to a high temperature within the resin. These facts are of concern for thick samples, being responsible for discoloration or degradation of the final material, cracking or crazing, or even distortion of the moulding [1–4].

The cure of resin deposited on a substrate, such as a coating, is also difficult to complete, because of the difficulty in heating the resin as well as the surroundings. The problem is how to achieve the cure of the resin coating at room temperature or at least at a temperature around 50 °C, and to determine the time required to obtain a perfect result. As the properties of the final material largely depend on the state of cure of the resin, various models have been built to describe the process by considering heat conduction and heat evolved by the reaction [4–8]. Another typical model was constructed and tested for the cure of coatings made of epoxy resin at low temperature, and it could predict the value of the state of cure in terms of parameters such as temperature and time. In addition, a mechanical property such as hardness was chosen to follow the cure, and correlations were established between this property and the state of cure [9].

The purpose of this chapter is to study the matter transfer taking place between an epoxy resin coating and a liquid, when they are in contact for a length of time. As the resin is generally also in contact with air at intervals, the subsequent process of desorption of the liquid which has entered the resin is examined. The liquid selected for these studies was water, and temperatures ranged from 20 to 100 °C.

The process of absorption is assumed to be controlled by transient diffusion, while the following stage of desorption is explained by diffusion of water through the film and evaporation from the surface.

The effect of the temperature of drying is especially considered, as temperature acts not only on the rate of evaporation but also on the diffusivity, diffusion of liquid through the coating playing the main role in the process of drying.

14.2 Drying Process

14.2.1 Theory

Some assumptions are made in order to explain the process, and an analytical solution as well as a numerical model are given.

Assumptions

The following assumptions are made:

1. The process of drying is controlled by transient diffusion of water through the coating and evaporation from the surface.
2. The diffusivity of water is constant during the stage of desorption, as found from experiments.
3. The initial profile of concentration of water at the beginning of the stage of drying may be uniform or not. This is a reason why an analytical solution is given, as well as a numerical model, the analytical solution being effective only in the case of initial uniform concentration.
4. The rate of evaporation is proportional to the difference between the actual concentration of water on the resin surface and the concentration which is at equilibrium with the surrounding atmosphere, the coefficient of proportionality being the rate of evaporation of the pure water under the same conditions.
5. The thickness of the coating remains constant during the stage of desorption, the amount of water located in the coating being very small.

Mathematical Treatment

The one-dimensional diffusion in the coating of thickness $2L$ is expressed by Fick's law with constant diffusivity:

$$\frac{\partial C}{\partial t} = D\frac{\partial^2 C}{\partial x^2} \tag{14.1}$$

where C is the water concentration in the resin at position x and time t.

The initial and boundary conditions are

$$t = 0, \quad -L < x < L, \qquad\qquad C_{in} \quad \text{(resin)} \tag{14.2}$$

$$t > 0, \quad -D\frac{\partial C}{\partial x} = \frac{F_0}{\rho}(C_s - C_{ext}) \quad \text{(surface)} \tag{14.3}$$

where F_0 is the rate of evaporation, C_s is the actual water concentration at the surface and C_{ext} the water concentration on the surface required to maintain equilibrium with the surrounding atmosphere. The left-hand side of this equation is the rate of diffusion to the surface.

When the initial concentration of water in the resin is uniform, and the liquid evaporates from both surfaces of the coating, the solution is given [10] as

$$\frac{C_{\text{ext}} - C_{x,t}}{C_{\text{ext}} - C_{\text{in}}} = \sum_{n=1}^{\infty} \frac{2S \cos \dfrac{\beta_n x}{L}}{(\beta_n^2 + S^2 + S) \cos \beta_n} \exp\left(-\frac{\beta_n^2}{L^2} Dt\right) \qquad (14.4)$$

where the β_ns are the positive roots of

$$\beta \tan \beta = S \qquad (14.5)$$

with the dimensionless number

$$S = \frac{LF_0}{D\rho} \qquad (14.6)$$

The total amount of water leaving the sheet up to time t, Q_t, is expressed as a fraction of Q_∞, the corresponding quantity after infinite time:

$$\frac{Q_\infty - Q_t}{Q_\infty} = \sum_{n=1}^{\infty} \frac{2S^2}{\beta_n^2 (\beta_n^2 + S^2 + S)} \exp\left(-\frac{\beta_n^2}{L^2} Dt\right) \qquad (14.7)$$

Numerical Model

A numerical model with finite differences is necessary when the initial concentration of water in the coating is not uniform, or when the diffusivity is not constant.

Within the Coating

The thickness of the coating is divided into equal slices of thickness Δx as shown in Fig. 14.1. The balance of water entering and leaving the slice at position n during the increment of time Δt is calculated, and the new concentration CN_n after time Δt is thus expressed in terms of the previous concentration at the same and adjacent places:

$$CN_n = \frac{1}{M} [C_{n-1} + (M - 2)C_n + C_{n+1}] \qquad (14.8)$$

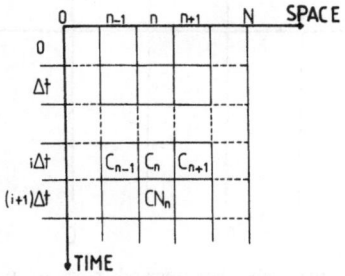

Fig. 14.1. Space–time diagram for numerical analysis.

where C_n is the concentration at position n and time t, CN_n the new concentration at position n and time $t + \Delta t$ and the dimensionless number M is

$$M = \frac{(\Delta x)^2}{D\Delta t} \qquad (14.9)$$

On the Surface
The slice of thickness $\Delta x/2$ next to the surface is shown in Fig. 14.2. From the matter balance during the time increment Δt calculated in this slice by considering the diffusion within the solid and evaporation from the surface, the new concentration at the surface is obtained:

$$CN_0 = \frac{1}{M} [2C_1 + (M - 2 - 2N)C_0 + 2NC_{ext}] \qquad (14.10)$$

where C_0 and CN_0 are the concentrations on the surface at times t and $(t + \Delta t)$, respectively, C_{ext} is the concentration of water on the surface which is necessary to maintain equilibrium with the surrounding atmosphere, and the dimensionless number N is

$$N = \frac{F_0\Delta x}{\rho D} \qquad (14.11)$$

Amount of Water Remaining
The amount of water remaining is obtained by integrating the water concentration with respect to space.

14.2.2 Experiment

Materials

The coating was obtained by mixing epoxy resin (Lopox 200, CDF Chimie, Orkem) and hardener (D 2605 – CDF Chimie, Orkem) containing an anhydride of Me THP and an accelerator with 0.75% of an aromatic amine. The binary

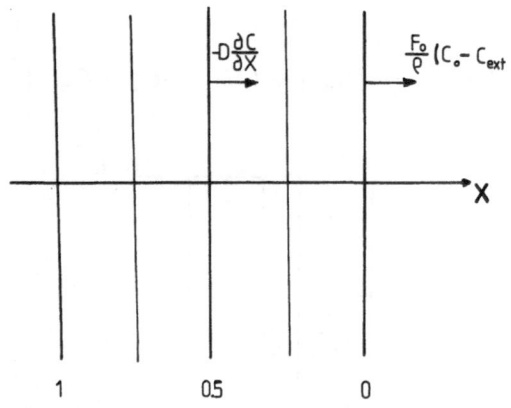

Fig. 14.2. Matter balance within the half slice located next to the surface, with diffusion and evaporation.

mixture was prepared at room temperature with continuous stirring. The weight percent composition was optimised in a previous work [11]: epoxy resin 56%, hardener 44%.

State of Cure of the Resin

The kinetics of cure were determined by calorimetry (DSC 111, Setaram). The resin was heated at a constant rate of $5\,°C\,min^{-1}$ from room temperature to the temperature at which the reaction was complete. Quite a large sample (150 mg) of resin was used, to ensure that the sample was representative of a large batch. The kinetics parameters were calculated using the Freeman and Caroll technique, by assuming that the cure reaction is expressed by an Arrhenius equation [12].

Preparation of the Coating

The resin mixture was deposited on steel sheets, the coating having an average thickness of 0.1 cm. After curing for 44 h, a part of the coating was studied by calorimetry, showing an extent of cure of 90% at this time [9].

Matter Transfer Studies

The resin coating was then removed from the steel plate after cure. The coating was immersed in water at various temperatures, and the kinetics of absorption followed by weighing the resin at intervals.

The resin coating was then removed and dried in air under the following conditions:

1. At 20 °C up to equilibrium, followed by drying at 80 and 100 °C.
2. At constant temperatures ranging from 20 to 100 °C.

14.2.3 Results

Results are presented for:

1. The values of the diffusivity, amount of water absorbed and desorbed and the rate of evaporation.
2. The validity of the model with the kinetics of drying.
3. The effect of temperature on the process of absorption and desorption.

Values of Parameters

Diffusivity
Generally, when the process is reversible, the diffusivity can be obtained either from absorption or desorption experiments, and it is often easier to work with the absorption stage.

In a few cases, as for instance with plasticised PVC, the process during the stage of absorption and desorption is quite different, and the diffusivity during the stage of desorption differs notably from the value measured during absorption. A similar pattern is observed in the case of thermosetting coatings.

The diffusivity during the stage of absorption is calculated by plotting the amount of water absorbed as a function of the square root of time and using the following equation:

$$\frac{Q_t}{Q_\infty} = \frac{2}{L}\left(\frac{Dt}{\pi}\right)^{0.5}$$

(14.12)

for a coating of thickness $2L$.

During the stage of desorption, the diffusivity is calculated by using Eq. (14.7) for long times, when only the first term of the series is significant:

$$\ln\left(\frac{Q_\infty - Q_t}{Q_\infty}\right) = -\frac{\beta_1^2}{L^2}Dt + \ln\left(\frac{2S^2}{\beta_1^2(\beta_1^2 + S^2 + S)}\right)$$

Rate of Evaporation
The rate of evaporation is determined either from the constant rate of loss in weight of pure water in a flask of given area, or from the slope of the kinetic curve obtained with the resin when time tends to zero [13].

Amount of Water Absorbed and Remaining
The amount of water absorbed by the coating, as well as the amount of water remaining in the coating after drying, were determined by weighing the resin at the beginning and the end of each stage.

Validity of the Model
The validity of the model was tested for the stage of desorption by comparing the experimental and calculated kinetics of loss in weight [14].

The sample was immersed in water at 80 °C, and the kinetics of absorption followed up to equilibrium.

This presaturated sample was then dried

• At 20 °C up to equilibrium, then at 80 °C and finally at 100 °C (Fig. 14.3)
• At 80 °C (Fig. 14.4)

The kinetics of drying were calculated by using either the numerical model or the analytical solution for constant temperatures of 20 °C (Fig. 14.5) and 80 °C (Fig. 14.4).

The data used for these calculations are shown in Table 14.1.

A good correlation is obtained for the kinetics of drying either at 20 °C or at 80 °C, proving the validity of the model.

The profiles of concentration of water within the coating are calculated by using the numerical model and the analytical solution. The proof of the validity of these profiles is given by the fact that the kinetics of desorption are obtained by integrating the concentration of water in the coating with respect to space. Some profiles are shown for the drying stage at 20 °C (Fig. 14.6) and at 80 °C (Fig. 14.7).

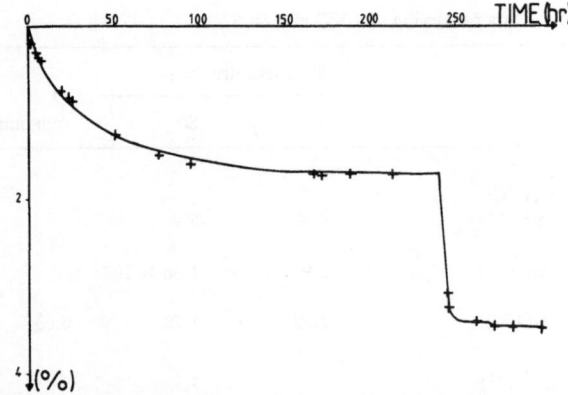

Fig. 14.3. Kinetics of drying for the coating (previously immersed in water at 80 °C) at 20, 80 and 100 °C. ($Q_i = 0.7564$ g, ΔQ absorbed = 4.04%, ΔQ desorbed = 3.46%.)

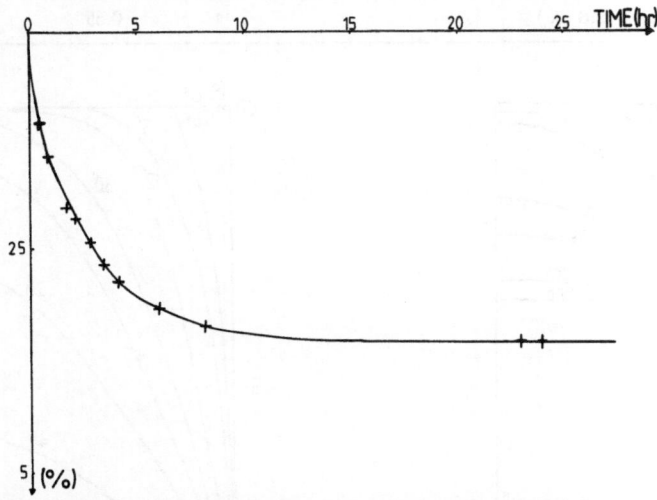

Fig. 14.4. Kinetics of drying of the coating (previously immersed in water at 80 °C) at 80 °C. ($Q_i = 0.7564$ g, ΔQ remaining = 0.635%.)

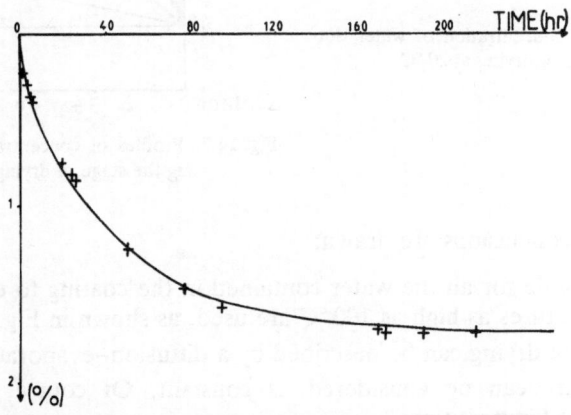

Fig. 14.5. Kinetics of drying of the coating (previously immersed in water at 80 °C) at 20 °C. ($Q_i = 0.7564$ g, ΔQ desorbed = 1.72%.)

Table 14.1. Data for drying at 20 °C and 80 °C

	Temperature (°C)		Q remaining
	20	80	
Drying at 20 °C followed by 80 °C:			
$D \times 10^8$ (cm^2 s^{-1})	0.52	8.9	
$\dfrac{F_0}{\rho} \times 10^6$ (cm s^{-1})	2.9	1.86×10^{-3}	
Q evaporated (%)	1.72	1.70	0.63
Drying at 80 °C:			
$D \times 10^8$ (cm^2 s^{-1})		7.3	
$\dfrac{F_0}{\rho} \times 10^6$ (cm s^{-1})		1.86×10^{-3}	
Q evaporated (%)		3.44	0.65

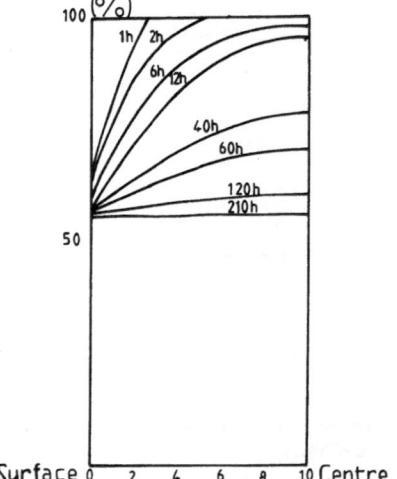

Fig. 14.6. Profiles of concentration of water during the stage of drying at 20 °C.

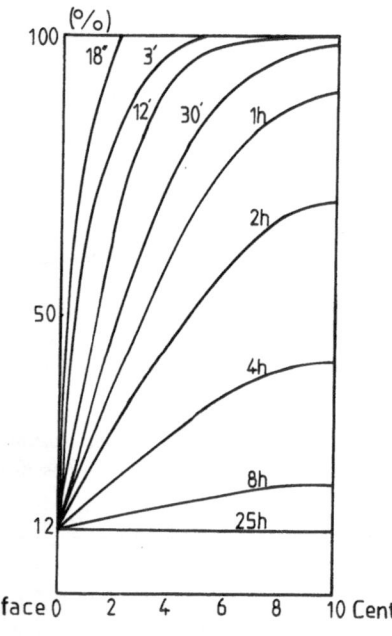

Fig. 14.7. Profiles of concentration of water during the stage of drying at 80 °C.

The following conclusions are drawn:

1. It is not possible for all the water contained in the coating to evaporate, even when temperatures as high as 100 °C are used, as shown in Fig. 14.3.
2. The process of drying can be described by a diffusion–evaporation process.
3. The diffusivity can be considered as constant. Of course, the diffusivity increases with temperature.
4. When an equilibrium is reached, a flat profile of concentration of water is

obtained. In fact, it is not real equilibrium but rather a pseudo equilibrium which is modified by increasing temperature.

5. The concentration of water on the coating surface decreases rather slowly during drying at 20 °C, because the rate of evaporation is not very high at this temperature. At 80 °C, this concentration on the surface drops very quickly, the rate of evaporation being considerably higher.

6. The water contained in the solid may perhaps be partly free and in part bound to the hydrophilic part of the resin, and high temperatures are thus necessary for this water to evaporate.

Effect of Temperature on Absorption and Desorption

Temperature is the main parameter acting not only on the kinetics of absorption and desorption but also on the amount of water absorbed during the absorption stage [15]. The kinetics of drying obtained from experiments and by using the numerical model with the data shown in Table 14.2 are shown for the following temperatures: 60 °C (Fig. 14.8), 80 °C (Fig. 14.4), 90 °C (Fig. 14.9) and 100 °C (Fig. 14.10).

In each case, the coating was immersed in water at the selected temperature

Table 14.2. Values of parameters at various temperatures

Temperature (°C)	ΔQ absorbed	ΔQ desorbed	D abs $\times 10^8$	D des $\times 10^8$	F_0 (cm s^{-1})
60	6.62	4.7	0.27	1.7	6.2×10^{-5}
80	4.06	3.57	1.7	8.1	1.4×10^{-4}
90	3.86	3.24	7.2	16	8.5×10^{-4}
100	3.57	3.57	26	17	1.2×10^{-3}

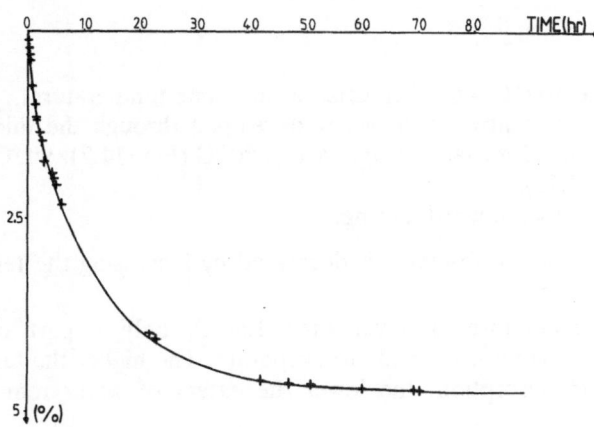

Fig. 14.8. Kinetics of drying for the coating (previously immersed in water up to equilibrium at 60 °C) at 60 °C.

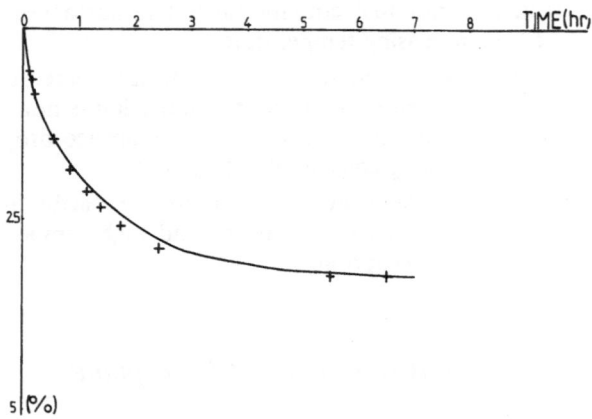

Fig. 14.9. Kinetics of drying for the coating (previously immersed in water at 90 °C up to equilibrium) at 90 °C.

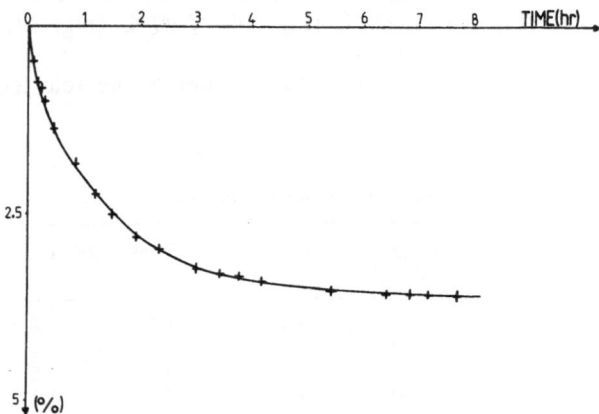

Fig. 14.10. Kinetics of drying for the coating (previously immersed in water at 100 °C up to equilibrium) at 100 °C.

ranging from 60 to 100 °C, and then dried at the same temperature.

The profiles of concentration of water developed through the thickness of the coating were calculated for 60 °C (Fig. 14.11), 80 °C (Fig. 14.7), 90 °C (Fig. 14.12) and 100 °C (Fig. 14.13).

Various conclusions are worth noting:

1. The amount of water absorbed is decreased by increasing the temperature of absorption.

2. When the temperature is lower than 100 °C, only a part of the water previously absorbed can be made to evaporate. The higher the temperature of absorption and desorption, the lower the extent of water remaining in the coating.

3. The process of drying is controlled by diffusion and evaporation, with constant diffusivity.

4. Diffusivity increases with temperature, following an Arrhenius law, as shown in Fig. 14.14.

5. At 100 °C, all the water which has previously entered the coating at the same temperature, can evaporate.

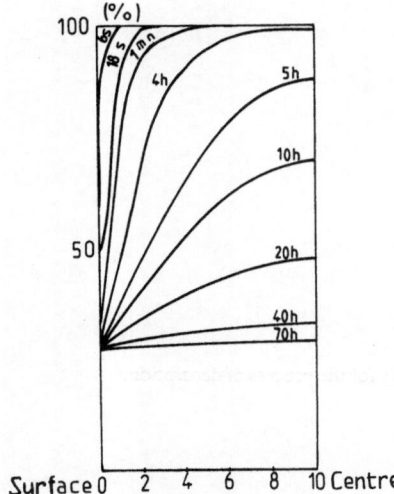

Fig. 14.11. Profiles of concentration of water developed within the coating thickness during the stage of drying at 60 °C, after absorption of water of 60 °C.

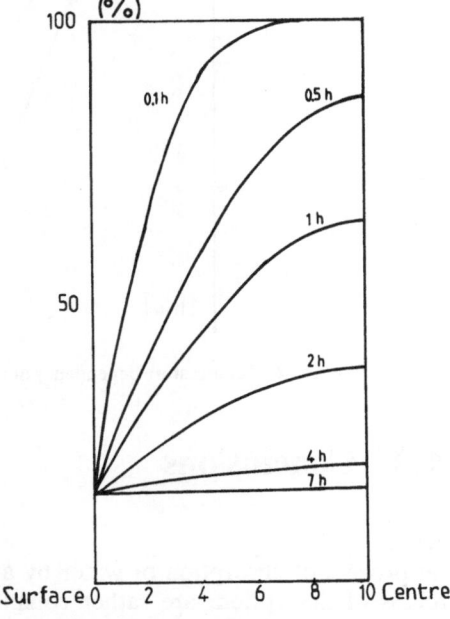

Fig. 14.12. Profiles of concentration of water developed within the coating thickness during the stage of drying at 90 °C, after absorption of water at 90 °C.

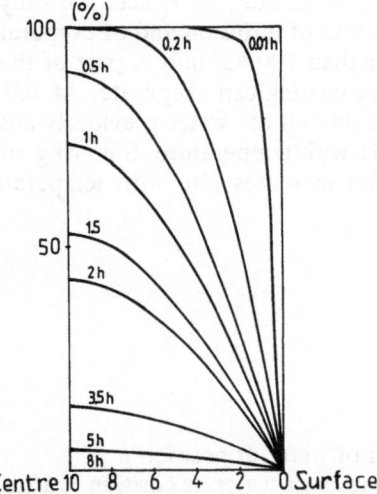

Fig. 14.13. Profiles of concentration of water developed within the coating thickness during the stage of drying at 100 °C, after absorption of water at 100 °C.

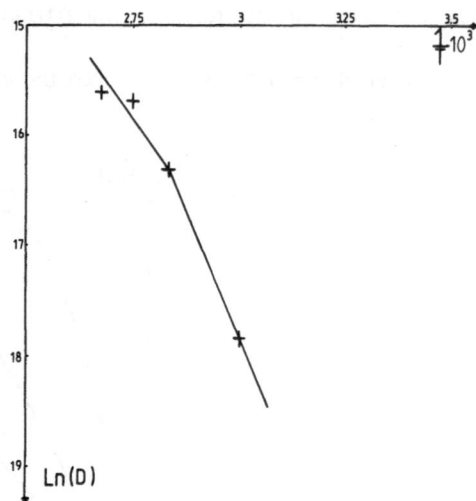

Fig. 14.14. Temperature dependency of the diffusivity for the process of desorption.

14.3   Conclusions

Process of Drying
The process of absorption of water by a thermosetting coating, and the following process of desorption, are rather complex, since only a part of the water which has previously entered the coating can evaporate.

However, the stage of drying can be described by a diffusion–evaporation process, with a constant diffusivity.

Effect of Temperature
Temperature is the main parameter, as it acts not only on the extent of water absorbed but also on the rate of diffusion and of evaporation.

At temperatures lower than 100 °C, only a part of the amount of water which has previously entered the coating can evaporate. At 100 °C, the process seems to be reversible in the sense that all the water previously absorbed can evaporate.

The diffusivity increases with temperature, following an Arrhenius law.

The rate of evaporation increases also with temperature, following the Clausius–Clapeyron law.

Symbols

C_n	Concentration of water at position n
CN_n	New concentration of water at position n after time Δt
D	Diffusivity ($cm^2\ s^{-1}$)
F_0	Rate of evaporation of water ($g\ cm^{-2}\ s^{-1}$)

$2L$	Thickness of the coating
M	Dimensionless number
N	Dimensionless number
Q_t, Q_∞	Amount of water leaving the sheet after time t and after infinite time, respectively
S	Dimensionless number
t	Time
x	Space coordinate
β_n	Positive roots of $\beta \tan \beta = S$
$\Delta t, \Delta x$	Increments of time and of space, respectively
ρ	Density of water ($g\,cm^{-3}$)

References

1. Progelhof RC, Throne JL. Non-isothermal curing of reactive plastics. Polym Engng Sci 1975; 15: 690–695
2. Mallik PK, Raghupathi N. Effect of cure cycle on mechanical properties of thick section fiber-reinforced poly thermoset mouldings. Polym Engng Sci 1979; 19: 774–778
3. Barone RM, Caulk DA. The effect of deformation and thermoset cure on heat conduction in a chopped-fibre reinforced polyester during compression moulding. Int J Heat Mass Transfer 1979; 22: 1021–1032
4. Pusatcioghu SY, Hassler JC, Frickle AL, McGee HA. Effect of temperature gradients on cure and stress gradients in thick-thermoset castings. J Appl Polym Sci 1980; 25: 381–393
5. Nixon JA, Hutchinson JM. Analysis of the cure of sheet moulding compounds. I. Development of the model. Plast Rubber Process Applic 1985; 5: 337–345
6. Nixon JA, Hutchinson JM. Analysis of the cure of sheet moulding compounds. II. Application of the model to various moulding geometries. Plast Rubber Process Applic 1985; 5: 359–363
7. Chater M, Vergnaud JM. Study of the process of cure of epoxide resin heated by liquid oil at constant temperature. Europ Polym J 1988; 24: 245–250
8. Chater M, Bouzon J, Vergnaud JM. Modelling of the cure of epoxide resin. Effect of the diameter of cylindrical moulds. Plast Rubber Process Applic 1987; 7: 199–205
9. Laoubi S, Vergnaud JM. Modelling of the cure of epoxy resin coating at low temperature (50 °C). Thermochim Acta, 1990; 162: 347–354
10. Crank J. The mathematics of diffusion, 2nd ed. Clarendon Press, Oxford, 1975, pp 60–62
11. Chater M, Lalart D, Michel-Dansac F, Vergnaud JM. Variation of enthalpy and kinetic parameters of the cure of epoxide resin with the composition of the binary system. Europ Polym J 1986; 10: 805–809
12. Lui H, Armand JY, Bouzon J, Vergnaud JM. Effect of sample size and heating rate on the DSC process for reactions of high enthalpy. Thermochim Acta 1988; 126: 81–92
13. Khatir Y, Bouzon J, Vergnaud JM. Liquid sorption by rubber sheets and evaporation, Models and experiments. Polym Test 1986; 6: 253–265
14. Laoubi S, Vergnaud JM. Modelling the process of absorption and desorption of water by coatings made of epoxy resin. Europ Polym J 1990; 26: 1359–1364
15. Laoubi S, Vergnaud JM. Effect of temperature on the transfer between epoxy coatings and water. Europ Polym J 1991; 27: 1425–1429

Chapter 15

Drying of Dosage Forms for Medical Applications

15.1 Introduction

All conventional dosage forms, except for continuous intravenous perfusion, do not release drugs at a constant rate, or with kinetics following a first-order reaction. With an oral form, the drug is usually very rapidly liberated, quickly builds up to a high concentration in the stomach, and then falls exponentially until the next dose. As a result, there is an undulating concentration pattern of the drug in the blood and tissues, in which high concentration alternates with low concentration. The optimal therapeutic level is only briefly present in this case.

Distribution of a drug in an organism can only be in equilibrium when its rate of continuous administration is the same as the rate of continuous elimination over a prolonged period; This has not been possible with conventional oral forms [1–3].

Recently, efforts have been directed to the development of methods for safer administration of drugs than the conventional methods. Special attention has been given to regulating the amount of drug released by means of monolithic devices where the drug is dispersed in a polymer matrix [4–7].

Various ways exist for preparing these monolithic devices, and three of them are worth noting:

1. By compression of the components in powder form either at room temperature or at higher temperature [8, 9].
2. By melting the mixture, or at least the polymer matrix [10, 11].
3. By transforming the polymer matrix into a paste with the help of a liquid in which the drug is not soluble. The paste is then easily shaped into beads [12–14].

There are some advantages and drawbacks for each of these above techniques. For instance, the last technique does not necessitate high pressure and high temperature for shaping the beads, but an inconvenience exists with the drying of the dosage form.

The main purpose of this chapter is to describe the process of drying of dosage forms made from Eudragit, a biocompatible polymer.

Spherical beads are considered, but the study can be applied to other shapes. In fact the process of drying a polymer is not simple. As shown in previous studies carried out with elastomers of various shapes, e.g., thin sheets [15], cylinders of finite length [16], tubes [17] or annuli [18], the process is controlled by diffusion through the polymer and evaporation from the surface. A mathematical and a numerical model are built up in order to describe the process. The first one can be used with constant diffusivity, while the numerical model can be used when the diffusivity is concentration dependent. With these two models, the rate of evaporation is proportional to the difference between the actual concentration of liquid on the surface and the concentration which is at equilibrium with the surrounding atmosphere.

Some emphasis is placed on determining the effect of parameters such as the radius of the bead and the temperature of drying. Temperature generally plays a leading role, because it acts not only on the rate of evaporation of the liquid but also on the rate of diffusion of the liquid within the polymer, the diffusivity being generally temperature dependent.

Moreover, the numerical model is of help to simulate the process. Some simulations are useful to determine the role played by the surrounding atmosphere in the process, and especially the volume of the atmosphere with regard to the amount of liquid to be evaporated.

15.2 Drying in an Infinite Atmosphere

The process of drying in the case of dosage forms made of a drug dispersed in a polymer, is examined. The transport of liquid is controlled by transient diffusion through the polymer and by evaporation from the surface. After testing the validity of the model, the effect of parameters such as temperature and radius of the beads on the process is studied.

15.2.1 Theory

Two models are described: one with analytical solutions when the diffusivity is constant and the conditions for the surrounding atmosphere are simple, the other using a numerical method with finite differences which can be used in more complex cases.

Assumptions

The following assumptions are made in order to build up the models:

1. The process of drying is controlled by transient diffusion of the liquid within the solid and evaporation from the surface.
2. The rate of evaporation is proportional to the difference between the

concentration of liquid at the surface and the concentration required to maintain equilibrium with the surrounding atmosphere, the coefficient of proportionality being the rate of evaporation of the pure liquid.

3. The bead is spherical in shape, and its dimensions do not change during the process.

4. The diffusivity is constant, as found from experiments.

5. The concentration of liquid in the bead is uniform at the beginning of the process.

6. The volume of the surrounding atmosphere is very large with regard to the volume of the bead, and the vapour pressure in air is constantly equal to zero.

Mathematical Treatment

The transfer of liquid within the bead is expressed by Fick's equation for a sphere:

$$\frac{\partial C}{\partial t} = D\left[\frac{\partial^2 C}{\partial r^2} + \frac{2}{r}\frac{\partial C}{\partial r}\right] \tag{15.1}$$

with a constant diffusivity.

The rate of evaporation from the surface is defined by the surface condition:

$$-D\frac{\partial C}{\partial r} = \frac{F_0}{\rho}(C_s - C_{eq}) \tag{15.2}$$

where C_s is the actual concentration of liquid at the surface, C_{eq} is the liquid concentration required to maintain equilibrium with the surrounding atmosphere and F_0 the rate of evaporation of pure liquid in g cm^{-2} s^{-1}.

As the sphere is initially at uniform concentration C_{in} and diffusivity is constant, the required solution of the above equations is [19]:

$$\frac{C_{r,t} - C_{eq}}{C_{in} - C_{eq}} = \frac{2SR}{r}\sum_{n=1}^{\infty}\frac{\sin\dfrac{\beta_n r}{R}}{\sin\beta_n[\beta_n^2 + S^2 - S]}\exp\left(-\frac{\beta_n^2 D}{R^2}t\right) \tag{15.3}$$

where the β_ns are the roots of

$$\beta_n \cot\beta_n + S - 1 = 0 \tag{15.4}$$

with the dimensionless number S

$$S = \frac{F_0 R}{\rho D} \tag{15.5}$$

The amount of liquid leaving the sphere after time t is expressed as a fraction of the corresponding quantity after infinite time by the relation

$$\frac{Q_t}{Q_\infty} = 1 - \sum_{n=1}^{\infty}\frac{6S^2}{\beta_n^2(\beta_n^2 + S^2 - S)}\exp\left(-\frac{\beta_n^2 D}{R^2}t\right) \tag{15.6}$$

Some roots of Eq. (15.4) are given in Table 4.2 for various values of S.

Numerical Model

A numerical model with finite differences, capable of solving the problem when the diffusivity is concentration dependent or when the initial concentration of liquid is not uniform, is used [20].

The radius of the sphere is divided into N spherical slices of constant thickness Δr. The matter balance during the time Δt is calculated within the solid, by considering the spherical membrane of radius r (Fig. 15.1):

$$\left[-D\frac{\partial C}{\partial r}S + D\frac{\partial C}{\partial r}S\right]\Delta t = 4\pi r^2 \Delta r[CN_r - C_r] \tag{15.7}$$

$$\text{at } r-\frac{\Delta r}{2} \qquad \text{at } r+\frac{\Delta r}{2} \qquad\qquad\qquad \text{at } r$$

On defining the function $G(r)$ as

$$G\left(r - \frac{\Delta r}{2}\right) = \left(r - \frac{\Delta r}{2}\right)^2 (C_{r-\Delta r} - C_r)D_{r-\Delta r/2} \tag{15.8}$$

the new concentration within the membrane of radius r after time Δt is expressed in terms of the previous concentrations at the same place and of the function G:

$$CN_r = C_r + \frac{\Delta t}{r^2(\Delta r)^2}\left[G\left(r - \frac{\Delta r}{2}\right) - G\left(r + \frac{\Delta r}{2}\right)\right] \tag{15.9}$$

This general equation cannot be used for the centre of the sphere. The matter balance is calculated within the sphere of radius $\Delta r/2$ during the time increment Δt, and the new concentration at the centre of the sphere after time Δt is obtained as a function of the previous concentration and the function G:

$$CN_0 = C_0 - \frac{24\Delta t}{(\Delta r)^4}G\left(\frac{\Delta r}{2}\right) \tag{15.10}$$

with

$$G\left(\frac{\Delta r}{2}\right) = \left(\frac{\Delta r}{2}\right)^2 D_{\Delta r/2}(C_0 - C_1) \tag{15.8'}$$

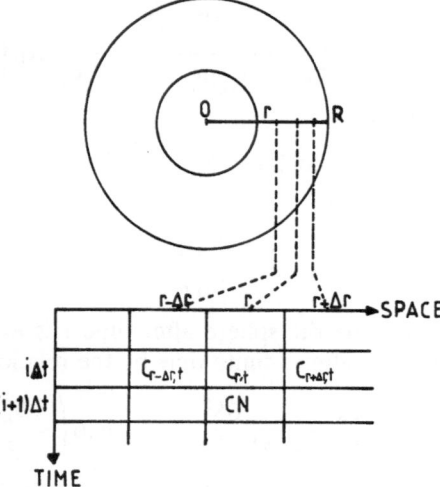

Fig. 15.1. Space–time diagram for numerical analysis for a spherical bead of radius R.

The matter balance is calculated during the increment of time Δt within the spherical membrane of thickness $\Delta r/2$ located next to the surface, by considering the diffusion within the solid and the evaporation from the surface:

$$\left[-D\frac{\partial C}{\partial r}S - \frac{F_0}{\rho}(C_R - C_{eq})S\right]\Delta t = 4\pi\left(R - \frac{\Delta r}{4}\right)^2\frac{\Delta r}{2}[CN_{R-\Delta r/4} - C_{R-\Delta r/4}]$$

at $\left(R - \dfrac{\Delta r}{2}\right)$ at R at $(R - \Delta r/4)$

(15.11)

With the simple assumption

$$CN_{R-\Delta r/4} - C_{R-\Delta r/4} = CN_R - C_R \tag{15.12}$$

the new concentration at the surface after time Δt is thus given as a function of the previous concentration and of the function G:

$$CN_R = C_R + \frac{2\Delta t}{\left(R - \dfrac{\Delta r}{4}\right)^2 (\Delta r)^2}G\left(R - \frac{\Delta r}{2}\right) - \frac{2R^2\Delta t F_0}{\left(R - \dfrac{\Delta r}{4}\right)^2 \Delta r \rho}(C_R - C_{eq})$$

(15.13)

The amount of liquid remaining in the bead at time t is calculated by integrating the concentrations at this time with respect to space:

$$Q_t = 4\pi \int_0^R C_r r^2 \, dr \tag{15.14}$$

which becomes with finite differences

$$Q_t = \frac{\pi}{6}(\Delta r)^3 C_0 + 4\pi(\Delta r)^3 \sum_{j=1}^{N-1} j^2 C_j + 2\pi(N - 0.25)^2(\Delta r)^3 C_{N-0.25} \tag{15.15}$$

by putting

$$r = j\Delta r \qquad R = N\Delta r \tag{15.16}$$

The value of the concentration at position $(N - 0.25)$ is obtained by

$$C_{N-0.25} = \tfrac{3}{4}C_N + \tfrac{1}{4}C_{N-1} \tag{15.17}$$

15.2.2 Experiment

Preparation of Dosage Forms

Eudragit RL, a copolymer of dimethylaminoethyl acrylate and ethylmethacrylate (Röhm Pharma) of MW = 150 000, in powder form, was the polymer matrix, and the drug used was sodium salicylate, also in powder form. The polymer matrix and drug were thoroughly mixed, and the mixture transformed into a viscous paste by adding ethanol. The paste was then pressed into spherical beads of different sizes. Their characteristics are given in Table 15.1.

Kinetics of Drying

The kinetics of drying were determined by weighing the beads at intervals. The beads were dried in open air at constant temperature (20 °C).

Table 15.1. Characteristics of the beads

	Radius (cm)		
	0.494	0.39	0.27
Total weight (mg)	697.2	352.3	251.5
Alcohol weight (mg)	120	65	50
% liquid (w/w)	17.2	18.5	19.9

The rate of evaporation of the liquid was obtained by following the weight of liquid evaporated from a flat flask of given area under the same conditions of temperature and surrounding atmosphere.

The effect of temperature on the process of drying was determined by drying the beads in a surrounding atmosphere of large volume, constantly stirred, at constant temperature. The selected values of temperature ranged from 20 to 60 °C.

15.2.3 Results

Results are presented for the determination of the diffusivity and rate of evaporation, and the validity of the model. Finally, an overview of all the parameters which affect the process of drying is given.

Determination of the Parameters

As the process of drying is controlled not only by the rate of evaporation but also by diffusion within the solid, the diffusivity and the rate of evaporation are parameters of interest.

Diffusivity

The diffusivity of the liquid was determined from the experimental kinetics of drying, by using the analytical solution Eq. (15.6). For long times, corresponding to a high value of Q_t, such as

$$0.7 < \frac{Q_t}{Q_\infty} < 1$$

the series in this equation converges very fast and the first term dominates. By neglecting the other terms, the equation can be written as follows:

$$\ln\left(\frac{Q_\infty - Q_t}{Q_\infty}\right) = -\frac{\beta_1^2 D}{R^2}t + \ln\left(\frac{6S^2}{\beta_1^2(\beta_1^2 + S^2 - S)}\right) \qquad (15.18)$$

By plotting the first term of this new equation as a function of time, a straight line is obtained from the slope of which the quantity $\beta_1^2 D/R^2$ is easily calculated.

The values of diffusivity can thus be evaluated by iterative calculus, because β_1 is also a function of diffusivity as shown in Eq. (15.5) (Fig. 15.2).

The diffusivity was found to be $4 \times 10^{-8} \pm 10^{-8}$ cm^2 s^{-1}.

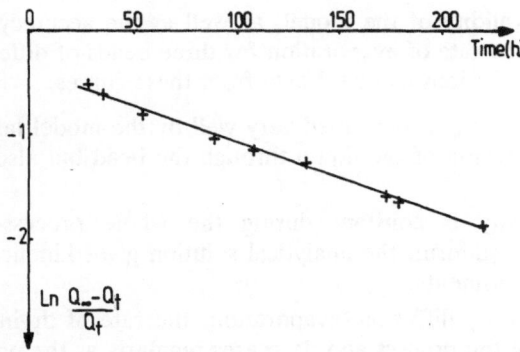

Fig. 15.2. $\ln(Q_\infty - Q_t/Q_\infty)$ as a function of the square root of time, for long drying time. Sample of weight 697.2 mg. (——, calculation; +, experiment.)

Rate of Evaporation

The rate of evaporation can be determined by two means:

- By evaporating the pure liquid under the same conditions of temperature and vapour pressure as those chosen for drying the bead. Special attention was given to the shape of the flask which was flat and full of liquid, as well as to the motion of air.

- By using the initial rate of drying of the beads, when the concentration of liquid is still uniform and constant, and therefore when the process may be assumed to be controlled by evaporation [15].

The rate of evaporation was found to be $2 \times 10^{-4}\ \mathrm{g\,cm^{-2}\,s^{-1}}$.

Validity of the Model

The validity of the model was tested by comparing the kinetics of drying of the different beads obtained either from experiments or by calculation. Figs 15.3 to

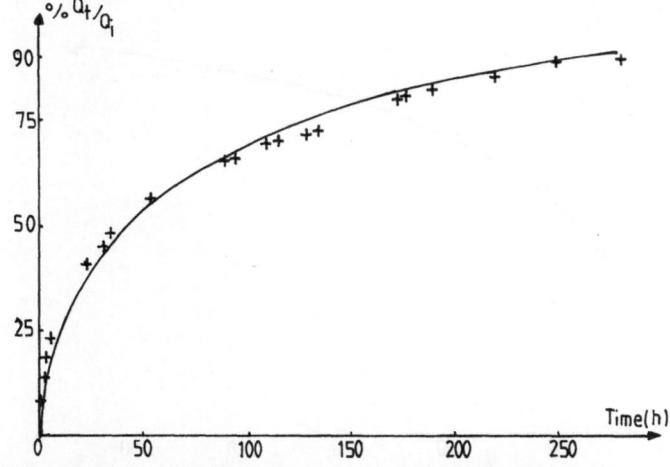

Fig. 15.3. Kinetics of drying at 20 °C for the bead of weight 697.2 mg and radius 0.494 cm. (——, calculation; +, experiment.)

15.5 illustrate the validity of the model, as well as the accuracy of the values of
the diffusivity and the rate of evaporation for three beads of different sizes.

The following conclusions can be drawn from these curves:

1.   The process of drying is described very well by the model taking into account
     not only the diffusion of the liquid through the bead but also the evaporation
     from the surface.
2.   As the diffusivity is constant during the whole process and the initial
     concentration is uniform, the analytical solution gives kinetics which correlate
     very well to experiments.
3.   Being controlled by diffusion–evaporation, the rate of drying is very high at
     the beginning of the process and decreases regularly as the process proceeds.
4.   Near the end of the process, the rate of drying becomes very low, and an
     asymptote is attained when the process of drying approaches completion.

Effect of the Parameters

Three parameters of interest are examined: the radius of the sphere, the
diffusivity and the rate of evaporation.

Radius

The size of the bead affects the evaporation as well as the diffusion. On the one
hand, a smaller radius means a larger external area per unit mass of solid, and
thus provokes faster drying. On the other hand, an increase in radius is followed
by an increase in the time of diffusion. This second fact is proved by the
dimensionless number Dt/R^2, showing that the time for diffusion is proportional
to the square of the radius.

Some profiles of concentration of liquid developed within the bead are shown
in Fig. 15.6 for various values of the radius ranging from 0.2 to 0.6 cm, at a
constant temperature of 20 °C.

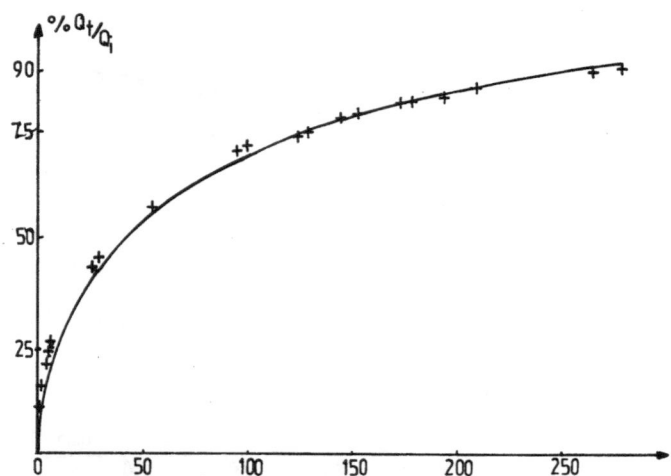

Fig. 15.4. Kinetics of drying at 20 °C for the bead of weight 352.3 mg and radius 0.39 cm. (——,
calculation; +, experiment.)

Diffusivity

The value of the diffusivity generally depends on the nature of the liquid and polymer [20]. It increases with temperature, often following an Arrhenius law with a constant energy of activation [21–23]. However, as it is not possible to predict by calculation the effect of temperature on the diffusivity, experiments are necessary to obtain this knowledge.

Rate of Evaporation

The rate of evaporation is a well-known parameter. In the case of a surrounding atmosphere of large volume, it is proportional to the vapour pressure of the liquid, this vapour pressure being expressed in terms of temperature by the Clausius–Clapeyron law.

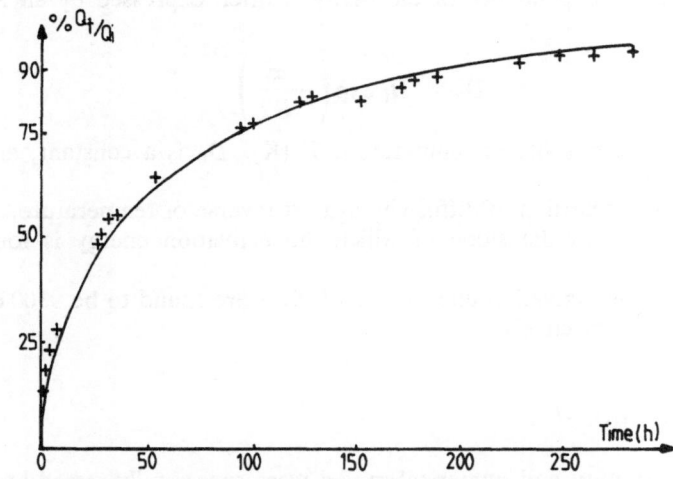

Fig. 15.5. Kinetics of drying at 20 °C for the bead of weight 251.5 mg and radius 0.27 cm. (——, calculation; +, experiment.)

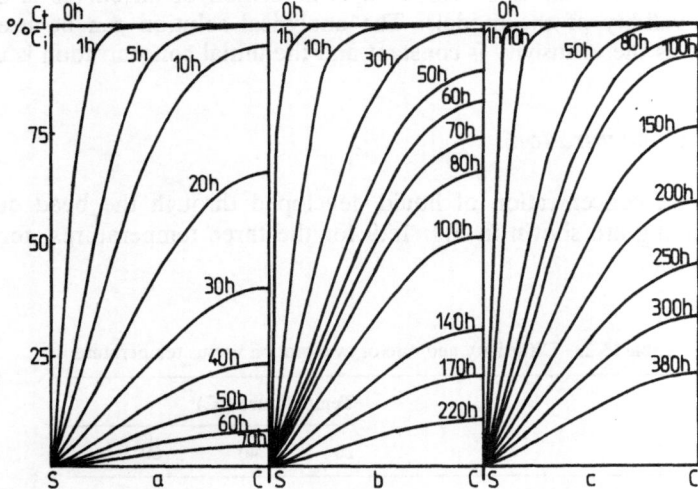

Fig. 15.6. Profiles of concentration developed at 20 °C within dosage forms of radii **a** 0.2 cm; **b** 0.4 cm; **c** 0.6 cm.

15.2.4 Effect of Temperature

Results are considered for the diffusivity as a function of temperature, the validity of the model for different temperatures and the effect of temperature on the profiles of concentration of liquid in the beads.

Diffusivity

The diffusivity was determined at various temperatures, by using the method described in the previous section and Eq. (15.18). The values of the diffusivity and rate of evaporation are shown in Table 15.2.

The temperature dependence of diffusivity is often expressed by an Arrhenius equation:

$$D_T = D_0 \exp\left(-\frac{E}{RT}\right) \tag{15.19}$$

where D_T is the diffusivity at temperature T (K), D_0 is a constant, and E the activation energy.

By plotting the logarithm of diffusivity against inverse of temperature, a straight line is obtained, from the slope of which the activation energy is found (Fig. 15.7).

The values of the activation energy and of D_0 were found to be 9300 cal mol^{-1} and 0.46 cm^2 s^{-1} respectively.

Validity of the Model

The analytical solution and numerical model were successfully tested by comparing the theoretical and experimental kinetics at the three temperatures ranging from 20 to 60 °C [24]. The effect of temperature on the kinetics of drying is clearly illustrated in Fig. 15.8. Good superimposition of all curves is observed, proving the validity of the models. The analytical solution can be used in the present case, as the diffusivity is constant and the initial concentration is uniform.

Profiles of Concentration

The profiles of concentration of liquid developed through the bead during the process of drying are shown in Fig. 15.9 for the three temperatures, for samples of radius 0.4 cm.

Table 15.2. Diffusivity and rate of evaporation versus temperature

	Temperature (°C)		
	20	40	60
$D \times 10^7$ (cm^2 s^{-1})	0.6	1.6	4.0
$F_0 \times 10^4$ (g cm^{-2} s^{-1})	2.7	7.8	19.8

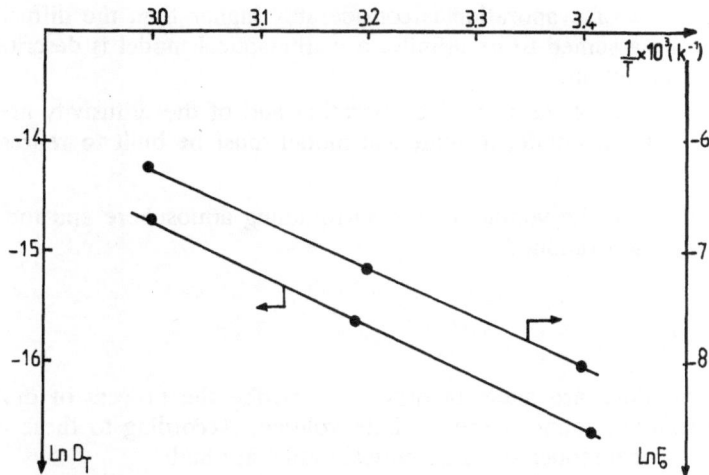

Fig. 15.7. ln D_T and ln F_0 versus the inverse of temperature.

Conclusions

The following conclusions are drawn from these results:

1. The effect of temperature on the kinetics of drying, illustrated in Fig. 15.8, is of concern. This parameter acts not only on the rate of evaporation but also on the rate of diffusion.
2. The effect of temperature on the profiles of concentration developed through the bead is very significant. These curves give complementary information, and a fuller insight into the process.
3. The diffusivity is expressed in terms of temperature by an Arrhenius law with constant activation energy. The diffusivity and the rate of evaporation vary with temperature according to an exponential law.

15.3   Drying in a Finite Atmosphere

The process of drying of solids, and especially of polymers, is very often controlled by diffusion of the liquid within the solid and evaporation from the surface. Thus the diffusivity and rate of evaporation are of great concern in the kinetics of drying. However, a third parameter, the volume of the surrounding atmosphere is also of interest. This parameter can play the role of a limiting factor in the sense that it defines the fraction of the liquid which is evaporated after infinite time, when equilibrium is reached.

The purpose of this section is to show how to build a mathematical and numerical model capable of describing the process of drying of a solid located in a surrounding atmosphere of finite volume. The following two cases are considered, according to the value of the rate of evaporation of the liquid as compared with the value of the diffusivity:

- Where the rate of evaporation is considerably higher than the diffusivity, so that it can be assumed to be infinite, a mathematical model is described with an analytical solution.
- Where the values of the rate of evaporation and of the diffusivity are of the same order of magnitude, a numerical model must be built to represent the process.

Thus the effects of the volume of the surrounding atmosphere and the rate of evaporation are also examined.

15.3.1 Theory

Different assumptions are made in order to describe the process of drying the solid in a surrounding atmosphere of finite volume. According to these assumptions, a mathematical model and a numerical model are built.

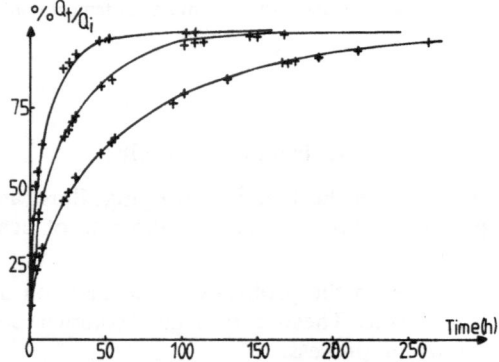

Fig. 15.8. Kinetics of drying for dosage forms at (*bottom to top:*) 20, 40 and 60 °C. (——, calculation; +, experiment.)

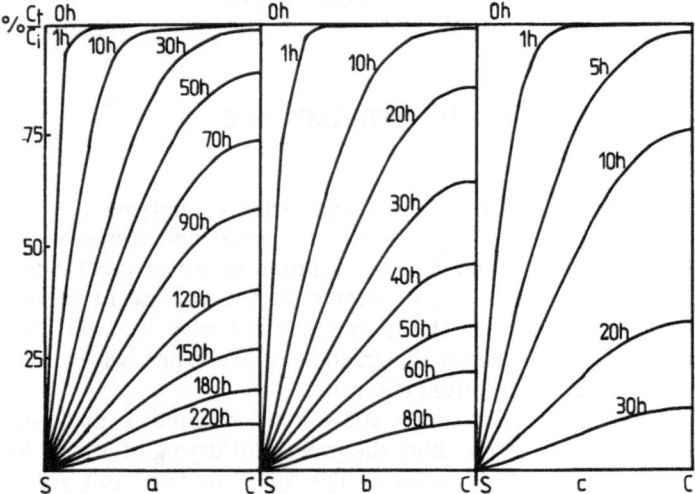

Fig. 15.9. Profiles of concentration developed within the dosage forms of radius 0.4 cm at **a** 20 °C; **b** 40 °C; **c** 60 °C.

Assumptions

The following assumptions are made:

1. The process of drying is controlled by diffusion of the liquid within the solid and evaporation from the surface.
2. The volume of the surrounding atmosphere is finite.
3. An analytical solution is found when:
   - The diffusivity is constant
   - The initial concentration in the solid is uniform
   - The surrounding atmosphere is initially free of vapour
   - The rate of evaporation is infinite
4. The numerical model must be used when the above assumptions shown in 3 do not hold.
5. The rate at which the liquid evaporates from the surface is constantly equal to the rate at which the liquid is brought by diffusion to the evaporating surface.

Mathematical Treatment

The general equation of diffusion in a sphere is given by

$$\frac{\partial C}{\partial t} = D\left[\frac{\partial^2 C}{\partial r^2} + \frac{2}{r}\frac{\partial C}{\partial r}\right] \tag{15.1}$$

where D is the constant diffusivity.
   The initial conditions are:
$$t = 0 \quad 0 < r < R, \quad C = C_{in} \quad \text{(sphere)}$$
$$V, \quad C = 0 \quad \text{(atmosphere)} \tag{15.20}$$

expressing that the initial concentration of liquid in the sphere is uniform, and that the volume V of the surrounding atmosphere is free of vapour.
   The boundary conditions express the fact that the rate at which the liquid evaporates from the surface of the sphere is always equal to the rate at which the vapour enters the surrounding atmosphere and to the rate at which the liquid is brought by diffusion to the evaporating surface:

$$t > 0, \quad V\frac{\partial C}{\partial t} = -4\pi R^2 D\frac{\partial C}{\partial r}, \quad r = R \quad \text{(surface)} \tag{15.21}$$

The total amount of liquid which has evaporated after time t, Q_t is expressed as a fraction of the corresponding quantity after infinite time, by the relation [19]

$$\frac{Q_\infty - Q_t}{Q_\infty} = \sum_{n=1}^{\infty} \frac{6\alpha(1 + \alpha)}{9 + 9\alpha + q_n^2\alpha^2} \exp\left(-\frac{q_n^2 D}{R^2}t\right) \tag{15.22}$$

where the q_ns are the non-zero roots of

$$\tan q_n = \frac{3q_n}{3 + \alpha q_n^2} \tag{15.23}$$

and α is the ratio of the volumes of the surrounding atmosphere and sphere:

$$\alpha = \frac{3V}{4\pi R^3} \tag{15.24}$$

This ratio is rewritten when there is a partition factor K between the evaporating substance in equilibrium in the sphere and the surrounding atmosphere:

$$\alpha = \frac{3V}{4\pi R^3 K} \qquad (15.24')$$

The ratio α is expressed in terms of the final fractional uptake of evaporating substance by the surrounding atmosphere by the relation

$$\frac{3Q_\infty}{4\pi R^3 C_{in}} = \frac{\alpha}{1 + \alpha} \qquad (15.25)$$

The concentration of the evaporating substance at the position r within the sphere is given by

$$\frac{C_{in} - C_{r,t}}{C_{in} - C_\infty} = 1 + \frac{R}{r} \sum_{n=1}^{\infty} \frac{q_n \frac{r}{R}}{\sin q_n} \frac{6(1 + \alpha)}{9 + 9\alpha + q_n^2 \alpha^2} \exp\left(-\frac{q_n^2 D}{R^2} t\right) \qquad (15.26)$$

where C_∞ is the concentration of the substance in the sphere after infinite time, at equilibrium.

The concentration of the evaporating substance at the external surface of the sphere is thus obtained by a more simple relation (with $r = R$ in Eq. (15.26)):

$$\frac{C_{in} - C_{r,t}}{C_{in} - C_\infty} = 1 + \sum_{n=1}^{\infty} \frac{6(1 + \alpha)}{9 + 9\alpha + q_n^2 \alpha^2} \frac{q_n}{\sin q_n} \exp\left(-\frac{q_n^2 D}{R^2} t\right) \qquad (15.27)$$

Some roots of Eq. (15.23) are given in Table 4.1, for various values of the ratio of the volumes α.

Numerical Analysis

There are many conditions for the analytical solution to be used: the diffusivity must be constant, the initial concentration in the solid uniform, the surrounding atmosphere initially free of vapour, and the rate of evaporation infinite [25].

In all other cases, the process of drying of a solid sphere in a surrounding atmosphere of finite volume can be described with the help of a numerical model which takes into account all the facts.

The radius of the sphere R is divided into N spherical membranes of equal thickness Δr. The matter balance is calculated at many places during the increment of time Δt, by considering the transport of matter by diffusion and by evaporation.

The new concentration after time Δt is thus determined as a function of the previous concentrations:

Within the Sphere

$$CN_r = C_r + \frac{\Delta t}{r^2 (\Delta r)^2} \left[G\left(r - \frac{\Delta r}{2}\right) - G\left(r + \frac{\Delta r}{2}\right) \right] \qquad (15.9)$$

with the function G

$$G\left(r - \frac{\Delta r}{2}\right) = \left(r - \frac{\Delta r}{2}\right)^2 (C_{r-\Delta r} - C_r) D_{r-\Delta r/2} \qquad (15.8)$$

Centre of the Sphere
The new concentration after time Δt is expressed in terms of the previous concentration at this place by

$$CN_0 = C_0 - \frac{24\Delta t}{(\Delta r)^4}G\left(\frac{\Delta r}{2}\right) \qquad (15.10)$$

where the function G is

$$G\left(\frac{\Delta r}{2}\right) = \left(\frac{\Delta r}{2}\right)^2(C_0 - C_1)D_{\Delta r/2} \qquad (15.8')$$

Surface
With the simple assumption made next to the surface:

$$CN_{R-\Delta r/4} - C_{R-\Delta r/4} = CN_R - C_R \qquad (15.12)$$

the new concentration at the surface after time Δt is expressed in terms of the previous concentration and function G:

$$CN_R = C_R + \frac{2\Delta t}{\left(R - \frac{\Delta r}{4}\right)^2(\Delta r)^2}G\left(R - \frac{\Delta r}{2}\right) - \frac{2R^2\Delta t F_0}{\left(R - \frac{\Delta r}{4}\right)^2\Delta r\rho}(C_R - C_{eq})$$

$$(15.13)$$

This equation is obtained by making the assumption that the rate of evaporation is constantly equal to the rate at which the liquid is brought by diffusion to the evaporating surface:

$$-D\frac{\partial C}{\partial r} = \frac{F_0}{\rho}(C_R - C_{eq}) \qquad (15.2)$$

where C_{eq} is the concentration on the surface required to maintain equilibrium with the surrounding atmosphere.

Equilibrium Concentration C_{eq}
Two assumptions are made:

- The vapour is an ideal gas.
- The concentration on the surface at equilibrium with the vapour can be related to the pressure of this vapour.

The ideal gas law is

$$PV = n_t RT \qquad (15.28)$$

where R is the universal gas constant, V is the volume of the surrounding atmosphere, P is the pressure of the vapour contained in the volume V at temperature T (kelvin) and n the number of moles of vapour at time t.

Various relationships are capable of describing the variation of the concentration on the surface C_{eq} with the pressure of vapour P. Two of them are considered.

One represents a linear absorption isotherm:

$$C_{eq} = kP \qquad (15.29)$$

The other when the absorption isotherm follows Langmuir's equation:

$$C_{eq} = k' \frac{bP}{1 + bP} \qquad (15.30)$$

where k, k' and b are constants.

The amount of liquid evaporated at time t is related to the number of moles of vapour n_t by

$$n_t = \frac{Q_t}{MW} \qquad (15.31)$$

where MW is the molecular weight of the evaporated substance.

Thus in the case of a linear absorption isotherm, the simple relationship is obtained between C_{eq} and Q_t:

$$C_{eq} = K_1 Q_t \qquad (15.32)$$

where K_1 is given by

$$K_1 = \frac{kRT}{VMW} \qquad (15.33)$$

In the case of the non-linear isotherm of absorption, the other relationship between C_{eq} and Q_t is obtained:

$$C_{eq} = k' \frac{K_2 Q_t}{1 + K_2 Q_t} \qquad (15.34)$$

where the constant K_2 is

$$K_2 = \frac{bRT}{VMW} \qquad (15.35)$$

Amount of Liquid Remaining in the Sphere
The amount of liquid remaining in the sphere can be determined by integrating the concentration of the liquid located in the sphere with respect to space, as shown in Eq. (15.15).

15.3.2 Experiment

Materials

The dosage form consisted of sodium salicylate as the drug in a polymer matrix made of Eudragit RL (Röhm Pharma). The beads had a radius of 0.38 cm and a empty weight of 222 mg. The initial liquid content (ethanol) was 49 mg.

Operational Conditions

The values of the diffusivity and the rate of evaporation were obtained at 60 °C with ethanol as the evaporating substance [24], and were found to be 3.5×10^{-7} cm^2 s^{-1} and 19.8×10^{-4} g cm^{-2} s^{-1} respectively.

15.3.3 Results

The process of drying was simulated by using either the mathematical model with an infinite rate of evaporation or the numerical model, which may be used in all cases. This approach is of interest, in the sense that it enables a thorough investigation throughout the various parts of the process, and it is thus capable of determining the role played by each factor [25].

Two parameters are especially considered:

- The value of the rate of evaporation F_0
- The volume of the surrounding atmosphere, defined by the value of the constant K_1

Effect of the Rate of Evaporation

The effect of the role played by the rate of evaporation F_0 on the process was explored by varying the value of this parameter and keeping constant the value of the other parameters, which were as obtained above, with $K_1 = 1$.

Three kinds of results obtained from calculation are given: the kinetics of drying, the change in concentration of liquid at the surface during the process, and the profiles of concentration of liquid developed through the dosage form.

Kinetics of Drying

The kinetics of drying are shown in Fig. 15.10 for various values of F_0 ranging from $19.8 \times 10^{-8} \, \text{g cm}^{-2} \text{s}^{-1}$ to infinity, while the other parameters are kept constant as stated above. The kinetics for an infinite rate of evaporation were determined with the help of the mathematical mode, and the kinetics with the finite rates of evaporation using the numerical model.

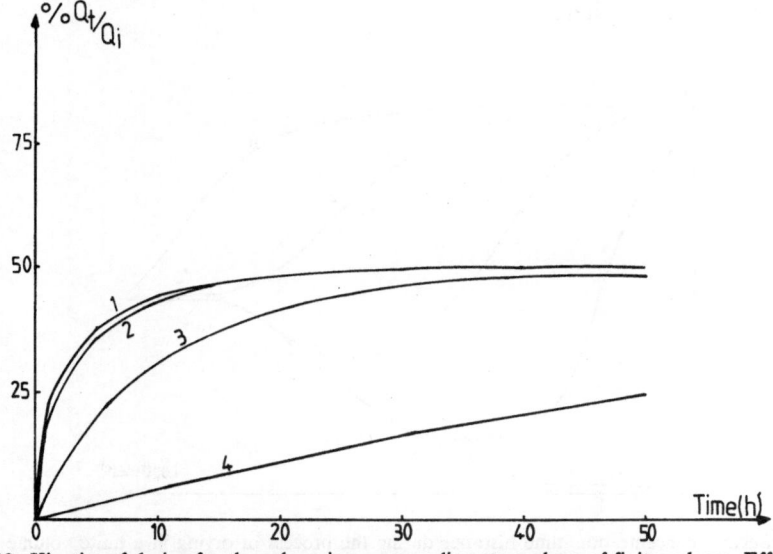

Fig. 15.10. Kinetics of drying for the sphere in a surrounding atmosphere of finite volume. Effect of the value given to the rate of evaporation of the liquid. ($D = 3.5 \times 10^{-7} \, \text{cm}^2 \text{s}^{-1}$, $R = 0.38 \, \text{cm}$, $K_1 = 1$.) 1: $F_0 = \infty$ or $19.8 \times 10^{-4} \, \text{g cm}^{-2} \text{s}^{-1}$ 2: $F_0 = 19.8 \times 10^{-6}$; 3: $F_0 = 19.8 \times 10^{-7}$; 4: $F_0 = 19.8 \times 10^{-8}$.

A fact of interest concerned with the rate of evaporation is observed: the kinetics calculated with infinite F_0 and $F_0 = 19.8 \times 10^{-4}$ superimposed very well, showing that the process of drying is controlled only by diffusion when the rate of evaporation is from 500 to 5000 times higher than the value of diffusivity.

Surface Concentration–Time Histories

The concentration of liquid at the surface provides interesting information on the process and can be easily obtained by calculation. It is very difficult to take measurements on the surface, as shown in a previous study on plasticised PVC using the attenuated total reflectance technique [26].

The values of the ratio of the concentration at the surface at time t to the corresponding value at time zero were plotted as a function of the logarithm of time (expressed in seconds) (Fig. 15.11).

Some results of interest are

1. For an infinite value of the rate of evaporation, the concentration at the surface falls abruptly to zero as soon as the process starts.
2. For a fairly high value of the rate of evaporation (19.8×10^{-4}), the concentration at the surface sharply declines during the first 10 s of the process.
3. For the values of the rate of evaporation which are higher than the value of the diffusivity, the surface concentration–time histories pass through a minimum. The time at which this minimum is obtained increases with decreasing the rate of evaporation.
4. This minimum is not observed when the value of the rate of evaporation is lower or is of the same order of magnitude as the value of diffusivity.
5. For times longer than 10 000 s, the surface concentration reaches the value at equilibrium. In the case of $K_1 = 1$, the value at equilibrium is half the initial value, since the liquid is distributed equally between the solid and the surrounding atmosphere.

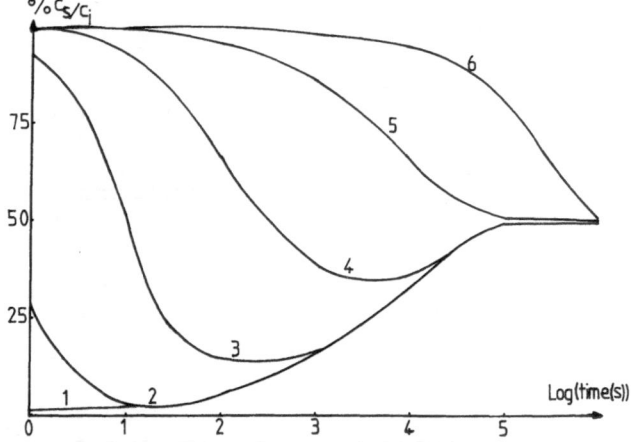

Fig. 15.11. Surface concentration–time histories during the process of drying in a finite volume of air. Effect of the value given to the rate of evaporation. ($D = 3.5 \times 10^{-7}$ cm^{2}s^{-1}, $R = 0.38$ cm, $K_1 = 1$. 1: $F_0 = \infty$; 2: $F_0 = 19.8 \times 10^{-4}$ g cm^{-2}s^{-1}; 3: $F_0 = 19.8 \times 10^{-5}$; 4: $F_0 = 19.8 \times 10^{-6}$; 5: $F_0 = 19.8 \times 10^{-7}$; 6: $F_0 = 19.8 \times 10^{-8}$.

Profiles of Concentration

The profiles of concentration developed within the spherical dosage form were also calculated for various values of the rate of evaporation using the numerical models (Figs 15.12 and 15.13). This additional information helps to understand the process of drying, and leads to the following conclusions:

1. Steep gradients of concentration are developed at the beginning of the process of drying, with a very low value of the concentration on the surface. The following statement can be made: the higher the rate of evaporation, the steeper the gradient of concentration during the early stage of the process.

2. These gradients of concentration then become less and less steep as drying proceeds, while the concentration at the surfaces increases slowly to the equilibrium value.

Effect of the Volume of the Surrounding Atmosphere

The value of the volume of the surrounding atmosphere with regard to the volume of the dosage form, is a parameter of prime importance. The effect of this parameter, expressed by the value of the constant K_1, is determined with the help of the numerical model, by varying the values given to K_1 and keeping the other parameters constant at the values initially determined.

Kinetics of Drying

The effect of the value given to K_1 on the process of drying is clearly shown in Fig. 15.14, where the kinetics of drying are plotted for various values of K_1 ranging from 1 to 0.

Some conclusions are worth noting:

1. The volume of the surrounding atmosphere affects the amount of liquid which evaporates at equilibrium.

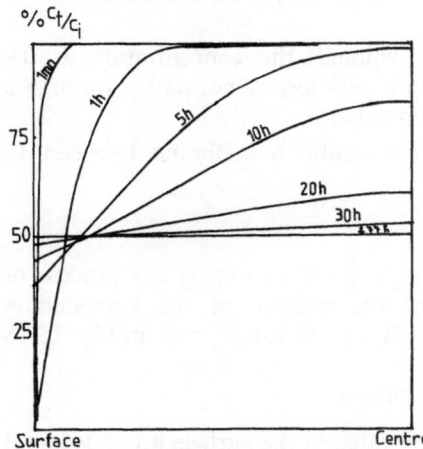

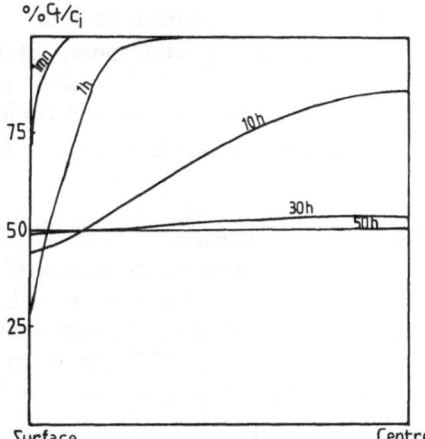

Fig. 15.12. Profiles of concentration developed through the dosage form during the process of drying, at various times (h). ($D = 3.5 \times 10^{-7}$ cm^2 s^{-1}, $R = 0.38$ cm, $K_1 = 1$, $F_0 = 19.8 \times 10^{-4}$ g cm^{-2} s^{-1}.)

Fig. 15.13. Profiles of concentration developed through the dosage form during the process of drying, at various times (h). ($D = 3.5 \times 10^{-7}$ cm^2 s^{-1}, $R = 0.38$ cm, $K_1 = 1$, $F_0 = 19.8 \times 10^{-6}$ g cm^{-2} s^{-1}.)

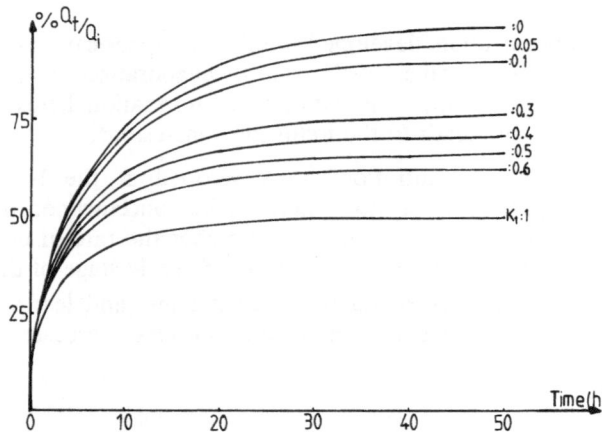

Fig. 15.14. Kinetics of drying of the sphere, for various volumes of the surrounding atmosphere (K_1). ($D = 3.5 \times 10^{-7}$ cm^2 s^{-1}, $R = 0.38$ cm, $F_0 = 19.8 \times 10^{-4}$ g cm^{-2} s^{-1}.)

2. When $K_1 = 0$, meaning that the volume of surrounding atmosphere is infinite, all the liquid evaporates from the sphere.
3. When $K_1 = 1$, the liquid at equilibrium is equally distributed between the solid and the surrounding atmosphere.

Surface Concentration–Time History
The surface concentration was calculated for various times and various values of the constant K_1. The surface concentration-time histories were then plotted for various values of K_1, for $F_0 = 19.8 \times 10^{-4}$ g cm^2 s^{-1}, using a logarithm scale for time (Fig. 15.15).
A few results are worth noting:

1. For a surrounding atmosphere of infinite volume ($K_1 = 0$), the concentration at the surface falls quickly to zero.
2. For a surrounding atmosphere of finite volume, the concentration at the surface drops first from the initial value to a very low value, passes through a minimum, and then rises to the equilibrium value.
3. These surface concentration–time histories exhibit a minimum between 10 and 100 s.

Profiles of Concentration
The profiles of concentration developed through the bead during the process of drying are calculated for various values of the volume of the surrounding atmosphere (K_1). These profiles are shown in Fig. 15.12 for $K_1 = 1$, in Fig. 15.16 for $K_1 = 0.5$, and in Fig. 15.17 for $K_1 = 0.1$.
A few conclusions can be drawn from these curves:

1. Steep gradients of concentration with a low value at the surface are developed during the drying process, especially in the early stage of the process.
2. As drying progresses, the gradients of concentration become less and less steep, until a flat gradient is reached at equilibrium.
3. Of course, the lower the value of K_1, the lower the value of the concentration at equilibrium in the dosage form.

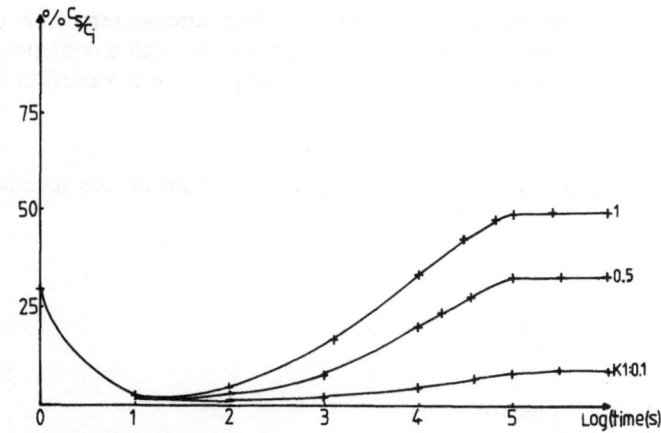

Fig. 15.15. Surface concentration–time history for various volumes of the surrounding atmosphere (K_1). $(D = 3.5 \times 10^{-7}\ cm^2\ s^{-1},\ R = 0.38\ cm,\ F_0 = 19.8 \times 10^{-4}\ g\,cm^{-2}\,s^{-1}.)$

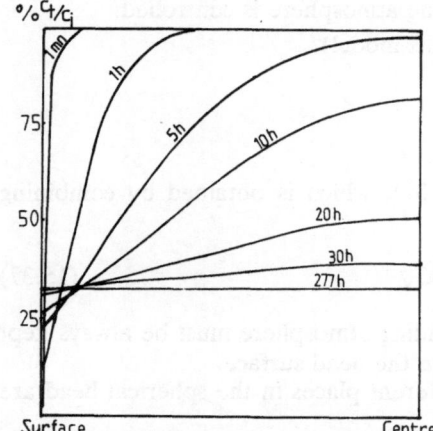

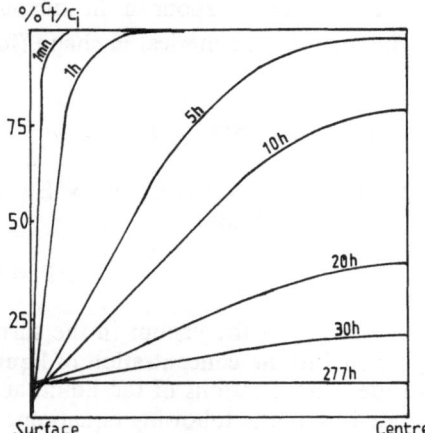

Fig. 15.16. Profiles of concentration developed through the dosage form during the process of drying, at various times (h). $(D = 3.5 \times 10^{-7}\ cm^2\,s^{-1},\ R = 0.38\ cm,\ K_1 = 0.5,\ F_0 = 19.8 \times 10^{-4}\ g\,cm^{-2}\,s^{-1}.)$

Fig. 15.17. Profiles of concentration developed through the dosage form during the process of drying, at various times (h). $(D = 3.5 \times 10^{-7}\ cm^2\,s^{-1},\ R = 0.32\ cm,\ K_1 = 0.1,\ F_0 = 19.8 \times 10^{-4}\ g\,cm^{-2}\,s^{-1}.)$

15.4 Drying Under Controlled Vapour Pressure

For a given liquid–polymer combination, the diffusivity and rate of evaporation depend on temperature. When the rate of evaporation is higher than the diffusivity, steep gradients of concentration are developed within the solid with a low concentration of liquid on the surface. This fact is a real drawback when the diffusivity is concentration dependent (diffusivity decreasing when the concentration is decreased), because the rate of transport by diffusion becomes so low next to the surface that part of the liquid is bound to be entrapped.

A solution in this difficult case can be given with the help of the numerical model, by acting upon the vapour pressure of the liquid in the surrounding atmosphere which is related to the surface concentration at equilibrium C_{eq}.

Instead of drying the material in a surrounding atmosphere free of vapour, drying must be carried out in a surrounding atmosphere with a controlled pressure of vapour. An example is given where the vapour pressure is varied in such a way that

$$C_{eq} = KIC_R \qquad (15.36)$$

where KI is a constant and C_R the concentration of liquid on the surface [27].

15.4.1 Theory

Assumptions

The following assumptions are made:

1. The process of drying is controlled by diffusion and evaporation.
2. The volume of the surrounding atmosphere is finite.
3. The pressure of vapour in the surrounding atmosphere is controlled.
4. The material is spherical in shape (for this model).

Numerical Analysis

The main relationship is given by Eq. (15.37), which is obtained by combining Eqs (15.29) and (15.36):

$$P = \frac{KI}{k}C_R \qquad (15.37)$$

The pressure of the vapour in the surrounding atmosphere must be always kept proportional to the concentration of liquid on the bead surface.

All the concentrations of the liquid at different places in the spherical bead are calculated using the following equations:

*   Centre   Eq. (15.10)
*   Within the sphere   Eq. (15.9)
*   Surface of the sphere   Eq. (15.13)
*   Vapour pressure   Eq. (15.37)

15.4.2 Simulation Using a Numerical Model

The process of drying was simulated by using the numerical model. The effect of the following two parameters on the process are considered: the value of the constant KI, and the value of the rate of evaporation.

Effect of KI

Calculations were made with the help of the numerical model, keeping the radius, diffusivity, rate of evaporation and initial amount of liquid present constant at the values initially determined.

The kinetics of drying were calculated, keeping these parameters constant, for each value given to the constant KI. The effect of the constant KI on the process is illustrated in Fig. 15.18, where KI ranges from 0.9 to 0.9999.

The surface concentration–time histories are also shown for various values of KI (Fig. 15.19).

The profiles of concentration developed throughout the bead were calculated for various times, when $KI = 0.9$ (Fig. 15.20) and when $KI = 0.999$ (Fig. 15.21).

The following conclusions can be drawn:

1. The effect of the value of the pressure of vapour in the surrounding atmosphere, throughout the range of KI, is of great concern.

2. A high value of the constant KI is responsible for the development of almost flat gradients of concentration.

3. The surface concentration decreases very slowly especially at the beginning of the process, when the constant KI is high (0.9999).

4. In the present case, the rate of evaporation is very high with regard to the diffusivity, provoking the formation of very steep gradients of concentration.

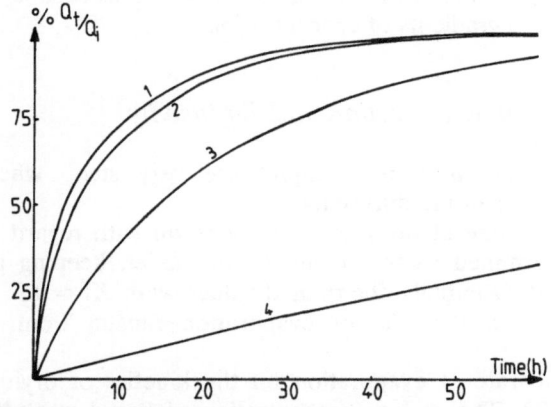

Fig.15.18. Kinetics of drying (calculated) for various values of the constant KI. 1: 0.9; 2: 0.99; 3: 0.999; 4: 0.9999. ($D = 3.5 \times 10^{-7}$ cm^2 s^{-1} $R = 0.38$ cm $F_0 = 19.8 \times 10^{-4}$ g cm^{-2} s^{-1}.)

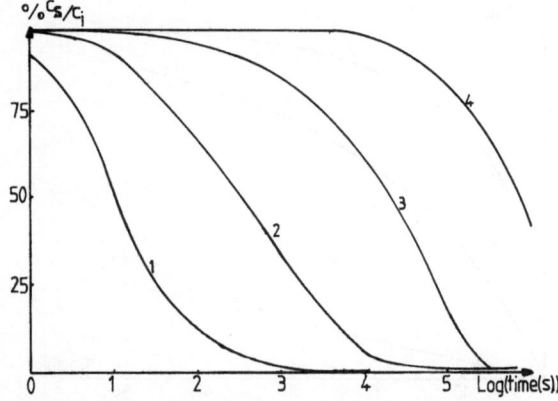

Fig. 15.19. Surface concentration (calculated) for various values of the constant KI. 1: 0.9; 2: 0.99; 3: 0.999; 4: 0.9999. ($D = 3.5 \times 10^{-7}$ cm^2 s^{-1} $R = 0.38$ cm $F_0 = 19.8 \times 10^{-4}$ g cm^{-2} s^{-1}.)

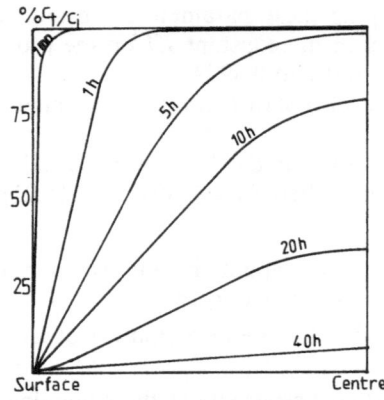

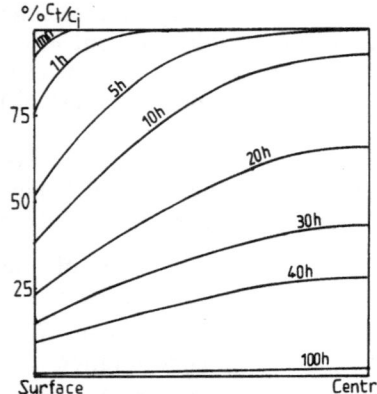

Fig. 15.20. Profiles of concentration (calculated) with controlled vapour pressure. ($D = 3.5 \times 10^{-7}$ cm^2 s^{-1}, $R = 0.38$ cm, $F_0 = 19.8 \times 10^{-4}$ g cm^{-2} s^{-1}, $KI = 0.9$.)

Fig. 15.21. Profiles of concentration (calculated) with controlled vapour pressure. ($D = 3.5 \times 10^{-7}$ cm^2 s^{-1}, $R = 0.38$ cm, $F_0 = 19.8 \times 10^{-4}$ g cm^{-2} s^{-1}, $KI = 0.999$.)

The use of a high value of the vapour pressure is thus capable of reducing the steepness of these gradients of concentration.

Effect of the Rate of Evaporation and Diffusivity

The gradients of concentration of liquid are very steep when the rate of evaporation is higher than the diffusivity.

The effect of the value of the rate of evaporation with regard to the value of diffusivity was determined using the numerical model, keeping the other parameters constant at their initially determined values, with $KI = 0.9$.

The values given to the rate of evaporation ranged from 19.8×10^{-4} to 19.8×10^{-7} g cm^{-2} s^{-1}.

The effect of the rate of evaporation on the kinetics of drying appears very clearly in Fig. 15.22. The surface concentration of liquid was also plotted as a function of log (time) for different values of the rate of evaporation (Fig. 15.23).

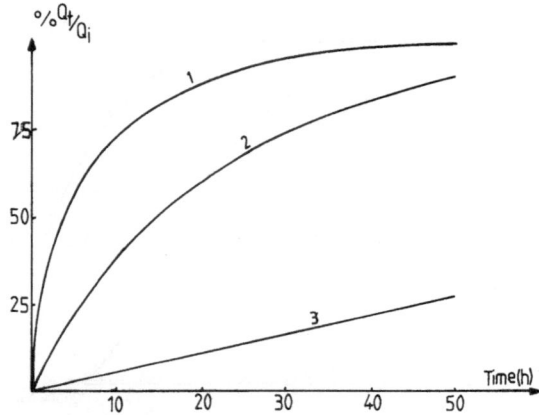

Fig. 15.22. Kinetics of drying (calculated) for various values of the rate of evaporation F_0. 1: 19.8×10^{-4}; 2: 19.8×10^{-6}; 3: 19.8×10^{-7}. ($D = 3.5 \times 10^{-7}$ cm^2 s^{-1}, $R = 0.38$ cm, $KI = 0.9$.)

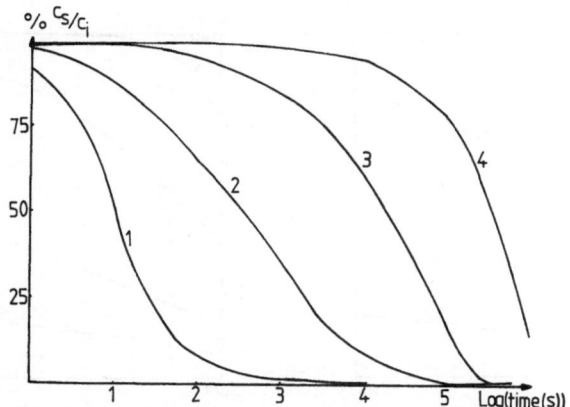

Fig. 15.23. Surface concentration (calculated) for various values of the rate of evaporation F_0. 1: 19.8×10^{-4}; 2: 19.8×10^{-5}; 3: 19.8×10^{-6}; 4: 19.8×10^{-7}. $(D = 3.5 \times 10^{-7} \text{ cm}^2\text{s}^{-1}, R = 0.38 \text{ cm}, KI = 0.9.)$

The profiles of concentration developed through the bead were calculated for two values of the rate of evaporation: 19.8×10^{-5} (Fig. 15.24), and 19.8×10^{-6} (Fig. 15.25).

The following results are worth noting:

1. Steep gradients of concentration are developed when the rate of evaporation is higher than the diffusivity, when the constant KI is equal to 0.9.

2. The gradients of concentration become very flat when the rate of evaporation is of the same order of magnitude as the diffusivity. In this case, it is not necessary to use a high value of KI, and the value of 0.9 is high enough.

3. Higher values of the constant KI are necessary to reduce the steepness of the high gradients developed when the rate of evaporation is higher than the diffusivity.

15.5 Conclusions

General conclusions on the drying process of dosage forms made of polymers can be drawn, either about the process or about the parameters.

Process of Drying
The process of drying is controlled by diffusion of liquid within the solid and by evaporation from the surface. The values of the diffusivity and of the rate of evaporation are of interest, as well as the value of the dimensionless number S.

Surrounding Atmosphere
The effect of the volume of the surrounding atmosphere is also of concern. It acts essentially upon the value of the amount of liquid distributed through the bead and the surrounding atmosphere, after infinite time when equilibrium is reached.

In the case of a surrounding atmosphere with finite volume, the vapour pressure is an interesting parameter. By controlling the value of this vapour

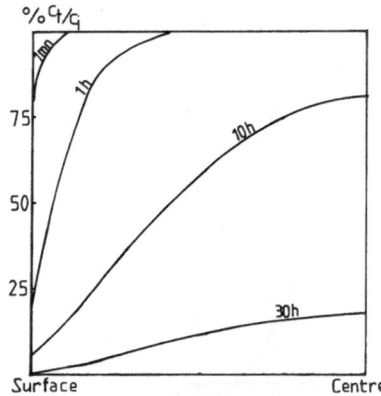

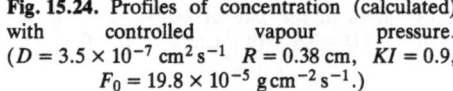

Fig. 15.24. Profiles of concentration (calculated) with controlled vapour pressure. ($D = 3.5 \times 10^{-7}$ cm^2 s^{-1} $R = 0.38$ cm, $KI = 0.9$, $F_0 = 19.8 \times 10^{-5}$ g cm^{-2} s^{-1}.)

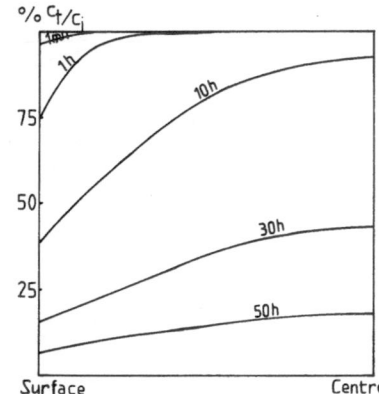

Fig. 15.25. Profiles of concentration (calculated) with controlled vapour pressure. ($D = 3.5 \times 10^{-7}$ cm^2 s^{-1} $R = 0.38$ cm, $KI = 0.9$, $F_0 = 19.8 \times 10^{-6}$ g cm^{-2} s^{-1}.)

pressure, for instance in such a way that it is proportional to the liquid concentration on the surface, it is possible to dry the sample with rather flat profiles of concentration of liquid within the solid. This method may be useful when the diffusivity is concentration dependent, in order to prevent the liquid from being entrapped in the solid.

Drying with a Programmed Temperature
High temperature is able to reduce the time of drying, as the diffusivity of the liquid through the polymer and the rate of evaporation increase exponentially with temperature. However, a drawback may appear with distortion of the shape of dosage forms because of the plasticity of humid beads at high temperature. A programmed temperature process is thus of value to achieve drying under the best conditions. Of course, three parameters are of concern: the temperature at the beginning and the end of the process of drying, and the rate of heating [28].

Symbols

C_{eq}	Concentration necessary to maintain equilibrium with the vapour pressure P in the surrounding atmosphere
$C_{r,t}$	Concentration of liquid at position r and time t
C_s, C_R	Concentration on the surface of the bead
CN_r	New concentration at position r, after time Δt
D_r	Diffusivity at position r (cm^2 s^{-1})
F_0	Rate of evaporation (g cm^{-2} s^{-1})
G	Function defined by Eq. (15.8)
N	Number of spherical membranes in the bead
q_n	Non-zero roots of Eq. (15.23)
Q_t, Q_∞	Amount of liquid evaporated after time t, and after infinite time, respectively

r	Position in the bead
R	Radius of the bead
S	Dimensionless number
t	Time
$\Delta r, \Delta t$	Increments of space and of time, respectively
ρ	Density of the liquid ($g\, cm^{-3}$)

References

1. Heilmann K. Therapeutic systems. Rate-controlled drug delivery; concept and development, Thieme Stratton, New York, 1984
2. Peppas NA, Gurny R, Doelker A, Buri P. Modelling of drug diffusion through swellable polymeric systems. J Membr Sci 1980; 7: 241–253
3. Peppas NA. Analysis of Fickian and non-Fickian drug release from polymers. Pharm Acta Helv 1985; 60: 110–111
4. Fessi H, Marty JP, Puisieux F, Carstensen JT. Square root of time dependence of matrix formulations with low drug content. J Pharm Sci 1982; 71: 749–752
5. Touitou E, Donbrow M. Drug release from non-disintegrating hydrophilic matrices: sodium salicylate as a model drug. Int J Pharm 1982; 11: 355–364
6. Heller J. Biodegradable polymers in controlled drug delivery. CRC Crit Rev Therm Drug Carrier Syst 1984; 1: 39–90
7. Focher B, Marzetti A, Sarto V, Baltrame PL, Carmetti P. Cellulosic materials: structure and enzymatic hydrolysis relationships. J Appl Polym Sci 1984; 29: 3329–3338
8. Droin A, Chaumat C, Rollet M, Taverdet JL, Vergnaud JM. Model of matter transfers between sodium salicylate–eudragit matrix and gastric liquid. Int J Pharm 1985; 27: 233–243
9. Malley I, Bardon J, Rollet M, Taverdet JL, Vergnaud JM. Modelling of controlled drug release in case of carbopol–sodium salicylate matrix in gastric liquid. Drug Dev Int Pharm 1987; 13: 67–81
10. Magron P, Rollet M, Taverdet JL, Vergnaud JM. Spherical oral polymer-drug device with two polymers for constant drug delivery. Int J Pharm 1987; 38: 91–97
11. Laghoueg N, Paulet J, Taverdet JL, Vergnaud JM. Oral polymer-drug devices with a core and an erodible shell for constant drug delivery. Int J Pharm 1989; 50: 133–139
12. Lui H, Magron P, Bouzon J, Vergnaud JM. Spherical dosage form with a core and shell. Experiments and modelling. Int J Pharm 1988; 45: 217–227
13. Armand JY, Magnard F, Bouzon J, Rollet M, Vergnaud JM. Modelling of the release of drug in gastric liquid from spheric forms with eudragit matrix. Int J Pharm 1987; 40: 33–41
14. Saber M, Magnard F, Bouzon J, Vergnaud JM. Modelling of matter transfers in drug-polymer devices used as galenic forms. J Polym Engng 1988; 8: 295–314
15. Khatir Y, Bouzon J, Vergnaud JM. Liquid sorption by rubber sheets and evaporation. Models and experiments. Polym Test 1986; 6: 253–265
16. Khatir Y, Bouzon J, Vergnaud JM. Non destructive testing of rubber for the sorption and desorption – evaporation of liquid by modelling (cylinders of finite length). J Polym Engng 1987; 7: 149–167
17. Khatir Y, Bouzon J, Vergnaud JM. Determination of processes of absorption and desorption of liquids with rubber tubings by using model and short tests. J Polym Engng 1987; 7: 275–299
18. Khatir Y, Bouzon J, Vergnaud JM. The kinetics of absorption and desorption of liquid by a rubber annulus using a model and short term tests. Plast Rubber Process Applic 1988; 9: 53–58
19. Crank J. The mathematics of diffusion, 2nd edn. Clarendon Press, Oxford, 1976, pp 96–97
20. Laghoueg-Derriche N, Vergnaud JM. Modelling the process of drying of dosage forms made of drug dispersed in a polymer. Int J Pharm 1991; 67: 51–57
21. Vergnaud JM. Scientific aspects of plasticizer migration from plasticized PVC into liquids. Polym Plast Technol Engng 1983; 20: 1–22
22. Blandin HP, David JP, Illien M, Malizewicz M, Vergnaud JM. Modelling of the drying process of coatings with various layers. J Coat Technol 1987; 59: 27–32
23. Koros WJ, Hopfenberg HB. Scientific aspects of migration of indirect additives from plastics to food. Food Technol 1979; 4: 56–60

24. Laghoueg-Derriche N, Bouzon J, Vergnaud JM. Effect of temperature on the drying process of dosage forms prepared by a humidity technique. Int J Pharm 1991; 67: 163–168
25. Laghoueg-Derriche N, Vergnaud JM. Effect of the volume of the surrounding atmosphere on the drying process of dosage forms made of polymer and drug. J Polym Engng 1991; 10(4): in press
26. Taverdet JL, Vergnaud JM. Surface analysis of plasticized PVC packagings by attenuated total reflectance. In: instrumental analysis of foods, vol 1. Academic Press, 1983, pp 367–377
27. Laghoueg-Derriche N, Vergnaud JM. Process of drying under controlled vapor pressure. Unpublished results
28. Laghoueg-Derriche N, Vergnaud JM. Drying of dosage forms prepared by a humidity technique using a programmed temperature system. Int J Pharm 1991; 71: 229–236

Chapter 16

Drying of a Polymer Sphere with Shrinkage

16.1 Introduction

During the process of absorption of a liquid by a polymer, swelling very often takes place, especially when the amount of liquid absorbed is large. Shrinkage of this polymer is thus observed during the process of desorption. The process is controlled by transient diffusion of the liquid within the polymer and evaporation from the surface. To obtain analytical solutions the mathematical treatment neglects the change in the dimensions of the polymer. Very often, numerical models with finite differences are built by considering a framework of reference with constant dimensions

A new model [1–4], based on a numerical method with finite differences, is built up, taking into account not only the transport of liquid through and out of the polymer, but also the shrinkage of the polymer during desorption.

Ethylene–vinyl acetate copolymers, available in beads, are used for testing the model. These beads are very useful for controlling the release of active agents in agriculture [5–9]. The absorption capacity of these copolymers is very large, and depends on the proportion of vinyl acetate present.

16.2 Theory

Assumptions

In order to set out the problem, the following assumptions are made:

1. The polymer is isotropic and homogeneous in composition.
2. The sample is spherical in shape and remains so during the process of drying.
3. The volume of the bead with the liquid in it is equal to the sum of the volumes of polymer and liquid (if the additivity of the volumes is not obtained, the law for the change must be known).

4. The transport of liquid through the polymer is controlled by transient diffusion, and by evaporation from the surface.
5. The rate of evaporation is proportional to the difference between the actual concentration of liquid at the surface and the concentration which is necessary to maintain equilibrium with the surrounding atmosphere.
6. The temperature remains constant during the process of evaporation, in spite of the endothermic enthalpy of vaporisation.

Mathematical Treatment

The radial transient diffusion through the polymer is described by Fick's law:

$$\frac{\partial C}{\partial t} = \frac{1}{r^2} \frac{\partial}{\partial r} \left[D r^2 \frac{\partial C}{\partial r} \right] \tag{16.1}$$

where C is the concentration of liquid in the polymer and D the diffusivity which may be constant or not.

Initial and boundary conditions for the stage of drying are

$$t = 0, 0 \leqslant r \leqslant R, C = C_{in} \text{ (bead)} \tag{16.2}$$

$$t > 0, -D \left(\frac{\partial C}{\partial r} \right)_R = \frac{F_0}{\rho} (C_R - C_{eq}) \text{ (surface)} \tag{16.3}$$

where R is the radius of the bead, C_R the actual concentration of liquid on the surface, C_{eq} the concentration of liquid necessary to maintain equilibrium with the surrounding atmosphere, F_0 the rate of evaporation of the pure liquid and ρ its density.

An analytical solution only exists when the diffusivity is constant and the radius of the sphere is kept constant. The problem of a change in dimension of the bead must be solved with the help of a numerical model.

Numerical Analysis

The radius of the bead R, free of liquid, is divided into N equal parts of thickness Δr, each of them being defined by the radial abscissa $j\Delta r$, j being an integer (Fig. 16.1).

$$R = N\Delta r, \qquad 0 \leqslant j \leqslant N \tag{16.4}$$

Three places are considered in the bead full of liquid, viz., within the bead, the centre and the surface.

Within the Bead, $1 \leqslant j \leqslant N - 1$

The spherical membrane located between the radial abscissae $r_{j-0.5}$ and $r_{j+0.5}$ is considered. When the bead is free of liquid, the volume of this membrane is

$$V_{j,\infty} = \frac{4\pi}{3} \left(\frac{R}{N} \right)^3 (3j^2 + \tfrac{1}{4}) \tag{16.5}$$

When the concentration of liquid in this membrane is $C_{j,t}$ at time t, this volume becomes

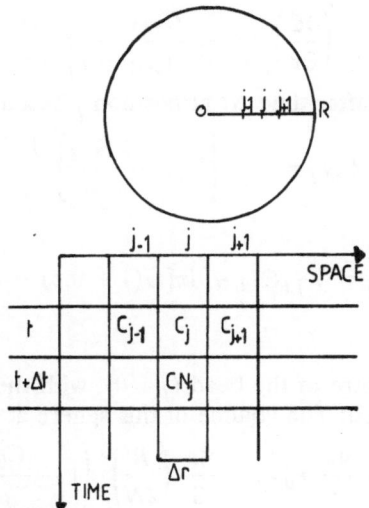

Fig. 16.1. Diagram for the numerical model.

$$V_{j,t} = \frac{4\pi}{3} (r_{j+0.5}^3 - r_{j-0.5}^3) \tag{16.6}$$

this volume $V_{j,t}$ being also equal to the sum of the volumes of the polymer and liquid (assumption 3)

$$V_{j,t} = V_{j,\infty} + V_{j,t} \frac{C_{j,t}}{\rho} \tag{16.7}$$

where ρ is the density of the diffusing liquid.

Eq. (16.7) can be rewritten as follows:

$$V_{j,t} = V_{j,\infty} \left[1 - \frac{C_{j,t}}{\rho}\right]^{-1} \tag{16.8}$$

The new abscissa at time t and position $(j + 0.5)\Delta r$, after uptake of liquid, can be obtained from the above equations:

$$r_{j+0.5,t}^3 - r_{j-0.5,t}^3 = \left(\frac{R}{N}\right)^3 (3j^2 + \tfrac{1}{4})\left(1 - \frac{C_{j,t}}{\rho}\right)^{-1} \tag{16.9}$$

The liquid balance during the time increment Δt is determined within the spherical membrane, by considering the diffusion. The new volume of this spherical membrane after time Δt, VN_j, can thus be expressed in terms of the previous volume at the same position and the diffusion:

$$VN_j CN_j = V_{j,t} C_{j,t} + 4\pi [G(j + 0.5) - G(j - 0.5)]\Delta t \tag{16.10}$$

with the function G defined by

$$G(j + 0.5) = r_{j+0.5}^2 D_{j+0.5} \left(\frac{\partial C}{\partial r}\right)_{j+0.5} \tag{16.11}$$

The gradient of concentration at the intermediate position $(j + 0.5)\Delta r$ can be approximated by the chord slope taken between j and $(j + 1)$:

$$\left(\frac{\partial C}{\partial r}\right)_{j+0.5} = \frac{C_{j+1} - C_j}{r_{j+1} - r_j} \tag{16.12}$$

The new concentration after time Δt at position j becomes

$$CN_j = A_{j,t}\left[V_{j,\infty} + \frac{A_{j,t}}{\rho}\right]^{-1} \tag{16.13}$$

on putting

$$A_{j,t} = VN_j CN_j = V_{j,t}C_{j,t} + 4\pi[G(j + 0.5) - G(j - 0.5)]\Delta t \tag{16.10'}$$

Centre of the Bead

A small sphere at the centre of the bead $(j = 0)$, with the final radius $\Delta r/2$ when free of liquid, is considered. The volume of this sphere at time t becomes

$$V_{0,t} = \frac{4\pi}{3} r_{0.5,t}^3 = \frac{4\pi}{3}\left(\frac{R}{2N}\right)^3\left(1 - \frac{C_{0,t}}{\rho}\right)^{-1} \tag{16.14}$$

and the radius of this small sphere, full of liquid, is

$$r_{0.5,t} = \frac{R}{2N}\left(1 - \frac{C_{0,t}}{\rho}\right)^{-1/3} \tag{16.15}$$

The liquid balance in this small sphere is calculated by considering the diffusion. The new concentration of liquid after time Δt is thus expressed in terms of the previous concentration and of the function G:

$$CN_0 = A_{0,t}\left[V_{0,\infty} + \frac{A_{0,t}}{\rho}\right]^{-1} \tag{16.16}$$

with the function A defined by

$$A_{0,t} = V_{0,t}C_{0,t} + 4\pi G(0.5)\Delta t \tag{16.10''}$$

Surface of the Bead

Infinite Rate of Evaporation. A spherical membrane next to the surface, located between the abscissae $r_{N-0.5}$ and r_N, is considered. The volume of this external membrane at infinite time, when free of liquid, is

$$V_{N,\infty} = \frac{4\pi}{3}\left(\frac{R}{N}\right)^3[N^3 - (N - 0.5)^3] \tag{16.17}$$

At time t, the volume of this membrane is

$$V_{N,t} = \frac{4\pi}{3}(r_N^3 - r_{N-0.5}^3) \tag{16.18}$$

From these two equations, a relationship is obtained between the two values of the volume of this membrane at time t and after infinite time when the liquid has evaporated:

$$V_{N,t} = V_{N,\infty}\left[1 - \frac{C'_{N,t}}{\rho}\right]^{-1} \tag{16.19}$$

The abscissa of the surface of the bead can thus be determined as a function of the concentration of liquid, and of the abscissa of the internal surface of the membrane:

$$r_{N,t}^3 - r_{N-0.5,t}^3 = \left(\frac{R}{N}\right)^3 [\tfrac{3}{2}N^2 - \tfrac{3}{4}N + \tfrac{1}{8}]\left[1 - \frac{C_{N,t}'}{\rho}\right]^{-1} \qquad (16.20)$$

The abscissa of the surface of the bead can thus be obtained by using Eqs (16.14), (16.9) and (16.20):

$$r_{N,t}^3 = \left(\frac{R}{N}\right)^3 \left[\tfrac{1}{8}\left(1 - \frac{C_{0,t}}{\rho}\right)^{-1} + \sum_{k=1}^{N-1} (3k^2 + \tfrac{1}{4})\left(1 - \frac{C_{k,t}}{\rho}\right)^{-1}\right.$$

$$\left. + (\tfrac{3}{2}N^2 - \tfrac{3}{4}N + \tfrac{1}{8})\left(1 - \frac{C_{N,t}'}{\rho}\right)^{-1}\right] \qquad (16.21)$$

where the contribution of the small sphere, of the spherical membranes within the bead and of the external membrane can be seen.

Finite Rate of Evaporation. The spherical membrane next to the surface of the bead, located between the radial abscissae $r_{N-0.5}$ and r_N is considered. From the matter balance during the increment of time Δt, the new mean concentration in the membrane is obtained in terms of the previous concentration, the diffusivity and the rate of evaporation:

$$A_N = VN_N CN_N' = V_N C_N' - 4\pi G(N - 0.5)\Delta t - 4\pi r_N^2 \frac{F_0}{\rho}(C_N - C_{eq})\Delta t \qquad (16.22)$$

with the function G defined in Eq. (16.11).

The new mean concentration in the external membrane is

$$CN_N' = A_N \left(V_{N,\infty} + \frac{A_N}{\rho}\right)^{-1} \qquad (16.23)$$

where $V_{N,\infty}$ is defined in Eq. (16.17).

The new concentration on the surface CN_N is obtained by linear extrapolation:

$$CN_N = \tfrac{4}{3}CN_N' - \tfrac{1}{3}CN_{N-1} \qquad (16.24)$$

Amount of Liquid in the Bead

The amount of liquid in the bead is obtained by integrating the concentration with respect to space:

$$Q_t = \sum_{j=0}^{N-1} V_{j,t} C_{j,t} + V_{N,t} C_{N,t}' \qquad (16.25)$$

The mean concentration in the external membrane, $C_{N,t}'$ can be calculated as follows:

$$C_{N,t}' = \tfrac{3}{4}C_{N,t} + \tfrac{1}{4}C_{N-1,t} = \tfrac{1}{4}C_{N-1,t} \qquad (16.26)$$

as the concentration on the surface is zero.

The total amount of liquid in the bead is thus expressed by the general relation

$$\frac{Q_t}{4\pi}\left(\frac{N}{R}\right)^3 = \frac{C_{0,t}}{24}\left(1 - \frac{C_{0,t}}{\rho}\right)^{-1} + \sum_{j=1}^{N-1}\left(j^2 + \tfrac{1}{12}\right)\left(1 - \frac{C_{j,t}}{\rho}\right)C_{j,t}$$

$$+ \left(\tfrac{1}{2}N^2 - \tfrac{1}{4}N + \tfrac{1}{24}\right)\left(1 - \frac{C'_{N,t}}{\rho}\right)^{-1} C'_{N,t} \qquad (16.27)$$

Conditions of Stability
The conditions of stability are given by

$$4\left[\frac{F_0}{\rho}\frac{N^2}{\Delta r} + \frac{D}{(\Delta r)^2}(N - 0.5)^2\right]\Delta t < N^3 - (N - 0.5)^3 \qquad (16.28)$$

16.3 Experiment

Materials

Beads of ethylene vinyl acetate (EVA) copolymers were used for the polymer. The characteristics of the polymer are shown in Table 16.1. n-hexane was used for the liquid.

Methods

The EVA beads were saturated by immersion in n-hexane. The kinetics of absorption were followed by weighing the beads at intervals.

During the process of drying, the saturated bead was exposed to the surrounding atmosphere at constant temperature. The volume of air was very large, and a slight agitation was maintained. The kinetics of drying were followed by weighing the beads at intervals.

16.4 Results

The essential objective in this chapter is to build a numerical model able to describe the process of drying by considering the liquid transport by transient diffusion within the polymer and evaporation from the surface, the change in

Table 16.1. Characteristics of the polymer and liquid

Material	% vinyl acetate	Melt index	Density (g cm^{-3})
EVA	40	55	0.96
n-hexane			0.66

Q empty bead (Q_i) = 0.02149 g.
Q liquid absorbed (Q_∞) = 0.01969 g.

dimension of the polymer resulting from this liquid transport. The validity of the model was tested by comparing the kinetics of evaporation obtained by experiments and calculation. Moreover, as up to now classical models without change in dimensions of the solid have been widely used, a relevant comparison between the results obtained with these two kinds of models can be instructive for the knowledge of the process.

Mathematical Model without Shrinkage and with Constant Diffusivity

As the dimension of the bead is kept constant, a problem appears very strongly with the selection of this value: is it the radius of the bead free of liquid or that of the bead presaturated with liquid?

This classical model has been used so far in various conditions [10], and especially in the case of desorption of the liquid previously absorbed by immersing the bead in water [5-9]. In the simple case of constant diffusivity, an analytical solution is found for the problem of diffusion–evaporation [10]. The main assumption is concerned with the boundary condition: the rate of evaporation is constantly equal to the rate at which the liquid is brought to the surface by diffusion (Eq. (16.3)).

An analytical solution is given under these conditions for the kinetics of drying:

$$\frac{Q'_\infty - Q'_t}{Q'_\infty} = \sum_{n=1}^{\infty} \frac{6S^2}{\beta_n^2(\beta_n^2 + S^2 - S)} \exp\left(-\frac{\beta_n^2}{R^2}Dt\right) \qquad (16.29)$$

the β_ns being the roots of

$$\beta_n \cot \beta_n + S - 1 = 0 \qquad (16.30)$$

with the dimensionless number S

$$S = \frac{RF_0}{D\rho} \qquad (16.31)$$

The constant diffusivity was determined from experiments carried out during the stage of absorption. It was then assumed that the diffusivity during desorption is the same as that measured during absorption.

The evaporation rate was determined from the experimental kinetics of drying, at the beginning of the process.

The kinetics of drying obtained by experiments and by calculation are compared in Fig. 16.2. Good agreement is observed between experiments and the results calculated with the mathematical model for the two values $R = 0.175$ cm (empty bead) and $R = 0.231$ cm (saturated bead).

Numerical Model with Shrinkage

The kinetics of drying were calculated with the numerical model, by considering the change in dimension of the bead as well as the diffusion–evaporation process.

The values of the parameters are the same as those used with the mathematical model.

- Diffusivity $D = 16 \times 10^{-7}$ cm^2 s^{-1}

- Rate of evaporation $F_0/\rho = 3 \times 10^{-4}$ cm s^{-1}

As shown in Fig. 16.3, a fairly good superimposition of the experimental and calculated kinetics is observed, proving the validity of the numerical model.

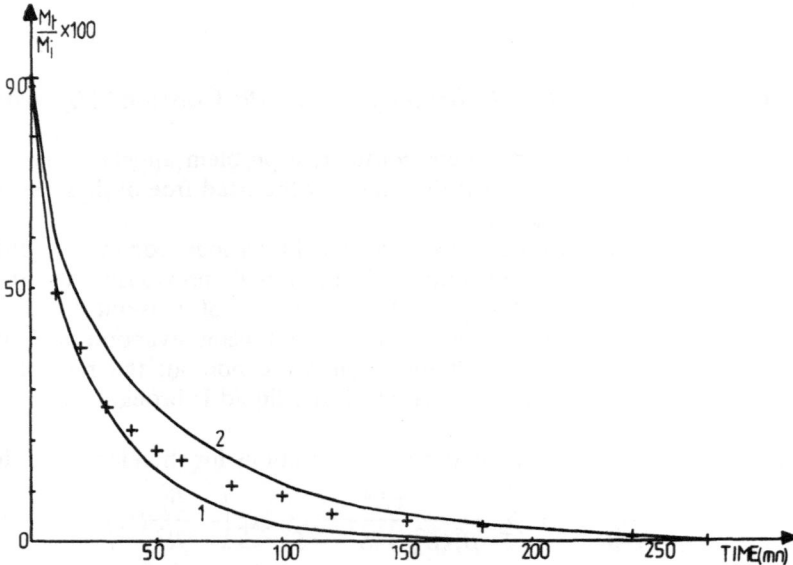

Fig. 16.2. Kinetics of drying obtained using the mathematical model with constant diffusivity and without change in dimension of the bead. +, experiments. 1: $R = 0.175$ cm. 2: $R = 0.231$ cm. ($D = 16 \times 10^{-7}$ cm^2 s^{-1}, $F_0/\rho = 3 \times 10^{-4}$ cm s^{-1}.)

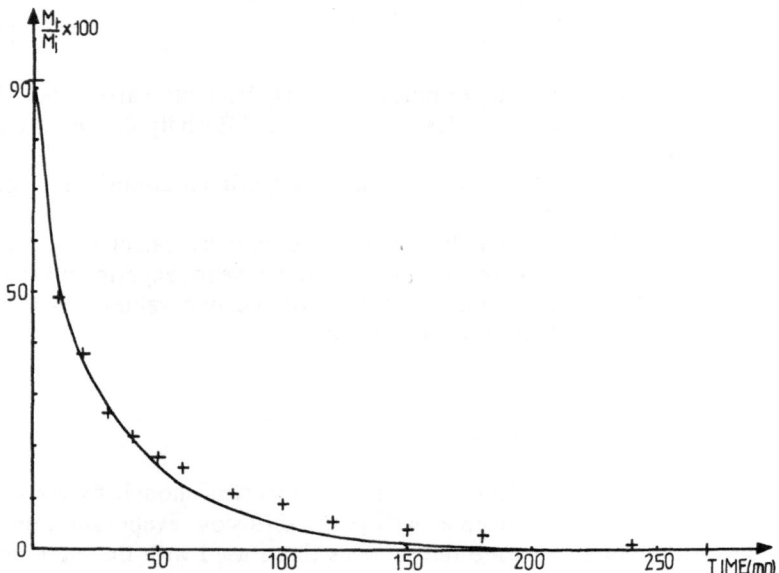

Fig. 16.3. Kinetics of drying obtained using the numerical model with diffusion–evaporation and change in dimension. (+, experiments; ——, calculation.)

Profiles of Concentration

The profiles of concentration of liquid developed through the bead were determined by using the two models. A comparison between these profiles affords further insight into the nature of the process.

Mathematical Model Without Shrinkage
The analytical solution with constant diffusivity is given for the concentration of the liquid in terms of position and time [10]:

$$\frac{C_{r,t}}{C_{in}} = \frac{2SR}{r} \sum_{n=1}^{\infty} [\beta_n^2 + S^2 - S]^{-1} \frac{\sin\dfrac{\beta_n r}{R}}{\sin\beta_n} \exp\left(-\frac{\beta_n^2}{R^2}Dt\right) \qquad (16.32)$$

$C_{r,t}$ being the concentration at position r and time t, and C_{in} the initial uniform concentration.

The profiles calculated with the help of the mathematical model are shown in Fig. 16.4, for each constant value given to the radius: the radius of the bead free of liquid (left) and of the saturated bead (right).

Numerical Model with Shrinkage
The profiles of concentration of liquid through the bead obtained by using the numerical model are shown in Fig. 16.5. The shrinkage of the bead can be observed in this figure, the position of the surface passing from around 52 at the beginning of the drying process to 40 which corresponds to the bead free of liquid.

Complementary results are given in Fig. 16.6 where the concentration of liquid at various places is depicted as a function of time: at the centre (1); at one-fifth

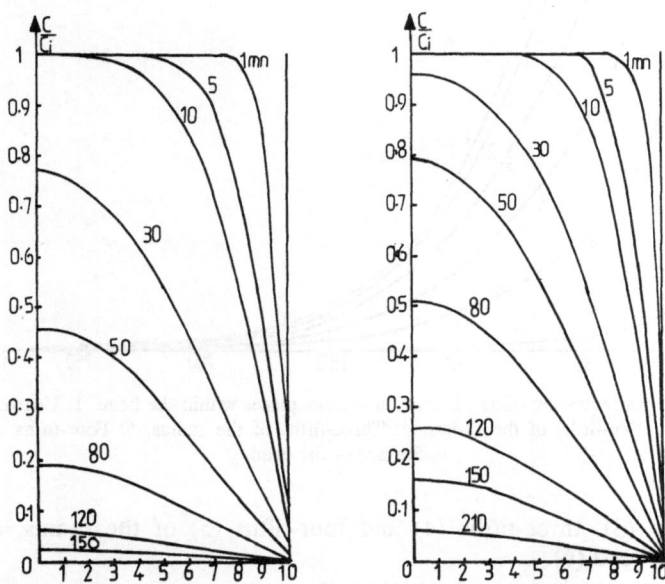

Fig. 16.4. Profiles of concentration within the bead during the process of drying, obtained using the mathematical model. Left: $R = 0.175$ cm. Right: 0.231 cm.

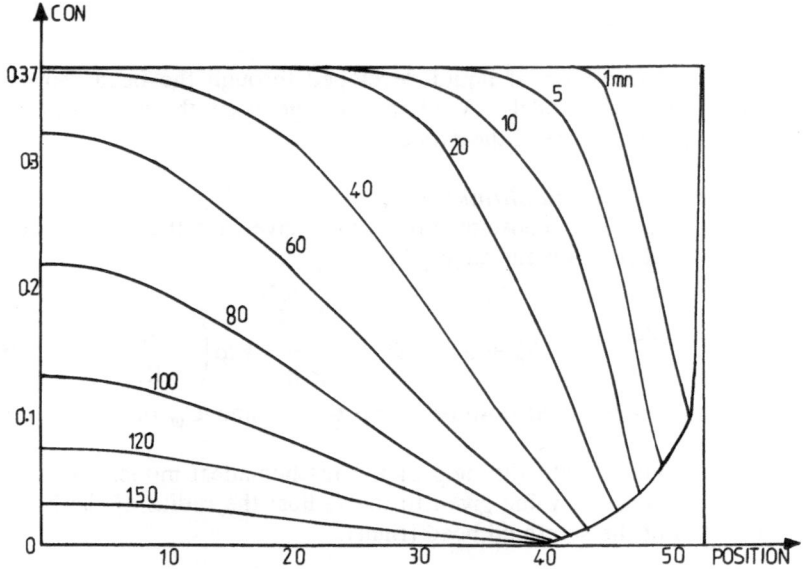

Fig. 16.5. Profiles of concentration within the bead during the process of drying, obtained using the numerical model.

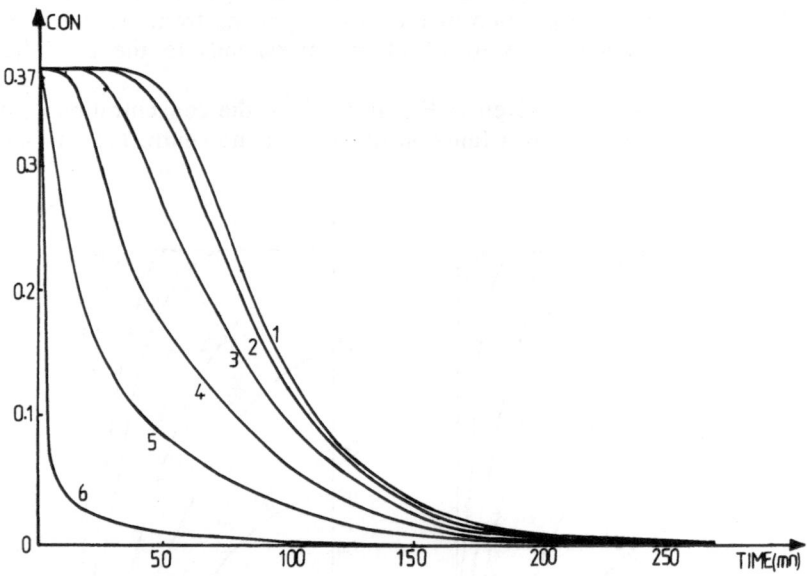

Fig. 16.6. Liquid concentration–time histories in various places within the bead. 1: Centre. 2: One-fifth of the radius. 3: Two-fifths of the radius. 4: Three-fifths of the radius. 5: Four-fifths of the radius. 6: Surface of the bead.

(2), two-fifths (3), three-fifths (4) and four-fifths (5) of the radius; and at the surface of the bead (6).

The following conclusions are of interest:

1. A significant change in dimensions of the bead is observed in Fig. 16.5. This

decrease in dimensions is not the same according to the position taken within the bead, and it follows the desorption of the liquid.

2. The concentration of liquid in the bead is higher when the swelling is neglected than when it is taken into account, when the radius considered is that of the bead free of liquid.

Change in Dimensions of the Bead

The numerical model is capable of providing the change in dimensions of the bead at various positions and times. Six positions are selected in Fig. 16.7, located between the centre and the surface of the bead.

Two conclusions are worth noting:

1. The rate at which shrinkage takes place is higher on the surface, at the beginning of the drying process.
2. Shrinkage on the surface occurs as soon as the process starts, and it takes some time for shrinkage to start at positions within the bead.

16.5 Conclusions

The new numerical model takes into account not only the diffusion of liquid within the polymer and the evaporation from the surface, but also the change in dimensions of the bead resulting from the drying process.

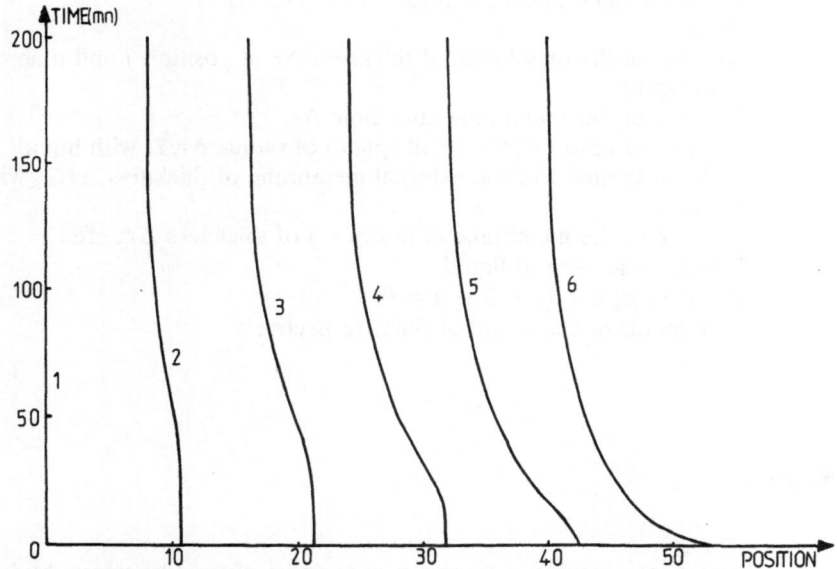

Fig. 16.7. Extent of shrinkage at various positions in the bead as a function of time. 1: Centre. 2: One-fifth of the radius. 3: Two-fifths of the radius. 4: Three-fifths of the radius. 5: Four-fifths of the radius. 6: Surface of the bead.

The results obtained with this numerical model are quite different from the results calculated with the mathematical model which neglects the change in dimensions of the bead. The liquid concentration is lower when both the swelling and shrinkage are considered.

Moreover, a difficulty arises with the value of the radius to be taken for the bead in the case of the mathematical model.

Symbols

C_{in}	Initial concentration of liquid in the bead (g cm^{-3})
$C_{j,t}$	Concentration of liquid at position j and time t
CN_j	Concentration at position j after time Δt
$C'_{N,t}$	Mean concentration in the membrane next to the surface of thickness $\Delta r/2$ when free of liquid
D	Diffusivity (cm^{-2} s^{-1})
$G(j)$	Function defined by Eq. (16.11)
j	Integer defining position
N	Number of equal intervals on the radius of the bead free of liquid
Q_t	Amount of liquid in the bead (Eq. 16.25)
Q'_t, Q'_∞	Amount of liquid evaporated after time t, after infinite time, respectively
r	Radial abscissa
$r_{j,t}$	Radial abscissa at position j and time t
R	Radius of the bead (free of liquid in the numerical model)
S	Dimensionless number defined by Eq. (16.31)
t	Time
$V_{j,t}$	Volume of the membrane of thickness Δr at position j and time t, with liquid
VN_j	Volume of this membrane after time Δt
$V_{0,t}$	Volume at time t of the small sphere of radius $\Delta r/2$, with liquid
$V_{N,t}$	Volume at time t of the external membrane of thickness $\Delta r/2$, with liquid
$V_{j,\infty}$	Volume of the membrane at position j of thickness Δr, after infinite time, free of liquid
β_n	Roots of $\beta_n \cot \beta_n + S - 1 = 0$
$\Delta r, \Delta t$	Increments of space and of time, respectively

References

1. Bouzon J, Vergnaud JM. Modelling of the process of desorption of liquid by polymer bead, by considering diffusion and shrinkage. Europ Polym J 1991; 27: 115–120
2. Senoune A, Bouzon J, Vergnaud JM. Modelling the process of liquid absorption by a polymer sphere, by considering diffusion and subsequent swelling. J Polym Engng 1990; 9 (3): 213–236

3. Bakhouya A, Bouzon J, Vergnaud JM. Modelling the process of diffusion–evaporation of a liquid from a polymer sphere, by considering the shrinkage of the sphere. Plast Rubber Proc Applic 1991; 15: 263–271

4. Mouffok B, Bouzon J, Vergnaud JM. Modelling the process of evaporation of hydrocarbons out of polymers by considering diffusion-evaporation and shrinkage. J Comput Polym Sci 1991; 1: 56–62

5. David H, Bouzon J, Vergnaud JM. Controlled absorption and release of an active agent by using EVAc beads. Effect of various parameters. Europ Polym J 1989; 25: 1007–1011

6. David H, Bouzon J, Vergnaud JM. Absorption of anilin by EVA copolymers and desorption into water. J Control Release 1988; 8: 151–156

7. David H, Bouzon J, Vergnaud JM. Modelling of desorption of liquid from an EVA polymer device composed of a core and shell. Plast Rubber Proc Applic 1989; 11: 9–16

8. David H, Bouzon J, Vergnaud JM. Modelling of absorption and desorption of a liquid by a polymer device made of a core and shell. Europ Polym J 1989; 25: 89–94

9. David H, Bouzon J, Vergnaud JM. Modelling of matter transfer with a polymer device made of a core and shell. Effect of the capacity of absorption by the silicone shell. Europ Polym J 1989; 25: 939–945

10. Crank J. The mathematics of diffusion, 2nd edn. Clarendon Press, Oxford, 1976, p. 96

Subject Index